Python for Accounting and Finance

Sunil Kumar

Python for Accounting and Finance

An Integrative Approach to Using Python for Research

Power lies not in the answers but in the tools, we use to find them

Sunil Kumar
Bristol, RI, USA

ISBN 978-3-031-54682-2   ISBN 978-3-031-54680-8 (eBook)
https://doi.org/10.1007/978-3-031-54680-8

© The Editor(s) (if applicable) and The Author(s), under exclusive license to Springer Nature Switzerland AG 2024

This work is subject to copyright. All rights are solely and exclusively licensed by the Publisher, whether the whole or part of the material is concerned, specifically the rights of translation, reprinting, reuse of illustrations, recitation, broadcasting, reproduction on microfilms or in any other physical way, and transmission or information storage and retrieval, electronic adaptation, computer software, or by similar or dissimilar methodology now known or hereafter developed.
The use of general descriptive names, registered names, trademarks, service marks, etc. in this publication does not imply, even in the absence of a specific statement, that such names are exempt from the relevant protective laws and regulations and therefore free for general use.
The publisher, the authors, and the editors are safe to assume that the advice and information in this book are believed to be true and accurate at the date of publication. Neither the publisher nor the authors or the editors give a warranty, expressed or implied, with respect to the material contained herein or for any errors or omissions that may have been made. The publisher remains neutral with regard to jurisdictional claims in published maps and institutional affiliations.

Cover illustration: imagenavi

This Palgrave Macmillan imprint is published by the registered company Springer Nature Switzerland AG
The registered company address is: Gewerbestrasse 11, 6330 Cham, Switzerland

Paper in this product is recyclable.

To my beloved wife Kavita, and our precious gems Nikshika and Anishka, who constantly inspire me to explore, learn, and share my knowledge. You are the true assets of my life.

Acknowledgements

Completing this book was made possible through the collective support and wisdom of many outstanding individuals. I would like to express my heartfelt gratitude to the University of Massachusetts Boston, its Accounting and Finance Department, and its PhD program, all of which provided an enriching environment that greatly contributed to my research and writing.

Special thanks are due to Professor Atreya Chakraborty, whose guidance and insights have been invaluable. I am also profoundly grateful to the faculty of the Accounting and Finance department at UMass Boston especially, Arindam Bandopadhyaya, Robert Kim, Sangwan Kim, Surit Tinaikar, and Lucia Silva Gao for their unwavering support and encouragement throughout this journey.

I also owe a deep sense of appreciation to Dr. Robert Taggart, Professor of Finance (Retired) at Boston College, whose teachings and mentorship have left an indelible mark on my professional life and academic pursuits. I also extend a special thanks to my friend Ankush Mohekar, who was always there when I needed him throughout this journey.

To all of you who supported me directly or indirectly during the writing of this book, thank you. Your collective wisdom has not only enlightened my path but has also enriched the pages of this work.

Prologue

In the swiftly evolving domains of business, accounting, and finance, harnessing the power of technology to enhance and expand upon traditional research methodologies has become increasingly vital. The advent of robust programming languages like Python has revolutionized the field of data analysis, enabling more sophisticated, nuanced, and efficient examination of complex data.

"Python for Accounting and Finance" is a comprehensive and illuminating exploration into the application of Python within the realm of accounting and finance research, as well as other business disciplines. Its contents are designed to serve as an indispensable guide for a diverse range of individuals engaged in these fields—PhD scholars, research faculty, industry professionals, and business researchers—who are keen to leverage the expansive capabilities of Python to elevate their research.

This book is predicated on the premise that Python has emerged as a programming language of choice in both academic research and applied research. Its open-source nature, combined with an extensive collection of libraries, gives it a distinct edge over many conventional, often proprietary, software programs. The book guides the reader through the comprehensive offerings of Python, from handling an array of data formats, including structured and unstructured data, to employing its advanced machine learning and artificial intelligence capabilities for predictive analytics.

The initial section of this book offers a solid foundation in Python, covering its fundamentals and key libraries, and regular expressions. The subsequent sections progressively delve into more specialized applications, beginning with data acquisition and cleaning, before moving on to exploratory data analysis and visualization, natural language processing, machine learning, and predictive analytics. Each section is meticulously crafted, presenting a judicious blend of theoretical knowledge and practical applications.

One of the standout features of "Python for Accounting and Finance" is its focus on real-world case studies and practical examples, supplemented with ready-to-use codes for most of the activities involved for research in these disciplines. This approach enables readers to contextualize and apply their learning immediately. The book is also replete with exercises that provide hands-on experience, reinforcing the concepts and techniques presented.

To derive maximum benefit from this book, it is imperative to implement the codes yourself and modify them as per your requirements. The learning journey through Python is one of active engagement and personal experimentation—getting your hands dirty, so to speak, is indeed the key to mastering this tool.

By the end of this comprehensive guide, readers will have developed a firm understanding of Python programming within the context of accounting, finance, and broader business research. They will be equipped with the skills to tackle real-world analytical problems in their professional pursuits. The journey through this book is not merely about learning a programming language; it is about embracing a powerful tool that unlocks a deeper understanding of research in these disciplines.

Welcome to a transformative journey into the world of Python for Accounting, Finance, and Business Research. Let the exploration begin!

Contents

Part I Introduction and Fundamentals

1 Introduction to Python for Accounting and Finance Research 3
  Benefits of Python in Accounting and Finance Research 4
  Overview of Python Programming Language 5
  Installing and Setting up Python Environment 8

2 Introduction to Python Language 11
  Data Types, Variables, and Operators 12
  Control Flow Statements 13
  Functions and Modules 20
  Data Structures in Python 21
  Input and Output 22
  File Handling in Python 23
  The os Module 25
  Object-Oriented Programming in Python 26

3 Regular Expressions for Python 31
  re Functions 32
  Building Blocks of Regex 33
    Literals 33
    Metacharacters 34
    Quantifiers 37

		Character Classes	39
		Escape Sequences	45
		Groups in Regex	47
		Substitution or Replacement Metacharacters	49
		Assertions	51
	Regular Expressions Cheat Sheet		56
4	**Important Python Libraries**		59
	Library		59
	Data Access Libraries		61
		BeautifulSoup	61
		Requests	63
		Scrapy	65
	Data Manipulation Libraries		67
		Pandas	67
		NumPy	70
		Dask	72
	Data Visualization Libraries		74
		Matplotlib	74
	Statistical Analysis Libraries		76
		SciPy	76
		StatModels	78
		PyMC3	80
	Machine Learning Libraries		82
		Scikit-Learn	82
		TensorFlow	84
		PyTorch	86
		Keras	87

Part II Data Acquisition and Cleaning

5	**Accessing Data from WRDS**		91
6	**Accessing Data from SEC EDGAR**		101
	Useful Modifications		104
		Limiting the Period	104
		Cleaning the HTML Tags	105
7	**Accessing Data from Other Sources**		109
	Data Contained in a Series of Webpages		110
	Data on Yahoo Finance		112
	Data on Cryptocurrency		114

	National Oceanic and Atmospheric Administration (NOAA) Data	116
	Twitter Data	118
	Google Trends Data	121
8	**Text Extraction and Cleaning**	125
	Extracting Useful Parts of Data	126
	Cleaning HTML Tags	129
9	**Text Normalization**	133
	Removal of Special Characters and Punctuation Marks	134
	Lowercasing	134
	Tokenizing	134
	Stop Word Removal	135
	Stemming	135
	Llematization	136
	Special Considerations in Accounting Data	142
10	**Corpus**	147
	Renaming Files	148
	Sorting Files	150
	Creating Corpus	152

Part III Exploratory Data Analysis and Visualization

11	**Data Visualization: Numerical Data**	157
	Matplotlib	158
	Heatmap	160
	3D Plot	160
	Box Plot	161
	Seaborn	163
	Plotly	167
	Bokeh	172
12	**Data Visualization: Text Data**	179
	Wordcloud	179
	Matplotlib	184
	Network Map	188
	Dimensionality Reduction Techniques	192
13	**Descriptive Statistics**	199
	Basic Descriptive Statistics	200
	Outlier Detection	204

Pearson's Correlation Coefficient	206
Time Series Descriptive Statistics	207
A Note on Sampling Techniques	209

Part IV Natural Language Processing and Text Analysis

14 Topic Modeling — 213
Latent Dirichlet Allocation (LDA) — 215
Non-negative Matrix Factorization (NMF) — 224
Probabilistic Latent Semantic Analysis (PLSA) — 229
Correlated Topic Model (CTM) — 231
Hierarchical Dirichlet Process (HDP) — 239

15 Word Embeddings — 243

16 Text Classification — 249
Naive Bayes — 250
Decision Trees — 259
Random Forests — 261
Deep Learning — 263

17 Sentiment Analysis — 265
Rule-Based Methods — 266
Lexicon-Based Methods — 278
Machine Learning Algorithms — 282

Part V Machine Learning and Predictive Analytics

18 Basic Regression — 303
Linear Regressions — 305
  Simple Linear Regression — 306
  Multiple Regression — 309
Regression Diagnostics — 314
  Linearity Check — 315
  Multicollinearity — 315
  Heteroscedasticity — 316
  Autocorrelation — 316
  Normality of Residuals — 317

19	**Logistic Regression**	319
	Implementing Logistic Regression in Python	320
	Confusion Matrix	323
	ROC Curve and AUC	323
	Precision, Recall, and F1-Score	325
20	**Probit and Logit Regression**	329
	Probit Regression	329
	Akaike Information Criterion (AIC) and Bayesian Information Criterion (BIC)	335
	Model Diagnostics	336
	Logit Model	339
21	**Polynomial Regression**	343
22	**Quantile Regression**	357
23	**Advanced Regressions**	365
	Tobit Regression	365
	Poisson Regression	373
	Negative Binomial Regression	374
	Instrumental Variables (IV) Regression	376
	Two-Stage Least Squares (2SLS) Regression	378
24	**Time Series Analysis**	381
	Autoregressive (AR) Model	388
	Moving Average (MA) Model	392
	ARMA, ARIMA, SARIMA, and SARIMAX	395
	Vector Autoregression (VAR) Model	398
	Vector Error Correction Model (VECM)	402
	Advantages of VECM	402
	GARCH Model	405
25	**Panel Data**	411
	Types of Panel Data Models	412
	Pooled OLS Models	414
	Fixed Effect Models	415
	Random Effect Models	417
	Dynamic Panel Data Models	419
	Panel Data Model Diagnostics	429

26	**Special Techniques in Multivariate Analysis**	435
	Principal Component Analysis	436
	Factor Analysis	441
	Cluster Analysis	444
	Canonical Correlation Analysis	451
	Discriminant Analysis	453

Part VI Advanced Topics

27	**Deep Learning**	459
	Neuron	463
	Deep Learning Techniques and Architectures	465
	Implementation of Deep Learning in Accounting and Finance	469
	Transformer Models	469
	FNN Models	481
	LSTM Models	486
	GRU Model	491
	CNN Model	492
	Autoencoder Model	497

Most Common Errors and Solutions	501
Index	505

Part I

Introduction and Fundamentals

1

Introduction to Python for Accounting and Finance Research

The disciplines of accounting and finance are inherently driven by data and heavily rely on analytical methodologies. As technological progress continues, the availability of data for analysis expands accordingly. To maintain competitiveness and ensure informed decision-making, researchers and practitioners in these fields must possess the ability to efficiently analyze large datasets. Python, a versatile programming language, has garnered significant attention in the realm of accounting and finance research. Its user-friendly syntax and robust libraries make it an optimal tool for performing tasks such as data analysis, machine learning, and data visualization. Acquiring proficiency in Python can yield substantial advantages in professional endeavors for doctoral candidates, researchers, and accounting professionals alike.

This chapter aims to introduce Python and elucidate its merits specifically within the context of accounting research. It commences with an overview of the language, encompassing its historical evolution, distinctive features, and wide-ranging applications. The discussion subsequently transitions to a step-by-step guide for installing and configuring a Python environment on the reader's personal computer, enabling a prompt initiation of coding activities. By the conclusion of this chapter, readers will have acquired a foundational understanding of Python and be adequately prepared to explore its extensive capabilities within the domains of accounting and finance research.

Benefits of Python in Accounting and Finance Research

Python has gained increasing popularity in accounting and finance research due to its numerous advantages. One notable benefit is its open-source nature, which extends to the majority of supporting libraries and tools. These open-source resources, accompanied by flexible and open licenses, not only make Python a cost-effective choice but also foster a collaborative and open community where code and ideas can be shared.

Moreover, Python's interpreted nature allows for runtime translation of code into executable byte code, resulting in fast and efficient execution. This characteristic makes Python particularly suitable for researchers working with large datasets. Python's multiparadigm nature further enhances its appeal, as it supports various programming and implementation paradigms, including object orientation, imperative, functional, and procedural programming. This flexibility empowers researchers to choose the most appropriate paradigm for their research, free from language constraints.

Python's versatility extends to its multipurpose nature, enabling its utilization for rapid interactive code development, building large applications, and performing low-level system operations as well as high-level analytics tasks. This adaptability enables researchers to employ Python across diverse research domains, ranging from data analysis to intricate modeling.

An additional advantage of Python is its dynamic typing feature, where types are inferred during runtime rather than statically declared as in compiled languages. This simplifies programming and alleviates distractions from technical details, allowing researchers to focus on their research objectives. Python's indentation awareness, which employs indentation for marking code blocks, enhances code readability and comprehension.

Python's cross-platform compatibility ensures its availability across major operating systems, including Windows, Linux, and Mac OS. It is applicable for desktop and web applications and can be employed on various hardware, ranging from powerful servers to smaller devices like the Raspberry Pi. This cross-platform capability ensures researchers can work with Python regardless of their hardware or operating system preferences.

The advantages of Python in accounting and finance research extend beyond its language features. Python bridges the realms of economics and data science by providing researchers with a language that abstracts technical programming aspects and is easily learnable, even for individuals without a technical background. Its concise and English-like syntax makes it an ideal tool for interdisciplinary research. Python's support for prototyping and rapid iterative development is facilitated by its interactive interpreter tools, allowing researchers to write and execute code line by line and immediately observe results.

Python's capability to handle both structured and unstructured data sets it apart from traditional tools like SAS and STATA, which primarily handle structured data. With Python, researchers can work with diverse data formats, including text, images, audio, and numerical data, facilitating comprehensive analyses and deeper insights. This empowers Python to be a powerful tool in accounting and finance research, for instance, enabling the analysis of text data in financial statements to uncover meaningful insights or measuring market sentiment for investment decision-making. Python also allows researchers to gauge management sentiment while drafting financial statements, aiding in identifying potential red flags. Moreover, Python facilitates research utilizing additional databases not available in structured form. Its capacity for creating self-learning models that adapt to new events enhances the accuracy of predictions and forecasts.

Overview of Python Programming Language

Python is a high-level, interpreted programming language that has become increasingly popular among accounting and finance researchers in recent years. The language was created in 1991 by Guido van Rossum with a focus on code readability and simplicity, making it an ideal choice for researchers with little to no programming experience.

Python's syntax is designed to be easy to read and write, with a structure that emphasizes code readability over traditional syntax elements such as semicolons or braces. This makes Python a popular choice for beginners and experts alike in the accounting and finance research fields. In addition, Python supports multiple programming paradigms, including object-oriented, imperative, and functional programming, providing researchers with flexibility in how they approach their research.

Python has a large standard library that provides a wide range of functionalities, including string manipulation, file I/O, web scraping, and database management. This extensive library of tools has made Python a popular choice for researchers who want to work with large datasets or perform complex data analysis tasks. In addition, Python has a large and active community that contributes to a wide range of open-source libraries and frameworks, making it easy to build complex applications in accounting and finance research.

Python's popularity in data science and machine learning is largely due to the availability of popular libraries such as NumPy, Pandas, Matplotlib, and Scikit-learn. These libraries provide tools for data manipulation, visualization, and statistical analysis, making Python a popular language for scientific computing. In the field of accounting and finance research, Python can be used to analyze financial data, perform sentiment analysis on financial statements, and build predictive models for forecasting financial trends.

Python is also known for its ease of use and versatility, making it a popular choice for rapid application development and prototyping. Python code is usually one-third to one-fifth the size of equivalent C++ or Java code, leading to less coding time. Additionally, Python's simple, easy-to-learn syntax emphasizes readability and therefore reduces the cost of program maintenance. Python is completely free to use and distribute, making it an affordable choice for researchers.

Python's dynamic semantics and automatic garbage collection make it a powerful tool for accounting and finance research. The language provides high-level dynamic data types and supports dynamic type checking, allowing researchers to work with a variety of data formats, including text, images, and audio, in addition to numerical data. Python is also ideal for prototyping and rapid, iterative development, as its interactive interpreter tools provide environments where researchers can write and execute each line of code in isolation and see the results immediately.

Python is an interpreted, object-oriented, high-level programming language that is cross-platform and can be used on a wide range of operating systems, including Windows, Linux, and Mac OS. Python can be used as a scripting language or can be compiled to byte code for building large applications. Python is easy to learn because of its simplicity and frequent usage of English keywords in the code, making it an ideal choice for interdisciplinary research in accounting and finance.

Python also provides easy integration with other programming languages such as C, C++, COM, ActiveX, CORBA, and Java. This allows researchers to use Python in conjunction with other programming languages to create hybrid solutions that leverage the strengths of each language. Additionally, Python's popularity in the field of data science and machine learning can be attributed to its powerful libraries such as TensorFlow, PyTorch, and Keras. These libraries provide advanced tools for building and training neural networks, deep learning models, and natural language processing applications. Python's high-level syntax makes it easy to represent complex mathematical concepts and algorithms, which is essential for creating machine learning models.

In addition to its strong community support, Python also offers a wealth of resources for learning and development. There are numerous online courses, tutorials, and documentation available for free that can help users get started with Python and become proficient in its use. Many universities and academic institutions also teach Python as a part of their curriculum, reflecting the growing importance of the language in various fields.

Python is widely used in the financial industry, where it is used to develop tools for risk management, pricing models, and algorithmic trading. Its versatility makes it an ideal tool for financial analysis, and it is often used to analyze large datasets, create visualizations, and build predictive models. Many financial institutions also use Python for backtesting trading strategies and building automated trading systems.

Python is also popular in the field of web development, where it is used to create dynamic and interactive web applications. Python's simplicity and ease of use make it a popular choice for web developers, and it is often used with popular web frameworks such as Django and Flask.

In the world of scientific computing, Python is used extensively for numerical computing, simulation, and visualization. Its powerful libraries such as NumPy, SciPy, and Matplotlib provide advanced tools for numerical computing and data visualization. Python is also used in the field of computational biology, where it is used for data analysis, genome sequencing, and gene expression analysis.

Python's versatility and ease of use have made it an increasingly popular choice for educators and students in the field of computer science. Its simple syntax and powerful libraries make it an ideal language for teaching programming concepts and techniques. Python is often used in introductory computer science courses, where students learn programming fundamentals and gain practical experience with coding.

Python is a versatile, powerful, and widely used programming language that has become increasingly popular in recent years. Its ease of use, flexibility, and strong community support have made it a popular choice for a wide range of applications, from web development and data science to scientific computing and machine learning. With its powerful libraries and tools, Python has become an essential tool for researchers, developers, and educators in various fields, making it an exciting language to learn and use.

Installing and Setting up Python Environment

The process of installing and setting up Python for use in accounting and finance research may appear daunting to newcomers, but it can be accomplished with relative ease. Initially, the first step entails downloading and installing Python itself, which can be obtained from the official Python website. Once downloaded, executing the installation file and following the prompts completes the installation process.

Alternatively, individuals may opt for Anaconda, a widely used Python distribution that comes pre-packaged with numerous popular libraries and tools for scientific computing and data analysis. Anaconda includes Jupyter Notebook, an interactive development environment (IDE) for Python. To install Anaconda, individuals can visit the Anaconda website, download the appropriate version suitable for their operating system, and then proceed with the installation process by executing the downloaded file and adhering to the prompts.

Following the installation of Anaconda, users can access the Anaconda Navigator, a graphical user interface that facilitates application and environment management within Anaconda. Launching the Navigator involves opening the Anaconda Prompt or Terminal and entering "anaconda-navigator" followed by the Enter key.

Once the Navigator interface is accessible, users can create a new environment specifically tailored to their accounting and finance research. This can be accomplished by navigating to the "Environments" tab and selecting the "Create" option. A suitable name for the environment should be assigned, and the desired version of Python should be chosen, with the recommendation being the latest version.

Upon the creation of the environment, users can proceed to install additional packages and libraries necessary for their research. This can be done by navigating to the "Home" tab and selecting the "Install" option. Users can then search for packages by name or explore available packages categorized accordingly. Once the desired package is located, a simple click followed by selecting "Apply" will initiate the installation into the designated environment.

To commence using Python for accounting and finance research, users can launch Jupyter Notebook from the Navigator interface. Jupyter Notebook provides an interactive web-based environment where users can develop and share documents containing live code, equations, visualizations, and narrative text. Launching Jupyter Notebook involves selecting the "Home" tab and clicking "Launch" under the Jupyter Notebook section.

After launching Jupyter Notebook, users can create a new notebook by selecting "New" and then "Python 3". This action will create a notebook where Python code can be written and executed by pressing Shift + Enter.

While Jupyter Notebook serves as a commendable environment for Python code development and execution, alternative Integrated Development Environments (IDEs) are available that users may prefer. Prominent IDEs for Python include PyCharm, Spyder, and Visual Studio Code. These IDEs offer additional features such as code highlighting, code completion, and debugging capabilities, which can enhance the Python development experience. Users are encouraged to explore these options and select the IDE that aligns best with their specific requirements.

Having successfully installed and configured Python along with the necessary tools, users can now embark on exploring the immense potential this powerful language holds for their accounting and finance research endeavors.

2

Introduction to Python Language

This chapter serves as an integral foundation for the application of Python in accounting and finance research, illuminating the rudimentary aspects of Python programming. This chapter delves into the fundamental components that constitute the Python programming language, thus establishing a robust groundwork for the intricate applications of Python in subsequent chapters. This exploration of Python's basics will equip readers with the necessary skills to manipulate data effectively and to create functional and efficient code for their research endeavors.

We begin our exploration with an examination of Python's Data Types, Variables, and Operators, providing a clear understanding of the building blocks used to structure and manipulate data. Following this, we delve into Control Flow Statements, paving the way for more complex programming logic. A detailed study of Functions, Modules, and Data Structures in Python then allows us to understand the abstraction and organization of code and data, respectively. With a firm grasp of these, we transition into File Handling in Python, an essential skill for any researcher dealing with data. The subsequent discussion of the **os** module provides an overview of how Python interacts with the operating system, a skill vital for automating and streamlining tasks. Finally, we introduce the paradigm of Object-Oriented Programming in Python, a powerful tool for encapsulating data and functionality into reusable and modular code. This comprehensive exploration of Python's basics thus forms a solid foundation for the advanced applications of Python in the realm of accounting and finance research.

Data Types, Variables, and Operators

Python is an object-oriented programming language, which means that all data is stored in objects. In Python, there are several built-in data types that can be used to store different kinds of data. The most commonly used data types in Python include:

- Integer: used to store whole numbers (e.g., 1, 2, 3)
- Float: used to store decimal numbers (e.g., 1.0, 2.5, 3.14)
- Boolean: used to store True or False values
- String: used to store text (e.g., "hello world")
- List: used to store multiple items in a single variable (e.g., [1, 2, 3])
- Tuple: used to store multiple items in a single variable (e.g., (1, 2, 3))
- Dictionary: used to store key-value pairs (e.g., {"name": "John", "age": 25})

To create a variable in Python, you simply need to assign a value to a name using the equals sign (=). For example:

```
x = 1
```

In this case, we have created a variable named *x* and assigned it the value of 1. Once a variable has been created, it can be used in expressions and calculations.

Python also provides several operators that can be used to perform calculations and manipulate data. These include:

- Arithmetic operators: used to perform basic arithmetic operations (e.g., + for addition, − for subtraction, * for multiplication, / for division)
- Comparison operators: used to compare two values and return a boolean value (e.g., == for equality, < for less than, > for greater than)
- Logical operators: used to combine two or more boolean values (e.g., and for logical AND, or for logical OR, not for logical NOT)
- Assignment operators: used to assign a value to a variable (e.g., = for simple assignment, += for addition assignment, −= for subtraction assignment)

Here's an example of how you can use variables and operators in Python:

```
x = 10
y = 5
z = x + y
print(z)
```

In this example, we have created two variables named x and y and assigned them the values of 10 and 5, respectively. We have then created a third variable named z and assigned it the value of $x + y$, which is 15. Finally, we have printed the value of z to the console using the print() function.

Control Flow Statements

Control flow statements in Python allow you to control the order in which statements are executed in a program. They are used to specify the sequence of execution of statements based on certain conditions.

a. If, Else, and Elif Statements

If, else, and elif statements are conditional statements that are used to execute certain code based on specific conditions. These statements allow the program to make decisions based on whether a certain condition is true or false.

- If Statements

The if statement is used to execute code if a certain condition is true. The syntax of an if statement is as follows:

```
if condition:
    # code to execute if condition is true
```

Here's an example:

```
x = 10
if x > 5:
    print("x is greater than 5")
```

In this example, the condition is x > 5. Since x is equal to 10, which is greater than 5, the code inside the if statement will be executed, and the output will be x is greater than 5.

- Else Statements

The else statement is used to execute code if the condition in the if statement is false. The syntax of an else statement is as follows:

```
if condition:
    # code to execute if condition is true
else:
    # code to execute if condition is false
```

Here's an example:

```
x = 2
if x > 5:
    print("x is greater than 5")
else:
    print("x is less than or equal to 5")
```

In this example, the condition is **x > 5**. Since **x** is equal to **2**, which is less than **5**, the code inside the else statement will be executed, and the output will be **x is less than or equal to 5**.

- Elif Statements

The elif statement is used to test multiple conditions and execute different code based on which condition is true. The syntax of an elif statement is as follows:

```
if condition1:
    # code to execute if condition1 is true
elif condition2:
    # code to execute if condition2 is true
else:
    # code to execute if all conditions are false
```

Here's an example:

```
x = 2
if x > 5:
    print("x is greater than 5")
elif x == 5:
    print("x is equal to 5")
else:
    print("x is less than 5")
```

In this example, the first condition (**x > 5**) is false, so the program moves on to the next condition (**x == 5**). Since **x** is not equal to **5**, the program executes the code inside the else statement, and the output is **x is less than 5**.

b. For and While Loops

Another key aspect of programming is the ability to perform iterative tasks. Two ways to do this in Python are with for loops and while loops.

- For Loops

A for loop is used to iterate over a sequence of elements. This sequence can be a list, tuple, string, or any other iterable object. The basic syntax for a for loop is:

```
for variable in sequence:
    # code to execute
```

The variable is assigned to each element in the sequence, and the code inside the loop is executed for each element.

For example, let's say we want to print each number in a list:

```
numbers = [1, 2, 3, 4, 5]

for num in numbers:
    print(num)
```

Output:

```
1
2
3
4
5
```

We can also use a for loop to iterate over a string:

```
my_string = "Hello, World!"

for char in my_string:
    print(char)
```

Output:

```
H
e
l
l
o
,

W
o
r
l
d
!
```

We can also use the range function to create a sequence of numbers to iterate over. The range function takes three arguments: start, stop, and step. The start argument is the first number in the sequence (inclusive), the stop argument is the last number in the sequence (exclusive), and the step argument is the amount by which to increment each number.

```
for i in range(1, 6):
    print(i)
```

Output:

```
1
2
3
4
5
```

- While Loops

A while loop is used to execute a block of code repeatedly as long as a condition is true. The basic syntax for a while loop is:

```
while condition:
    # code to execute
```

The condition is evaluated at the beginning of each iteration. If the condition is true, the code inside the loop is executed. This continues until the condition becomes false.

For example, let's say we want to print the numbers 1 to 5 using a while loop:

```
    i = 1
while i <= 5:
    print(i)
    i += 1
```

Output:

```
1
2
3
4
5
```

We can also use the break and continue statements in for and while loops, just like in if statements. The break statement is used to exit the loop completely, while the continue statement is used to skip the current iteration and move on to the next one.

```
# Example of using break and continue in a for loop
for i in range(1, 6):
    if i == 3:
        break
    elif i == 2:
        continue
    print(i)
```

Output:

```
1
```

```
# Example of using break and continue in a while loop
i = 1
while i <= 5:
    if i == 3:
        break
    elif i == 2:
        i += 1
        continue
    print(i)
    i += 1
```

Output:

```
1
4
5
```

c. Break, Continue, and Pass Statements

Python also provides two other important control flow statements: **break** and **continue**. These statements allow you to alter the flow of a loop based on certain conditions.

- Break Statement

The **break** statement is used to terminate a loop prematurely. When the interpreter encounters a **break** statement within a loop, it immediately exits the loop and resumes execution at the next statement after the loop. This is useful when you need to exit a loop early based on some condition.

Here's an example of using the **break** statement in a **for** loop:

```
for i in range(1, 11):
    if i == 5:
        break
    print(i)
```

In this example, the loop will iterate over the numbers 1 through 10. However, when the loop variable i is equal to 5, the break statement is executed, and the loop is terminated prematurely. As a result, only the numbers 1 through 4 are printed.

- Continue Statement

The continue statement is used to skip the current iteration of a loop and move on to the next iteration. When the interpreter encounters a continue statement within a loop, it immediately skips to the next iteration of the loop without executing any further statements in the current iteration. This is useful when you need to skip over certain iterations of a loop based on some condition.

Here's an example of using the continue statement in a while loop:

```
i = 0
while i < 10:
    i += 1
    if i % 2 == 0:
        continue
    print(i)
```

In this example, the loop will iterate over the numbers 1 through 10. However, when the loop variable **i** is even, the **continue** statement is executed, and the current iteration of the loop is skipped. As a result, only the odd numbers are printed.

- Pass Statement

In Python, the **pass** statement is used as a placeholder for code that is not yet implemented. It is an empty statement that does nothing and is used to avoid syntax errors when you have an empty block of code.

The **pass** statement is particularly useful when writing code that requires a function or loop structure to be defined, but you haven't yet written the code that goes inside. Rather than leaving the block empty, which will generate an error, you can use the **pass** statement to fill the block with something that does nothing.

Here is an example of the **pass** statement in a loop:

```
for i in range(10):
    if i % 2 == 0:
        pass
    else:
        print(i)
```

In this code, the **pass** statement is used to fill the block of code that runs when **i** is an even number. Since we don't want to do anything in this case, we simply use **pass** to fill the block.

Without the **pass** statement, the code would generate a syntax error because the **if** statement requires a block of code to be executed when the condition is true.

The **pass** statement can also be used in functions, classes, and other code blocks where a block of code is required but you don't want to execute any code.

For example, in a function that you are still designing, you can use the **pass** statement to fill the function body until you're ready to write the actual code:

```
def my_function():
    pass
```

This code creates a function **my_function** that does nothing. However, it allows you to define the function without generating an error due to an empty body.

In summary, the **pass** statement is used as a placeholder for code that is not yet implemented. It can be used in loops, functions, and other code blocks where a block of code is required but you don't want to execute any code.

Functions and Modules

Functions and modules are fundamental concepts in programming that allow code to be organized, modularized, and reused. In Python, functions and modules are easy to define and use, making them powerful tools for developers and researchers in a wide range of fields, including accounting and finance.

A function is a block of code that performs a specific task and can be called multiple times throughout a program. Functions in Python are defined using the **def** keyword, followed by the function name and any arguments that the function takes. The function body is indented and contains the code that the function executes when called.

Here's an example of a simple function that adds two numbers:

```python
def add_numbers(a, b):
    return a + b
```

This function takes two arguments, **a** and **b**, and returns their sum using the **return** keyword. The **return** statement is used to return a value from the function to the caller.

Functions in Python can also have default arguments, which are used when the caller does not provide a value for a particular argument. Here's an example of a function with default arguments:

```python
def greet(name, greeting="Hello"):
    print(greeting + ", " + name + "!")
```

This function takes two arguments, **name** and **greeting**, with **greeting** having a default value of "Hello". If the caller does not provide a value for **greeting**, the function will use the default value.

A module is a file that contains Python code, such as functions, variables, and classes. Modules allow code to be organized into separate files, making it easier to manage and reuse code across different projects. In Python, modules are imported using the **import** keyword, followed by the name of the module.

Here's an example of how to import the **math** module in Python:

```python
import math

print(math.pi)
```

This code imports the **math** module and prints the value of pi, which is defined in the **math** module.

In addition to built-in modules like **math**, Python allows users to create their own modules by defining functions and variables in a separate Python file. To use functions and variables from a custom module, the module must be imported using the **import** keyword.

Here's an example of how to create and use a custom module in Python:

1. Create a new file called **my_module.py** with the following code:

```
def greet(name):
    print("Hello, " + name + "!")
```

2. In a separate file, import the **greet** function from the **my_module** module:

```
from my_module import greet

greet("John")
```

This code imports the **greet** function from the **my_module** module and calls it with the argument **"John"**, which will print the greeting **"Hello, John!"**.

Data Structures in Python

Data structures are fundamental building blocks for any programming language, and Python has a rich set of data structures that allow for efficient storage and manipulation of data. These data structures provide a way to organize and manage data in a logical and structured way. Here are some of the most commonly used data structures in Python:

a. Lists: A list is a collection of items, where each item is assigned a unique index number. Lists are mutable, which means that they can be modified after they are created. Lists are created using square brackets, and items in the list are separated by commas. Here's an example:

```
fruits = ['apple', 'banana', 'orange', 'kiwi']
```

b. Tuples: Tuples are similar to lists, but they are immutable, which means that they cannot be modified after they are created. Tuples are created using parentheses, and items in the tuple are separated by commas. Here's an example:

```
coordinates = (10, 20)
```

c. **Sets:** A set is an unordered collection of unique items. Sets are created using curly braces, and items in the set are separated by commas. Here's an example:

```
numbers = {1, 2, 3, 4, 5}
```

d. **Dictionaries:** A dictionary is a collection of key-value pairs, where each key is associated with a value. Dictionaries are created using curly braces, and key-value pairs are separated by colons. Here's an example:

```
ages = {'John': 30, 'Alice': 25, 'Bob': 40}
```

e. **Dictionaries:** A dictionary is a collection of key-value pairs, where each key is associated with a value. Dictionaries are created using curly braces, and key-value pairs are separated by colons. Here's an example:

```
greeting = "Hello, world!"
```

These data structures provide a way to store and manipulate data in Python. They can be used in various ways, such as sorting, filtering, and searching data. By understanding the different data structures available in Python, you can choose the best one for your specific needs and optimize your code for performance.

Input and Output

Input and output (I/O) operations are essential for any programming language, and Python is no exception. In Python, there are several ways to perform I/O operations, including console-based (standard) input/output and file-based input/output.

Console-based input is the most common method of accepting user input in Python. It allows a program to receive data entered by the user through the command line. In Python, console-based input is performed using the **input()** function. The **input()** function prompts the user to enter data and waits for a response before proceeding. Here is an example:

```
name = input("Enter your name: ")
print("Hello, " + name + "!")
```

In this example, the **input()** function prompts the user to enter their name, and the response is stored in the **name** variable. The program then uses the **print()** function to output a personalized greeting.

Non-standard input methods, such as reading from a file, are also common in Python.

File Handling in Python

File handling is a crucial aspect of any programming language, and Python provides an easy-to-use and powerful interface for file operations. File handling is used to perform various operations on files such as reading, writing, and appending data. In Python, file handling is done through the built-in **open()** function, which returns a file object that can be used to read, write, or append data to a file.

a. Opening a File

To open a file in Python, we use the **open()** function. The basic syntax for opening a file in Python is:

```
file_object = open("filename", "mode")
```

The **open()** function takes two arguments:

- **filename:** The name of the file (including the path, if necessary) that we want to open.
- **mode:** The mode in which we want to open the file. The mode can be either read mode ('r'), write mode ('w'), append mode ('a'), or a combination of these modes.

Here are the different modes that we can use:

- 'r': This mode is used to read data from an existing file. It is the default mode when we open a file. If the file does not exist, we will get an error.
- 'w': This mode is used to write data to a file. If the file exists, it will be overwritten. If the file does not exist, a new file will be created.
- 'a': This mode is used to append data to an existing file. If the file does not exist, a new file will be created.
- 'x': This mode is used to create a new file. If the file already exists, we will get an error.

- **'b'**: This mode is used to open a file in binary mode.
- **'t'**: This mode is used to open a file in text mode. It is the default mode.

b. Closing a File

After we have finished using a file, it is important to close it to free up the resources that were used. To close a file in Python, we use the **close()** method of the file object. Here is an example:

```
file_object = open("filename", "mode")
# do something with the file
file_object.close()
```

c. Reading from a File

To read from a file in Python, we use the **read()** method of the file object. Here is an example:

```
file_object = open("filename", "r")
data = file_object.read()
print(data)
file_object.close()
```

This will read the entire contents of the file into the **data** variable.

d. Writing to a File

To write to a file in Python, we use the **write()** method of the file object. Here is an example:

```
file_object = open("filename", "w")
file_object.write("This is some text that we want to write to the file.")
file_object.close()
```

This will write the specified text to the file. If the file already exists, it will be overwritten.

e. Appending to a File

To append to a file in Python, we use the **write()** method of the file object with the mode 'a'. Here is an example:

```
file_object = open("filename", "a")
file_object.write("This is some text that we want to append to
the file.")
file_object.close()
```

In this example, the **open()** function is used to open the **example.txt** file in append mode. The **write()** method is then used to append the string "This is some additional text." to the end of the file. Finally, the **close()** method is used to close the file.

f. Reading and Writing to CSV Files

Python provides the **csv** module for working with CSV files. You can use the **csv.reader()** method to read data from a CSV file and the **csv.writer()** method to write data to a CSV file. For example:

```
import csv

# Reading from a CSV file
with open('example.csv', 'r') as file:
    reader = csv.reader(file)
    for row in reader:
        print(row)

# Writing to a CSV file
```

The os Module

The **os** module in Python provides a way to interact with the file system. This includes creating, deleting, and renaming files and directories, as well as navigating the file system and changing file permissions.

To use the **os** module, it must be imported at the beginning of the Python script using the following command:

```
import os
```

Once the **os** module has been imported, there are several functions that can be used for file operations. One of these is the **is** function, which can be used to check whether a file or directory exists. The syntax for using the **is** function is as follows:

```
os.path.exists(path)
```

The **path** argument is a string representing the path to the file or directory that you want to check. The **os.path** module provides a way to construct file paths that is platform-independent, so you can use the same code on Windows, Mac, and Linux.

Here is an example of how to use the **is** function to check whether a file exists:

```
import os

if os.path.exists("example.txt"):
    print("The file exists.")
else:
    print("The file does not exist.")
```

If the file "example.txt" exists in the current directory, the output will be "The file exists." Otherwise, the output will be "The file does not exist."

Another useful function in the **os** module is **os.listdir()**, which returns a list of all the files and directories in a given directory. The syntax for using **os.listdir()** is as follows:

```
os.listdir(path)
```

The **path** argument is a string representing the path to the directory you want to list. Here is an example of how to use **os.listdir()** to list all the files in the current directory:

```
import os

files = os.listdir(".")
for file in files:
    print(file)
```

This code will print the names of all the files in the current directory.

In addition to these functions, the **os** module provides many other functions for file operations, such as creating directories, renaming files, and changing file permissions. With the **os** module, you can perform all the file operations you need to manage files and directories in your Python scripts.

Object-Oriented Programming in Python

Object-Oriented Programming (OOP) is a programming paradigm that involves using objects and their interactions to design and implement applications. Python is an object-oriented programming language, which means it supports OOP concepts such as encapsulation, inheritance, and polymorphism.

In Python, everything is an object. This means that you can create objects of various types, such as integers, strings, and lists, and use them in your programs. However, you can also create your own custom objects that represent the entities in your application. These custom objects can have attributes and methods that represent the data and behavior of the entity they represent.

Here are some important concepts in OOP in Python:

1. Classes: A class is a blueprint for creating objects. It defines the attributes and methods that an object of that class will have. You can create multiple objects from the same class, each with its own set of attributes.
2. Objects: An object is an instance of a class. It represents a specific entity in your application and has its own set of attributes and methods.
3. Attributes: Attributes are data members of an object that represent its state. They can be accessed and modified using dot notation.
4. Methods: Methods are functions that are defined in a class and can be called on an object of that class. They can operate on the object's attributes and change its state.
5. Inheritance: Inheritance allows you to create a new class that is a modified version of an existing class. The new class inherits all the attributes and methods of the parent class and can override or add new methods and attributes.
6. Polymorphism: Polymorphism is the ability of objects of different classes to be treated as if they were objects of the same class. This is achieved through the use of inheritance and method overriding.

In Python, you can define a class using the **class** keyword followed by the name of the class. Here's an example of a simple class definition:

```
class Person:
    def __init__(self, name, age):
        self.name = name
        self.age = age

    def say_hello(self):
        print(f"Hello, my name is {self.name} and I'm {self.age} years old.")
```

This class defines a **Person** object with two attributes (**name** and **age**) and one method (**say_hello**). The **__init__** method is a special method that gets called when an object of the class is created.

To create an object of this class, you can use the following code:

```
person = Person("Alice", 30)
```

This creates a **Person** object with the name "Alice" and age 30. You can access the object's attributes and methods using dot notation:

```
print(person.name)  # Output: Alice
person.say_hello()  # Output: Hello, my name is Alice and I'm 30 years old.
```

Inheritance is a powerful feature of object-oriented programming that allows you to define a new class based on an existing class. The new class is called the derived class or subclass, while the existing class is called the base class or superclass. The derived class inherits all the properties and methods of the base class, and can add new properties and methods or override existing ones.

To create a subclass in Python, you can simply include the name of the base class in parentheses after the name of the new class. For example, let's say we have a base class called Animal, and we want to create a subclass called Dog:

```
class Animal:
    def __init__(self, name):
        self.name = name

    def speak(self):
        raise NotImplementedError("Subclass must implement abstract method")

class Dog(Animal):
    def speak(self):
        return "Woof"
```

In this example, we define the Animal class with an __init__ method that initializes the name attribute, and a **speak** method that raises a **NotImplementedError** because it is an abstract method that must be implemented by any subclass.

We then define the Dog class as a subclass of Animal and override the **speak** method to return "Woof". Since the Dog class inherits from the Animal class, it also inherits the __init__ method and the name attribute.

Polymorphism is another important feature of object-oriented programming that allows you to use objects of different classes interchangeably, as long as they have the same interface. In other words, if two classes have the same methods and attributes, you can use objects of either class interchangeably.

For example, let's say we have two classes called Circle and Rectangle, both of which have a **area** method that calculates the area of the shape:

```
class Circle:
    def __init__(self, radius):
        self.radius = radius

    def area(self):
        return 3.14 * self.radius ** 2

class Rectangle:
    def __init__(self, width, height):
        self.width = width
        self.height = height

    def area(self):
        return self.width * self.height
```

In this example, we define two classes, Circle and Rectangle, both of which have an **__init__** method that initializes the attributes of the shape, and an **area** method that calculates the area of the shape.

Because both classes have the same interface (i.e., the same methods and attributes), we can use objects of either class interchangeably in a function that expects a shape object:

```
def print_area(shape):
    print(f"The area of the shape is {shape.area()}")

c = Circle(5)
r = Rectangle(4, 6)

print_area(c)
print_area(r)
```

In this example, we define a function called **print_area** that takes a shape object as an argument and prints the area of the shape using the **area** method. We then create objects of the Circle and Rectangle classes, and pass them to the **print_area** function. Since both classes have the same interface, we can use them interchangeably in the function.

In this chapter, we have covered the basics of Python programming language, including data types, control flow statements, functions and modules, input and output operations, object-oriented programming, and more. While this chapter provides a solid foundation in Python programming, there is much more to learn. If you are interested in learning more about Python, there are many great resources available.

Some resources for learning Python include:

- Python official website: https://www.python.org/
- Codecademy Python Course: https://www.codecademy.com/learn/learn-python-3
- Coursera Python for Everybody: https://www.coursera.org/specializations/python

Additionally, there are many books available on Python programming, including:

- Python Crash Course, 2nd Edition by Eric Matthes
- Learning Python, 5th Edition by Mark Lutz
- Python for Data Analysis by Wes McKinney

In the next chapters, we will explore the applications of Python specific to accounting and finance research. These applications include data cleaning and manipulation, financial analysis, and more. By the end of this book, you will have a strong foundation in Python programming and be able to apply it to various accounting and finance research projects.

3

Regular Expressions for Python

Regular expressions (regex) are a powerful tool used for pattern matching and text manipulation. A regular expression is a sequence of characters that define a search pattern. It is widely used in programming languages such as Python to search and manipulate strings. In other words, regex allows us to specify a pattern of text that we want to match against, and then search for that pattern within a larger body of text.

Regex finds its application in many fields, including computer science, linguistics, and natural language processing. In the context of Python, regular expressions are used extensively for text processing and data cleaning. For example, if you have a large amount of text data that needs to be analyzed or processed, regular expressions can be used to search for specific patterns or structures in the text, such as finding all occurrences of a particular word or phrase, or identifying patterns of text that indicate the presence of certain types of data.

Understanding regular expressions is an important skill for any data scientist or analyst. It is a valuable tool that can be used to clean and preprocess text data, extract relevant information from large datasets, and perform advanced natural language processing tasks. In the context of academic accounting and finance research, regular expressions can be a powerful tool for text preprocessing and data analysis. With the increasing availability of unstructured data, such as financial reports, news articles, and social media posts, regular expressions can help extract relevant information and identify patterns in the data. This can lead to a deeper understanding of market trends, sentiment analysis, and risk management. Moreover, regular expressions can be used in conjunction with other data analysis techniques, such as machine learning and natural language processing, to improve the accuracy and efficiency of predictive models. Regular expressions are a valuable tool for researchers looking to extract valuable insights from textual data in accounting and finance.

In the following sections, we will discuss the basics of regular expressions and how to use them in Python. We will cover the syntax and rules for constructing regular expressions, as well as some of the most common use cases for regex in data analysis and text processing.

re Functions

In Python, the **re** module provides support for regular expressions. This module allows you to use regex functions and methods to search, find, and manipulate text data. The **re** module is a standard library in Python, so you don't need to install any external libraries to use regular expressions.

The **re** module provides several functions to work with regular expressions. These functions can be used to perform a wide range of tasks such as searching for a pattern in a string, replacing a pattern with another string, and splitting a string based on a pattern. Some of the commonly used **re** functions include:

- **re.search()**: Searches for a pattern in a string and returns the first match.
- **re.findall()**: Finds all the occurrences of a pattern in a string and returns them as a list.
- **re.sub()**: Replaces all the occurrences of a pattern in a string with another string.
- **re.split()**: Splits a string into a list of substrings based on a pattern.
- These functions can be used to perform various text processing tasks, such as data cleaning, text classification, and sentiment analysis, among others. These functions will be used in Python as follows:
- **re.search(pattern, string, flags=0)**: searches for the first occurrence of the pattern in the string and returns a match object.
- **re.match(pattern, string, flags=0)**: searches for the pattern at the beginning of the string and returns a match object.
- **re.findall(pattern, string, flags=0)**: returns all non-overlapping occurrences of the pattern in the string as a list of strings.
- **re.finditer(pattern, string, flags=0)**: returns an iterator of all non-overlapping occurrences of the pattern in the string as match objects.

- **re.split(pattern, string, maxsplit=0, flags=0)**: splits the string by the occurrences of the pattern and returns a list of strings.
- **re.sub(pattern, repl, string, count=0, flags=0)**: replaces all occurrences of the pattern in the string with the replacement string and returns the modified string.

Building Blocks of Regex

A discussion of regex can be done in terms of literals, metacharacters, quantifiers, character classes, escape sequences, groups.

Literals

Literals are the simplest form of regular expressions. They are characters that match themselves, with no special meaning. For example, the regular expression "hello" would match the string "hello" and only "hello". Literals can be any alphanumeric character, as well as special characters like punctuation and whitespace.

To use a literal in a regular expression, simply include the character or sequence of characters that you want to match. For example, to match the string "cat", you would use the regular expression "cat". This will match only the exact sequence of characters "cat".

It is important to note that literals are case-sensitive by default. This means that the regular expression "cat" will not match the string "Cat". To match both upper and lower case versions of a letter, you can use a character class, which we will cover in a later subsection. An example of using **re** using only literals is as follows:

```
import re

# Define a string to search
string = "The quick brown fox jumps over the lazy dog."

# Define a literal to search for
literal = "fox"

# Use the literal in a regular expression search
matches = re.findall(literal, string)

# Print the matches
print(matches)
```

This code searches for the literal string "fox" in the larger string "The quick brown fox jumps over the lazy dog." using the **re.findall()** function. The output of the code will be **['fox']**, indicating that the search found one match for the literal "fox" in the string.

Metacharacters

Metacharacters are special characters used in regular expressions that have a special meaning and are used to match specific patterns in a string. Here are the most common metacharacters used in regular expressions:

- **.** : Matches any single character except newline character.
- **^** : Matches the start of the string.
- **$** : Matches the end of the string.
- **** : Used to escape a special character.
- ***** : Matches 0 or more occurrences of the preceding character.
- **+** : Matches 1 or more occurrences of the preceding character.
- **?** : Matches 0 or 1 occurrence of the preceding character.
- **{m}** : Matches exactly m occurrences of the preceding character.
- **{m,n}** : Matches between m and n occurrences of the preceding character.
- **[]** : Matches any character within the brackets.
- **|** : Matches either the expression before or after the pipe symbol.
- **()** : Groups a series of expressions together.

For example, the regular expression **r'^\d{3}-\d{2}-\d{4}$'** uses the metacharacters ^, $, {}, and - to match a Social Security number in the format of ###-##-####.

In Python, we use the **re** module to work with regular expressions. We can use the **re.search()** function to search for a regular expression pattern within a string. Here's an example:

```
import re

string = 'The quick brown fox jumps over the lazy dog'
pattern = r'fox'

result = re.search(pattern, string)

if result:
    print('Pattern found')
else:
    print('Pattern not found')
```

3 Regular Expressions for Python 35

This code searches for the pattern **fox** within the string **'The quick brown fox jumps over the lazy dog'** using the **re.search()** function. If the pattern is found, it prints **Pattern found**. If the pattern is not found, it prints **Pattern not found**.

Other examples of using a combination of literals and quantifiers are as follows:

This code searches for the pattern **fox** within the string **'The quick brown fox jumps over the lazy dog'** using the **re.search()** function. If the pattern is found, it prints **Pattern found**. If the pattern is not found, it prints **Pattern not found**.

- Matching a specific word:

```
import re

text = "The quick brown fox jumps over the lazy dog."
pattern = r"brown"

match = re.search(pattern, text)

if match:
    print("Match found!")
else:
    print("No match found.")
```

This code searches for the string "brown" within the given text "The quick brown fox jumps over the lazy dog" using the regular expression module in Python. It uses the **re.search()** function to look for a match and if a match is found, it prints "Match found!". If no match is found, it prints "No match found."

- Matching a specific character set:

```
import re

text = "The quick brown fox jumps over the lazy dog."
pattern = r"[aeiou]"

match = re.search(pattern, text)

if match:
    print("Match found!")
else:
    print("No match found.")
```

This code uses the **re** module in Python to search for a match of any vowel character in a given text string. The regular expression pattern **[aeiou]** matches any single character that is either **a**, **e**, **i**, **o**, or **u**. The **search()** function of the **re** module is used to search the **text** string for a match of this pattern. If a match is found, the code prints "Match found!", otherwise it prints "No match found."

- Matching any non-whitespace character:

```
import re

text = "The quick brown fox jumps over the lazy dog."
pattern = r"\S"

match = re.search(pattern, text)

if match:
    print("Match found!")
else:
    print("No match found.")
```

This code searches for a non-whitespace character in the given text string. The regular expression pattern **\S** matches any character that is not a whitespace character. The **re.search()** function searches for the first occurrence of the pattern in the text string. If a match is found, the code prints "Match found!" to the console, otherwise it prints "No match found."

- Matching any digit character:

```
import re

text = "The year is 2022."
pattern = r"\d"

match = re.search(pattern, text)

if match:
    print("Match found!")
else:
    print("No match found.")
```

This code uses regular expressions in Python to search for a digit in a string. The string "The year is 2022" is assigned to the variable **text**, and the regular expression pattern **\d** is assigned to the variable **pattern**. The **\d** pattern matches any digit. The **re.search()** function searches for a match between the pattern and the text, and returns a match object if it finds a match. If a match is found, the code prints "Match found!". Otherwise, it prints "No match found."

- Matching any whitespace character:

```
import re

text = "The quick brown fox jumps over the lazy dog."
pattern = r"\s"

match = re.search(pattern, text)

if match:
    print("Match found!")
else:
    print("No match found.")
```

The code uses the re module to search for whitespace characters in the given string "text". It defines a pattern using the escape character " followed by the lowercase letter 's', which represents whitespace. It then uses the search function to look for a match with the pattern in the text string. If a match is found, it prints "Match found!" otherwise it prints "No match found."

Quantifiers

Quantifiers in regex are used to specify how many times a character or group of characters can occur in a pattern. They help in making the regex more flexible and powerful. Some commonly used quantifiers include:

: Matches zero or more occurrences of the preceding character.
: Matches one or more occurrences of the preceding character.
? : Matches zero or one occurrence of the preceding character.
{n} : Matches exactly n occurrences of the preceding character.
{n,m} : Matches between n and m occurrences of the preceding character.

Quantifiers can be applied to literals, character classes, groups, and other regular expressions. They can help in simplifying the regex and making it more efficient. However, excessive use of quantifiers can lead to performance issues and overfitting. It is important to use them judiciously and with proper testing. Some examples of using quantifiers in Python are:

- Matching repeated characters using + quantifier:

```
import re

text = "aaabbbccc"
pattern = r"a+"

match = re.search(pattern, text)

if match:
    print("Match found:", match.group())
else:
    print("No match found.")
```

This Python code uses the **re** module to search for a pattern of one or more 'a' characters in the string **text**. The pattern is specified using the regular expression syntax and the + quantifier, which matches one or more occurrences of the preceding character or group. If a match is found, the **search()** function returns a match object that contains information about the match, including the matching substring. The code then prints the matched substring.

- Matching optional characters using ? quantifier:

```
import re

text = "colour"
pattern = r"colo?ur"

match = re.search(pattern, text)

if match:
    print("Match found:", match.group())
else:
    print("No match found.")
```

This Python code uses the **re** module to search for a pattern in a string. The pattern is defined as **colo?ur**, which uses the **?** quantifier to indicate that the **u** character is optional. This means that the pattern will match both "color" and "colour". The **search()** method is used to search for the pattern in the **text** string. If a match is found, the **group()** method is used to return the matched text. In this case, the output would be "colour". If no match is found, the output would be "No match found."

- Matching a specific number of characters using {} quantifier:

```
import re

text = "The quick brown fox jumps over the lazy dog."
pattern = r"\b\w{4}\b"

match = re.search(pattern, text)

if match:
    print("Match found:", match.group())
else:
    print("No match found.")
```

In this Python code, the regular expression pattern **\b\w{4}\b** is used to find a four-letter word that occurs at the boundary of a word in the input text string. **\b** matches a word boundary, and **\w** matches a word character (letters, digits, or underscore). The **{4}** quantifier specifies that the preceding **\w** should match exactly four times. When run, the program searches for a match to this pattern in the input string "The quick brown fox jumps over the lazy dog", and prints "Match found: jumps", since "jumps" is a four-letter word at a word boundary.

Character Classes

In this Python code, the regular expression pattern **\b\w{4}\b** is used to find a four-letter word that occurs at the boundary of a word in the input text string. **\b** matches a word boundary, and **\w** matches a word character (letters, digits, or underscore). The **{4}** quantifier specifies that the preceding **\w** should match exactly four times. When run, the program searches for a match to this pattern in the input string "The quick brown fox jumps over the lazy dog", and prints "Match found: jumps", since "jumps" is a four-letter word at a word boundary.

Character classes in regex are a powerful tool for specifying the exact set of characters you want to match in a search pattern. They can be used to match any type of character, such as digits, letters, whitespace, or punctuation.

For example, the character class [0-9] matches any single digit from 0 to 9, while the character class [a-z] matches any single lowercase letter from a to z. You can also use negation to specify a set of characters that should not be matched, such as [^aeiou] to match any consonant.

Character classes can also be used to match specific types of characters, such as digits, letters, or whitespace. For example, \d matches any single digit, \w matches any alphanumeric character, and \s matches any whitespace character.

Here is a list of some common character classes in regex:

\d: Matches any digit character (0-9).
\D: Matches any non-digit character.
\w: Matches any word character (a-z, A-Z, 0-9, or _).
\W: Matches any non-word character.
\s: Matches any whitespace character (space, tab, newline, etc.).
\S: Matches any non-whitespace character.
.: Matches any character except newline (\n).

There are also some other character classes that can be used, depending on the regex flavor being used, such as:

[:alpha:]: Matches any alphabetical character.
[:digit:]: Matches any digit character.
[:alnum:]: Matches any alphanumeric character.
[:space:]: Matches any whitespace character.
[:punct:]: Matches any punctuation character.

Some examples of character classes in Python are:

```
import re

# regex pattern to match a phone number
pattern = r'\d{3}-\d{3}-\d{4}'

# sample phone number
phone_number = '123-456-7890'

# match the phone number using the regex pattern
match = re.search(pattern, phone_number)

if match:
    print('Phone number found:', match.group())
else:
    print('Phone number not found')
```

The code defines a regular expression pattern **r'\d{3}-\d{3}-\d{4}'** to match a phone number in the format of ###-###-####. It then defines a phone number string **phone_number = '123-456-7890'**. The **re.search()** function is used to search for the pattern within the phone number string, and if a match is found, it prints "Phone number found:" followed by the matched phone number. In this case, the match is found and the output will be "Phone number found: 123-456-7890". If there is no match, the output will be "Phone number not found".

- Matching an Email Address:

```python
import re

# regex pattern to match an email address
pattern = r'\b[A-Za-z0-9._%+-]+@[A-Za-z0-9.-]+\.[A-Z|a-z]{2,}\b'

# sample email address
email_address = 'test@example.com'

# match the email address using the regex pattern
match = re.search(pattern, email_address)

if match:
    print('Email address found:', match.group())
else:
    print('Email address not found')
```

This code is searching for an email address using regular expression in Python. The regex pattern used here is **r'\b[A-Za-z0-9._%+-]+@[A-Za-z0-9.-]+\.[A-Z|a-z]{2,}\b'**. It starts with the **\b** anchor which matches at a word boundary. Then it matches one or more occurrences of characters that can be letters (both uppercase and lowercase), digits, period, underscore, percent, plus, or hyphen. After that, it matches the @ symbol followed by one or more occurrences of letters (both uppercase and lowercase), digits, period, or hyphen. Finally, it matches a period followed by two or more letters (uppercase or lowercase) to match the top-level domain. The **if-else** statement checks if a match is found or not and prints the appropriate message.

- Extracting Words from a String:

```python
import re

# regex pattern to match words
pattern = r'\b\w+\b'

# sample string
text = 'This is a sample string.'

# extract words from the string using the regex pattern
words = re.findall(pattern, text)

# print the extracted words
print('Words found:', words)
```

The code uses the regular expression module **re** to find all words in a given string. The regular expression pattern used is **r'\b\w+\b'**, which matches one or more word characters (**\w+**) that are surrounded by word boundaries (**\b**). Word characters include alphanumeric characters and underscores. The **findall()** function of the **re** module is used to find all non-overlapping matches of the pattern in the input string. In this case, the input string is **'This is a sample string.'**, and the output will be a list of words found in the string, which will be ['This', 'is', 'a', 'sample', 'string'].

- Matching a URL:

```
import re

# regex pattern to match a URL
pattern = r'https?://(?:[-\w.]|(?:%[\da-fA-F]{2}))+'

# sample URL
url = 'http://www.example.com'

# match the URL using the regex pattern
match = re.search(pattern, url)

if match:
    print('URL found:', match.group())
else:
    print('URL not found')
```

This Python code uses the **re** module to search for a URL in a given string. The regular expression used to find a URL is stored in the variable **pattern**, which matches a string that starts with either "http://" or "https://" and is followed by one or more characters that can be a combination of letters, digits, and special characters like dots and hyphens. The **re.search()** function is used to search for a match for the pattern in the given URL. If a match is found, the code prints the message "URL found" along with the matched URL, otherwise, it prints "URL not found".

- Matching a Date:

```
import re

# regex pattern to match a date in YYYY-MM-DD format
pattern = r'\d{4}-\d{2}-\d{2}'

# sample date
date = '2023-04-08'

# match the date using the regex pattern
match = re.search(pattern, date)

if match:
    print('Date found:', match.group())
else:
    print('Date not found')
```

This code uses the **re** module to search for a date string in **YYYY-MM-DD** format in a given input string. The regex pattern used is **r'\d{4}-\d{2}-\d{2}'**, which matches four digits for the year, two digits for the month, and two digits for the day, separated by hyphens. The input string **date** is searched for a match using the **re.search()** method, which returns a match object if a match is found. The **if** condition checks if a match was found, and if so, it prints a message with the matched date using the **match.group()** method. If no match is found, it prints a message indicating the date was not found.

- Matching a Zip Code:

```
import re

# regex pattern to match a date in YYYY-MM-DD format
pattern = r'\d{4}-\d{2}-\d{2}'

# sample date
date = '2023-04-08'

# match the date using the regex pattern
match = re.search(pattern, date)

if match:
    print('Date found:', match.group())
else:
    print('Date not found')
```

This code uses the Python **re** module to search for a date in a specific format (**YYYY-MM-DD**) within a string. The regular expression pattern used is **\d{4}-\d{2}-\d{2}**, which matches any sequence of four digits, followed by a hyphen, followed by two digits, another hyphen, and finally two more digits. The **search** function is used to search for a match of this pattern within the **date** string. If a match is found, the code prints a message saying that the date was found and prints the matched string. Otherwise, it prints a message saying that the date was not found.

- Matching an IP Address:

```
import re

# regex pattern to match an IP address
pattern = r'\b(?:[0-9]{1,3}\.){3}[0-9]{1,3}\b'

# sample IP address
ip_address = '192.168.0.1'

# match the IP address using the regex pattern
match = re.search(pattern, ip_address)

if match:
    print('IP address found:', match.group())
else:
    print('IP address not found')
```

This code uses regular expressions in Python to match an IP address. The regex pattern used is **r'\b(?:[0-9]{1,3}\.){3}[0-9]{1,3}\b'**, which matches a group of three numbers separated by a period, repeated three times, followed by a single number, all enclosed in word boundaries. The sample IP address used is '192.168.0.1'. The **re.search()** function is used to search for the pattern within the string. If a match is found, it prints 'IP address found:' along with the matched string. If no match is found, it prints 'IP address not found'.

- Matching a Social Security Number:

```
import re

# regex pattern to match a social security number
pattern = r'\b\d{3}-\d{2}-\d{4}\b'

# sample social security number
ssn = '123-45-6789'

# match the social security number using the regex pattern
match = re.search(pattern, ssn)

if match:
    print('Social security number found:', match.group())
else:
    print('Social security number not found')
```

This Python code demonstrates how to use regular expressions to search for a social security number in a given string. The regular expression pattern used is **\b\d{3}-\d{2}-\d{4}\b**, which matches a sequence of three digits, a hyphen, two digits, another hyphen, and four digits. In this code, the social security number to be searched is provided in the **ssn** variable, and the **re.search()** function is used to find a match for the regular expression pattern.

If a match is found, the message "Social security number found" along with the matched string is printed to the console, otherwise the message "Social security number not found" is printed.

- Matching a Credit Card Number:

```
import re

# regex pattern to match a credit card number
pattern = r'\b(?:\d{4}[- ]){3}\d{4}\b'

# sample credit card number
credit_card = '1234-5678-9012-3456'

# match the credit card number using the regex pattern
match = re.search(pattern, credit_card)

if match:
    print('Credit card number found:', match.group())
else:
    print('Credit card number not found')
```

The code uses the **re** module to find and validate a credit card number. The regex pattern used is **r'\b(?:\d{4}[-]){3}\d{4}\b'**, which matches a credit card number with a format of four-digit groups separated by hyphens or spaces. The **\b** at the beginning and end of the pattern indicates word boundaries. The **(?:...)** syntax is used for a non-capturing group, and the **\d{4}[-]** matches four digits followed by a hyphen or a space. The **{3}** quantifier indicates that this group should be repeated three times, and the final **\d{4}** matches the last four digits of the credit card number. If the credit card number is found, the program prints a message saying so, otherwise it prints a message saying it was not found.

Escape Sequences

Escape sequences are special characters that are used to match specific characters or classes of characters in regular expressions. They are usually represented by a backslash () followed by a letter or sequence of letters.

Some common escape sequences in regular expressions include:

\n: Matches a newline character.
\t: Matches a tab character.
\r: Matches a carriage return character.
\d: Matches any digit character (0-9).

\w: Matches any word character (letters, digits, or underscore).
\s: Matches any whitespace character (spaces, tabs, newlines, etc.).
\b: Matches a word boundary.
.: Matches any character (except for a newline).

Escape sequences can also be used to match specific characters that have special meaning in regular expressions, such as the period (.) or the backslash () itself. To match these characters literally, you need to use an escape sequence.

For example, if you want to match a period character in a regular expression, you need to use the escape sequence . because the period has a special meaning in regular expressions (matches any character). Similarly, if you want to match a backslash character, you need to use the escape sequence \. Here are some examples of using escape sequences in Python:

- Matching a Tab Character (\t)

```
import re

text = "Name\tAge\tGender\nJohn\t25\tMale\n"
pattern = r"\t"

matches = re.findall(pattern, text)
print("Number of matches:", len(matches))
```

This code uses the regex pattern **r"\t"** to match a tab character (**\t**) in the **text** string. It then uses the **re.findall()** function to find all matches of the pattern in the text, and returns the number of matches found. In this case, the output will be "Number of matches: 2", as there are two tab characters in the text.

- Matching a Newline Character (\n)

```
import re

text = "This is the first line.\nThis is the second line."
pattern = r"\n"

matches = re.findall(pattern, text)
print("Number of matches:", len(matches))
```

This code uses the regex pattern **r"\n"** to match a newline character (**\n**) in the **text** string. It then uses the **re.findall()** function to find all matches of the pattern in the text, and returns the number of matches found. In this case, the output will be "Number of matches: 1", as there is one newline character in the text.

- Matching a Null Character (\0)

```
import re

text = "This is a null character: \x00"
pattern = r"\x00"

matches = re.findall(pattern, text)
print("Number of matches:", len(matches))
```

This code uses the regex pattern **r"\x00"** to match a null character (\0) in the **text** string. It then uses the **re.findall()** function to find all matches of the pattern in the text, and returns the number of matches found. In this case, the output will be "Number of matches: 1", as there is one null character in the text.

Groups in Regex

Groups in regular expressions allow you to match multiple characters as a single unit. This can be useful in cases where you need to apply a quantifier or a modifier to a specific set of characters within a larger string.

To create a group in regex, you can use parentheses () around the characters you want to group together. For example, the regex pattern **(ab)+** would match one or more occurrences of the sequence "ab".

Groups can also be used to capture the matched text for later use. Capturing groups are created by surrounding the group with parentheses, and the captured text can be referred to by its group number or name. For example, the regex pattern **(\d{3})-(\d{3})-(\d{4})** would capture a phone number in the format of XXX-XXX-XXXX, with the area code in group 1, the prefix in group 2, and the line number in group 3.

Groups can also be made non-capturing by using the **?:** syntax. For example, the regex pattern **(?:ab)+** would match one or more occurrences of the sequence "ab", but the matched text would not be captured as a group.

Some of the examples of groups usage are as follows:

- Extracting Groups from a Matched Pattern:

```python
import re

# regex pattern to match a date in YYYY-MM-DD format
pattern = r'(\d{4})-(\d{2})-(\d{2})'

# sample date
date = '2023-04-08'

# match the date using the regex pattern
match = re.search(pattern, date)

if match:
    # access groups using group method and group index
    year = match.group(1)
    month = match.group(2)
    day = match.group(3)
    print('Year:', year)
    print('Month:', month)
    print('Day:', day)
else:
    print('Date not found')
```

In this example, the regular expression **(\d{4})-(\d{2})-(\d{2})** matches a date string in the format of **YYYY-MM-DD**, where **(\d{4})** represents the group for the year, **(\d{2})** represents the group for the month, and **(\d{2})** represents the group for the day. The **search** function of the **re** module is used to find the first occurrence of the pattern in the **date** string. The **group** method of the resulting **Match** object is used to access the matched groups by their index positions.

- Using Named Groups for Easier Access:

```python
import re

# regex pattern to match a name in first name-last name format
pattern = r'(?P<first>\w+)\s(?P<last>\w+)'

# sample name
name = 'John Doe'

# match the name using the regex pattern
match = re.search(pattern, name)

if match:
    # access named groups using group method and group name
    first_name = match.group('first')
    last_name = match.group('last')
    print('First Name:', first_name)
    print('Last Name:', last_name)
else:
    print('Name not found')
```

In this example, the regular expression **(?P<first>\w+)\s(?P<last>\w+)** matches a name string in the format of **first name-last name**, where **(?P<first>\w+)** represents the named group for the first name and **(?P<last>\w+)** represents the named group for the last name. The **search** function of the **re** module is used to find the first occurrence of the pattern in the **name** string. The **group** method of the resulting **Match** object is used to access the matched named groups by their group names.

- Using Group References for Matching Repeated Patterns:

```
import re

# regex pattern to match a repeated word
pattern = r'(\b\w+\b)\s+\1'

# sample string
text = 'hello hello world world'

# find repeated words using the regex pattern
matches = re.findall(pattern, text)

if matches:
    print('Repeated words found:', matches)
else:
    print('No repeated words found')
```

In this example, the regular expression **(\b\w+\b)\s+\1** matches a repeated word in a string, where **(\b\w+\b)** represents the group for a word and **\s+\1** represents a group reference to the first group, ensuring that the matched word is repeated. The **findall** function of the **re** module is used to find all non-overlapping occurrences of the pattern in the **text** string, returning a list of matched groups.

Substitution or Replacement Metacharacters

Substitution or replacement metacharacters in regular expressions allow you to perform search and replace operations on text. These metacharacters are used in conjunction with the **re.sub()** function in Python, which replaces all occurrences of a pattern in a string with a new substring.

Here are some commonly used substitution or replacement metacharacters in regex:

$&: Inserts the entire matched string.

$n: Inserts the nth captured group from the pattern. For example, **$1** will insert the first captured group, **$2** will insert the second captured group, and so on.

$': Inserts the part of the string after the matched substring.

'$": Inserts the part of the string before the matched substring.

$+: Inserts the last captured group.

\g: This is used to specify the group number to be replaced. For example, \g<1> represents the first group in the pattern.

\g<name>: This is used to specify the group name to be replaced. For example, \g<name> represents the group named 'name' in the pattern.

Here are some examples of usage of replacement metacharacters:

- Replacing a Word with Another Word:

```
import re

text = 'The quick brown fox jumps over the lazy dog.'
pattern = r'brown'
replacement = 'red'

new_text = re.sub(pattern, replacement, text)

print(new_text)
```

In this example, we're using the **sub** function to reverse the order of two words ("brown" and "fox") in the given text. The regular expression used here matches two words separated by whitespace and captures them in separate groups. The replacement string **\2 \1** swaps the order of the captured groups.

- Replacing with a Dynamic Value Based on a Captured Group:

```
import re

text = 'Today is 2023-04-09'
pattern = r'(\d{4})-(\d{2})-(\d{2})'
replacement = r'\2/\3/\1'

new_text = re.sub(pattern, replacement, text)

print(new_text)
```

In this example, we're using the **sub** function to replace a date in the format "YYYY-MM-DD" with a date in the format "MM/DD/YYYY". The regular expression used here matches the date in the first format and captures the year, month, and day in separate groups. The replacement string \2/\3/\1 rearranges the captured groups to match the desired format.

Assertions

In regular expressions, assertions are zero-width patterns that match a position in the input string rather than a character. They are used to impose certain conditions on the match without including the actual characters in the match result. Following are the types of assertions in regex:

i. Positive Lookahead: This assertion matches a pattern only if it is followed by another pattern. It is represented by the syntax **(?=pattern)**. For example, the regex pattern **foo(?=bar)** would match the substring **foo** only if it is followed by **bar**. An example of positive lookahead assertion would be:

```
import re

# regex pattern with a positive lookahead assertion
pattern = r'\d+(?= dollars)'

# sample text to search for the pattern
text = 'I have 100 dollars in my wallet.'

# search for the pattern using the regex
match = re.search(pattern, text)

# print the match
if match:
    print('Match found:', match.group())
else:
    print('Match not found')
```

In this example, the pattern **\d+** matches one or more digits, and **(?= dollars)** is a positive lookahead assertion that matches only if the pattern is followed by the string " dollars". Therefore, the code should output **Match found: 100**.

ii. Negative Lookahead: This assertion matches a pattern only if it is not followed by another pattern. It is represented by the syntax **(?!pattern)**. For example, the regex pattern **foo(?!bar)** would match the substring **foo** only if it is not followed by **bar**. An example of negative lookahead assertion would be:

```python
import re

# regex pattern with a negative lookahead assertion
pattern = r'\d+(?! dollars)'

# sample text to search for the pattern
text = 'I have 100 euros in my wallet.'

# search for the pattern using the regex
match = re.search(pattern, text)

# print the match
if match:
    print('Match found:', match.group())
else:
    print('Match not found')
```

In this example, the pattern \d+ matches one or more digits, and (?! dollars) is a negative lookahead assertion that matches only if the pattern is not followed by the string " dollars". Therefore, the code should output Match found: 100.

iii. **Positive Lookbehind:** This assertion matches a pattern only if it is preceded by another pattern. It is represented by the syntax (?<=pattern). For example, the regex pattern (?<=foo)bar would match the substring bar only if it is preceded by foo. An example of positive lookbehind assertion would be:

```python
import re

# regex pattern with a positive lookbehind assertion
pattern = r'(?<=\$\d{3}\.)\d{2}'

# sample text to search for the pattern
text = 'The price is $123.45.'

# search for the pattern using the regex
match = re.search(pattern, text)

# print the match
if match:
    print('Match found:', match.group())
else:
    print('Match not found')
```

In this example, **(?<=\$\d{3}\.)** is a positive lookbehind assertion that matches only if the pattern is preceded by a dollar sign followed by exactly three digits and a dot. Therefore, the pattern **\d{2}** matches only the two digits after the dot. Therefore, the code should output **Match found: 45**.

iv. **Negative Lookbehind:** This assertion matches a pattern only if it is not preceded by another pattern. It is represented by the syntax (?<!pattern). For example, the regex pattern (?<!foo)bar would match the substring

bar only if it is not preceded by foo. An example of negative lookbehind assertion would be:

```python
import re

# regex pattern with a negative lookbehind assertion
pattern = r'(?<!\$)\d{2}'

# sample text to search for the pattern
text = 'The price is 123.45 dollars.'

# search for the pattern using the regex
match = re.search(pattern, text)

# print the match
if match:
    print('Match found:', match.group())
else:
    print('Match not found')
```

In this example, **(?<!\$)** is a negative lookbehind assertion that matches only if the pattern is not preceded by a dollar sign. Therefore, the pattern **\d{2}** matches only the two digits in the price (45). Therefore, the code should output **Match found: 45**.

v. Word boundary in regex matches the boundary between a word character (as defined by the **\w** metacharacter) and a non-word character. It can be denoted by the **\b** metacharacter. An example of word boundary would be:

```python
import re

# regex pattern to match the word 'cat' only when it appears as
a standalone word
pattern = r'\bcat\b'

# sample text
text = 'The cat is sitting on the mat.'

# search for the word 'cat' using the regex pattern
match = re.search(pattern, text)

if match:
    print('Match found:', match.group())
else:
    print('Match not found')
```

In the above example, we use the **\b** metacharacter to specify a word boundary. This ensures that the word 'cat' is matched only when it appears as a standalone word and not as part of a larger word (e.g., 'catastrophe' or 'category'). The regex pattern **\bcat\b** matches the string 'cat' only when it is surrounded by word boundaries. When we run the code, the **re.search()** function finds a match for the pattern in the sample text and returns a match object. The **if** statement checks if a match was found and prints the result accordingly.

A negative word boundary is used very rarely, but an example here would be relevant.

```python
import re

# regex pattern to match the word 'cat' only if it's not
preceded by a letter or digit
pattern = r'\Bcat\b'

# sample text to search for the pattern
text = 'The cat in the cathedral is not a caterpillar.'

# find all occurrences of the pattern in the text
matches = re.findall(pattern, text)

# print the matches
print(matches)
```

In this example, the regex pattern **\Bcat\b** uses the negative word boundary **\B** to match the word 'cat' only if it's not preceded or followed by a letter or digit. The **re.findall()** function is used to find all occurrences of the pattern in the given **text**. When this code is executed, it will output the following result:

```
['cat']
```

Since there is only one occurrence of the word 'cat' in the text that matches the pattern, the output is a list with only one element.

vi. Comment in regex is denoted by the **(?#comment)** syntax. It allows adding comments to the regex pattern without affecting the pattern matching. The text inside the **(?#)** will be ignored by the regex engine. In Python, the triple-quoted strings (''' or """) are often used to create multiline string literals. While these can be used to include comments in code, they are not considered as an official commenting syntax in regular expressions. However, they can be used to add comments in regular expressions if the regex is compiled with the **re.VERBOSE** flag. In this case, the comments are ignored by the regex engine. An example would be:

```
import re

# regex pattern to match a valid email address
pattern = r'''
    ^                              # start of string
    [a-zA-Z0-9._%+-]+              # username
    @                              # @ symbol
    [a-zA-Z0-9.-]+\.[a-zA-Z]{2,}   # domain name
    $                              # end of string
'''

# sample email address
email = 'test@example.com'

# match the email address using the regex pattern
match = re.search(pattern, email, re.VERBOSE)

if match:
    print('Valid email address found:', match.group())
else:
    print('Invalid email address')
```

In this example, we're using the **re.VERBOSE** flag to enable comments in the regular expression. The **pattern** variable contains the regular expression for matching a valid email address, which includes comments to explain each part of the pattern.

The regular expression starts with ^ to indicate the beginning of the string, followed by a pattern to match the username part of the email address. The @ symbol is matched literally, followed by a pattern to match the domain name part of the email address. The regular expression ends with $ to indicate the end of the string.

By using comments in the regular expression, we can make it easier to read and understand the pattern. The **re.VERBOSE** flag allows us to include comments and whitespace in the pattern without affecting the matching of the regular expression.

Regular expressions, often abbreviated as regex, serve as an instrumental apparatus for investigating and manipulating textual data within the Python programming environment. This mechanism enables rapid and efficient extraction of data from text files, the ability to comb through extensive documents, and the alteration of strings to comply with specific patterns. While the syntax of regular expressions may initially appear complex and intimidating, it ultimately evolves into an intuitive tool for text processing, given sufficient practice and exposure.

To facilitate the initiation into Python's implementation of regular expressions, this document provides an array of examples elucidating common regular expression patterns and functions. These instances span diverse text processing endeavors such as pattern matching, grouping, substitution, and assertions. Moreover, the text delves into pivotal concepts within regular expressions, including metacharacters, quantifiers, character classes, and word boundaries.

The concluding portion of this chapter includes a comprehensive cheat sheet of regular expression functions as a resource. This serves as a handy reference guide for the most prevalently employed regular expression functions within Python, complemented with examples of their application. This resource is intended to enhance the reader's comfort with regular expressions, thereby fostering proficiency in their application.

Regular Expressions Cheat Sheet

REGEX	DESCRIPTION
Literals and Metacharacters	
^	The start of a string
$	The end of a string
.	Wildcard which matches any character, except newline (\n)
|	Matches a specific character or group of characters on either side (e.g., a|b corresponds to a or b)
\	Used to escape a special character
a	The character "a"
ab	The string "ab"
Quantifiers	
*	Used to match 0 or more of the previous (e.g., xy*z could correspond to "xz", "xyz", "xyyz", etc.)
?	Matches 0 or 1 of the previous
+	Matches 1 or more of the previous
{5}	Matches exactly 5
{5, 10}	Matches everything between 5–10
Character Classes	
\s	Matches a whitespace character
\S	Matches a non-whitespace character
\w	Matches a word character
\W	Matches a non-word character
\d	Matches one digit
\D	Matches one non-digit
[\b]	A backspace character

(continued)

(continued)

REGEX	DESCRIPTION
\c	A control character
Escape Sequences	
\n	Matches a newline
\t	Matches a tab
\r	Matches a carriage return
\ZZZ	Matches octal character ZZZ
\xZZ	Matches hex character ZZ
\0	A null character
\v	A vertical tab
Groups	
(xyz)	Grouping of characters
(?:xyz)	Non-capturing group of characters
[xyz]	Matches a range of characters (e.g., x or y or z)
[^xyz]	Matches a character other than x or y or z
[a-q]	Matches a character from within a specified range
[0-7]	Matches a digit from within a specified range
Substitution or Replacement Metacharacters	
$'	Insert before matched string
$'	Insert after matched string
$+	Insert last matched
$&	Insert entire match
$n	Insert nth captured group
Assertions	
(?=xyz)	Positive lookahead
(?!xyz)	Negative lookahead
?!=or ?<!	Negative lookbehind
\b	Word Boundary (usually a position between \w and \W)
?#	Comment

4

Important Python Libraries

In recent years, Python has gained significant popularity in the programming world, primarily due to its extensive and versatile ecosystem of libraries. These libraries have enabled Python's use across a wide range of industries, such as software engineering, data science, and machine learning. Consequently, an increasing number of aspiring programmers and industry professionals are adopting Python as their primary programming language.

In this chapter, we will examine some of the most prominent and widely used Python libraries employed by academic researchers in the fields of accounting and finance. The focus will be on providing a comprehensive understanding of the benefits these libraries offer while the practical applications of these libraries in addressing real-world research challenges will be covered throughout the rest of the book.

Library

In programming, a library is a pre-written collection of utility methods, classes, and modules that can be used to perform specific tasks without having to create functionalities from scratch. The APIs of libraries have a narrow scope, such as Strings, Input/Output, and Sockets, and require fewer dependencies, making them highly reusable. Code reusability is the primary reason for using libraries, as they allow developers to avoid reinventing the wheel and instead focus on solving the actual problem at hand. Libraries are like software installed on a computer, and their utilities can be utilized in multiple applications, enabling developers to achieve more in less time and with fewer dependencies.

Installing a library in Python is a simple process that can be done using the pip command, which is a package installer for Python. Here are the steps to install a library in Python:

1. Open your command prompt or terminal and navigate to the directory where your Python environment is installed.
2. Type the following command to install the library:

```
pip install library_name
```

Replace 'library_name' with the name of the library you want to install.

3. Press Enter to execute the command. This will download and install the library along with its dependencies.

Once the installation is complete, you can start using the library in your Python code. To import the library, simply add the following line at the beginning of your Python script:

```
import library_name
```

Replace 'library_name' with the name of the library you have installed.

Note that some libraries may require additional dependencies or installation steps. You can find more information about a specific library's installation process in its documentation. If you encounter any issues installing a library using pip, you can try running the command prompt as an administrator as a troubleshooting step. If the issue persists, you can seek help from the library's documentation or online forums to resolve the issue.

For the purpose of this discussion, I divide Python libraries that are useful for academic researchers in accounting and finance into five functional categories. It is pertinent to mention that this categorization is only for the purpose of discussion. More often a library does more than one function and, therefore, may fall into more than one category.

Data Access Libraries

Data access libraries play a critical role in enabling academic researchers in accounting and finance to collect, access, and process large amounts of data from various sources. These libraries provide researchers with the tools and functionalities they need to access data from sources such as databases, web pages, and APIs. By using data access libraries, researchers can automate the process of collecting data, which can save them time and effort. Moreover, data access libraries allow researchers to collect data from a wide range of sources, which can help them gain insights into different aspects of their research topics. Overall, data access libraries are an essential component of any academic researcher's toolkit. Here are some important libraries for data access that could be useful for academic research in accounting and finance:

BeautifulSoup

BeautifulSoup is a widely used Python library designed to facilitate the process of parsing, navigating, and extracting information from HTML and XML documents. Developed by Leonard Richardson and maintained by the community, BeautifulSoup simplifies web scraping tasks by providing intuitive and easy-to-use methods to parse and manipulate HTML structures. It can handle both well-formed and poorly-structured HTML, making it a robust tool for web scraping. By providing a friendly interface to navigate and search the parsed document, BeautifulSoup allows users to efficiently extract relevant data.

To install BeautifulSoup, use the following command in your terminal or command prompt:

```
pip install beautifulsoup4
```

Additionally, BeautifulSoup requires a parser to function. The most commonly used parser is 'lxml', which can be installed using the following command:

```
pip install lxml
```

To begin using BeautifulSoup, import the necessary libraries and create a new BeautifulSoup object:

```python
from bs4 import BeautifulSoup
import requests

url = 'https://example.com'
response = requests.get(url)
soup = BeautifulSoup(response.content, 'lxml')
```

The **BeautifulSoup** constructor takes two arguments: the HTML content (in this case, the content of the response object) and the name of the parser ('lxml').

BeautifulSoup provides various methods to navigate and search the HTML tree:

- Accessing tags: To access a tag, simply treat it like an attribute:

```
title = soup.title
```

- Accessing tag attributes: To access a tag's attributes, treat them like a dictionary:

```
link = soup.a['href']
```

- Searching for tags: The **find()** and **find_all()** methods allow you to search for tags based on their name, attributes, or content:

```
all_links = soup.find_all('a')
specific_links = soup.find_all('a', class_='example_class')
```

Once you've located the desired tags, you can extract their text content using the **.text** or **.get_text()** methods:

```
title_text = soup.title.text
```

BeautifulSoup offers numerous advantages that make it a popular choice for web scraping tasks. Its ability to handle both well-formed and poorly-structured HTML allows developers to work with a wide range of websites, even those that may not adhere to best coding practices. Additionally, the library is easy to learn and use, with a user-friendly interface and intuitive methods for parsing, navigating, and extracting information from HTML documents. Another notable advantage of BeautifulSoup is its flexibility. It works seamlessly with various HTML parsers, such as 'lxml' and Python's built-in 'html.parser', providing users with multiple options based on their specific needs. This flexibility extends to its searching capabilities, which allow users to search for tags based on their name, attributes, or content, enabling the efficient extraction of relevant information. BeautifulSoup is also compatible with a wide range of Python versions, ensuring its accessibility to developers working with different iterations of the language. Furthermore, as an open-source project, BeautifulSoup benefits from an active community that continually contributes to its development, maintenance, and improvement, ensuring that the library remains up-to-date and reliable.

One of the most significant drawbacks of this library is its performance. The library is known to be slower than other alternatives, such as lxml, when it comes to parsing large or complex HTML documents. This can be particularly problematic for projects that require the parsing of massive amounts of data or real-time processing. Another limitation is that BeautifulSoup solely focuses on parsing HTML and XML documents. It does not provide built-in support for executing JavaScript or handling AJAX requests, which are common features in modern web applications. As a result, developers may need to utilize additional libraries, such as Selenium or Pyppeteer, to handle websites that rely heavily on JavaScript for their content. BeautifulSoup's simplicity, while an advantage in many cases, can also be considered a limitation. The library does not include advanced features or optimization techniques that more specialized scraping tools might offer. For example, BeautifulSoup does not provide built-in support for handling CAPTCHAs, managing cookies, or handling sessions, which may be necessary for some web scraping tasks. In these cases, developers may need to combine BeautifulSoup with other libraries or tools to achieve their desired results.

Requests

Requests is a popular and user-friendly Python library designed to simplify the process of making HTTP requests. It enables developers to send HTTP/1.1 requests and effortlessly interact with web services, making it a crucial component in the modern web development ecosystem. Developed by Kenneth Reitz, Requests has been widely adopted for its ease of use, reliability, and extensibility. Requests is not a part of the Python standard library, but it can be easily installed using pip, the Python package installer. To install Requests, simply run the following command:

```
pip install requests
```

Once installed, Requests enables you to send various types of HTTP requests, such as GET, POST, PUT, DELETE, and others. The most common request, the GET request, can be performed using the **get()** function:

```python
import requests

response = requests.get('https://api.example.com/data')
```

Requests returns a **Response** object containing server's response data. This object includes useful attributes, such as **status_code**, **headers**, and **text** (or **content** for binary data), as well as methods to decode JSON data:

```python
import requests

response = requests.get('https://api.example.com/data')

if response.status_code == 200:
    data = response.json()
    print("Data received:", data)
else:
    print("Request failed with status code:", response.status_code)
```

The Requests library makes it easy to send data along with your requests, specify custom headers, and control timeouts. For example, sending a JSON payload with a POST request:

```python
import requests
import json

headers = {'Content-Type': 'application/json'}
data = {'key': 'value'}
url = 'https://api.example.com/data'

response = requests.post(url, headers=headers, data=json.dumps(data), timeout=5)

if response.status_code == 201:
    print("Data created successfully")
else:
    print("Request failed with status code:", response.status_code)
```

Requests raises exceptions for certain types of network errors, such as timeouts, DNS resolution failures, or SSL certificate issues. To handle these exceptions gracefully, you can use a try-except block:

```
import requests
from requests.exceptions import RequestException

try:
    response = requests.get('https://api.example.com/data', timeout=3)
except RequestException as e:
    print("An error occurred:", e)
```

The Requests library has several advantages that make it a popular choice among Python developers. Its simplicity and ease of use are arguably its most significant benefits. The intuitive interface abstracts the complexities of handling HTTP requests, allowing developers to quickly send and receive data from web services. This user-friendly design makes it easy for beginners to get started with web development in Python. Another advantage is its reliability, as the library is well-maintained and regularly updated, ensuring it remains compatible with current web standards. Moreover, Requests is highly extensible, with a wide range of built-in features and support for custom middleware, making it suitable for a variety of web development tasks.

Despite its many benefits, there are a few disadvantages to using the Requests library. One potential drawback is that Requests only supports synchronous operations, meaning that it blocks the execution of the program while waiting for a response from the server. This can lead to performance issues, particularly in applications that require high levels of concurrency or deal with numerous simultaneous requests. In such scenarios, using an asynchronous library like aiohttp or httpx might be a better choice. Another disadvantage is that Requests, being an external library, adds an additional dependency to a project. While this is generally not a significant concern, it can be problematic in certain contexts where minimizing dependencies is crucial, such as in a resource-constrained environment or when deploying a serverless application. Also, the Requests library is limited to the HTTP/1.1 protocol, meaning it does not support newer protocols like HTTP/2 or HTTP/3, which offer improved performance and features.

Scrapy

Scrapy is an open-source Python library that allows users to extract and process structured data from websites. It was developed by Scrapinghub in

2008 and has since become one of the most popular web scraping frameworks among data scientists and developers alike. Scrapy offers a robust and scalable solution to web scraping, allowing users to navigate and collect data from even the most complex websites.

Before diving into Scrapy, it is essential to install the library and its dependencies. The recommended way to do this is through the Python Package Installer (pip):

```
pip install scrapy
```

Once installed, you can start building your first Scrapy project. To create a new Scrapy project, run the following command:

```
scrapy startproject my_project
```

This command will generate a project structure with essential files and directories for your web scraping project.

Scrapy's Components:

- Spiders: Spiders are the core component of a Scrapy project, responsible for defining how a website should be scraped and how the data should be extracted. Users write custom spider classes, inheriting from Scrapy's Spider class, to define the scraping behavior.
- Items: Items are data containers used to store the extracted data. They are Python classes with fields defined to store specific data points from the target website.
- Middlewares: Middlewares are hooks that allow you to process requests and responses. They are useful for handling redirects, user-agent spoofing, and other customizations.
- Pipelines: Pipelines are responsible for processing the data after it has been extracted. They can be used to perform tasks like data validation, cleaning, and storage.
- Settings: The settings file is where you can configure various aspects of your Scrapy project, such as the user agent, download delay, and other settings.

Scrapy has numerous advantages that make it a popular choice among developers and data scientists. One of its primary benefits is its speed and efficiency. Scrapy's asynchronous networking library, Twisted, allows it to handle multiple requests simultaneously, enabling the rapid extraction of data from multiple web pages. Additionally, Scrapy's modular architecture encourages code reusability, making it easier to maintain and share code across multiple projects or within a team. Another advantage of Scrapy is its extensibility. Users can extend the library's functionality by creating custom middlewares, pipelines, and extensions to meet the specific requirements of various web scraping scenarios. Furthermore, Scrapy comes with built-in CSS and XPath selectors, which simplifies the process of targeting and extracting specific data elements from web pages with precision.

Despite its many advantages, Scrapy also has some drawbacks. One potential issue is its relatively steep learning curve for those new to web scraping or Python. Users may need to invest time in understanding Scrapy's components, such as spiders, items, middlewares, and pipelines, before being able to efficiently build their projects. Another disadvantage is that Scrapy may be considered overkill for small-scale or straightforward web scraping tasks. In such cases, simpler libraries like BeautifulSoup or Requests might be more suitable and easier to use. Moreover, Scrapy's reliance on the Twisted networking library might also pose a challenge for users unfamiliar with asynchronous programming, potentially increasing the complexity of certain projects. Moreover, while Scrapy is excellent for structured data extraction, it may not be the best option for dealing with JavaScript-heavy websites or those that require rendering. In these cases, using a headless browser such as Selenium or integrating Scrapy with Splash can be necessary, adding an extra layer of complexity to the web scraping process.

Data Manipulation Libraries

Data manipulation is a critical step in academic research, particularly in accounting and finance, as researchers often work with large and complex datasets. Data manipulation libraries provide researchers with tools to access, clean, filter, transform, and aggregate data, making it easier to draw insights and conclusions. These libraries enable researchers to perform complex data manipulations efficiently, freeing up their time to focus on more important aspects of their research. By using data manipulation libraries, researchers can reduce the likelihood of errors and produce more accurate and meaningful results. Some of key data manipulation libraries are as follows:

Pandas

The Pandas library is an open-source data analysis and data manipulation library for Python. Developed by Wes McKinney in 2008, it has since become an indispensable tool for data scientists, analysts, and programmers who work with data in Python. Pandas provides an easy-to-use interface for handling data in various formats, such as CSV, Excel, JSON, and SQL databases. With its powerful data structures and functions, the library greatly simplifies data processing and analysis tasks.

Key features of Pandas are:

- Data Structures: Pandas introduces two primary data structures, namely Series and DataFrame. A Series is a one-dimensional labeled array, whereas a DataFrame is a two-dimensional labeled data structure with columns of potentially different types. These data structures are designed to handle a wide variety of data types and provide a range of functionalities for data manipulation.
- Handling Missing Data: Pandas offers robust support for handling missing data. It represents missing values as NaN (Not a Number) and provides methods to detect, drop, and fill these missing values.
- Data Alignment and Merging: Pandas facilitates data alignment and merging by providing functions such as 'merge', 'concat', and 'join'. These functions allow users to combine multiple DataFrames or Series based on various conditions, including indices, columns, or values.
- Reshaping and Pivoting: With Pandas, you can easily reshape your data by stacking, unstacking, or pivoting your DataFrame. This makes it simple to transform data into the desired format for further analysis or visualization.
- Time Series Functionality: Pandas provides extensive support for working with time series data. It offers time-based indexing, resampling, and frequency conversion, making it an ideal library for time series analysis.
- Performance: Pandas is built on top of NumPy, a powerful numerical library for Python, which significantly enhances its performance. The library also supports Cython, allowing for further performance improvements when needed.

Most recent Anaconda distribution of Python comes with Pandas preinstalled. If you don't have Pandas preinstalled, you can use the pip package manager as follows:

```
pip install pandas
```

To start using Pandas, import the library and create a DataFrame or Series:

```python
import pandas as pd

# Creating a Series
data = pd.Series([1, 2, 3, 4])

# Creating a DataFrame
data = {'Column1': [1, 2, 3, 4],
        'Column2': ['A', 'B', 'C', 'D']}
df = pd.DataFrame(data)
```

Pandas provide for easy handling of missing values. The missing data can be filled using interpolation. The relevant code would be:

```
# Interpolate missing values using linear interpolation
interpolated_data = original_data.interpolate(method='linear')
print(interpolated_data.head())
```

Pandas offers following methods of interpolation:

- **linear**: Fill missing values using linear interpolation, which connects two points with a straight line. This is the default method and is suitable for data with a linear trend.
- **time**: Fill missing values based on the time index, taking into account the time distance between points. This method is appropriate for time series data with irregular intervals.
- **index, values**: Use the index or values, respectively, to perform interpolation.
- **nearest**: Fill missing values with the nearest available data point.
- **polynomial**: Fill missing values using a polynomial function of a specified order.
- **spline**: Fill missing values using a spline interpolation of a specified order.
- **pad, ffill**: Fill missing values using the previous value in the time series (forward fill).
- **bfill, backfill**: Fill missing values using the next value in the time series (backward fill).
- **from_derivatives**: Interpolate based on the first derivative of the data points.
- **piecewise_polynomial**: Fill missing values using a piecewise polynomial of a specified order.
- **cubicspline**: Fill missing values using a cubic spline interpolation.
- **akima**: Fill missing values using the Akima interpolator.
- **pchip**: Fill missing values using the PCHIP (Piecewise Cubic Hermite Interpolating Polynomial) method.
- **quadratic**: Fill missing values using a quadratic function.

- **cubic**: Fill missing values using a cubic function.

Pandas offers several advantages that make it a popular choice for data manipulation and analysis in Python. One of its key strengths is the provision of powerful data structures like Series and DataFrame, which simplify handling complex data types and operations. These data structures come with a wide range of built-in functions for data cleaning, filtering, sorting, and aggregation, resulting in more efficient and readable code. Another advantage of Pandas is its ability to seamlessly handle missing data. It represents missing values as NaN and offers multiple functions to detect, drop, or fill them, making data cleaning and preprocessing easier. Furthermore, Pandas boasts strong support for time series data, which is crucial for various applications in finance, economics, and other fields. The library's time-based indexing, resampling, and frequency conversion features make it an ideal choice for time series analysis.

On the other hand, there are some disadvantages to using Pandas. One primary concern is its memory consumption, as Pandas DataFrames can consume a significant amount of memory, especially when working with large datasets. This can lead to performance issues and may require users to resort to other libraries or techniques for handling big data, such as Dask or PySpark. Another drawback is that, while Pandas is highly versatile, it might not be the best choice for certain specialized tasks. For instance, when working with multidimensional arrays or tensors, users may find NumPy or TensorFlow more suitable. Additionally, for highly performance-sensitive applications, Pandas may not be the optimal choice due to its inherent overhead compared to lower-level libraries like NumPy. In such cases, users might need to resort to Cython or Numba for further performance optimization.

NumPy

NumPy, short for Numerical Python, is a powerful open-source library that serves as the backbone for a wide array of scientific computing and data analysis tasks in Python. Developed by Travis Oliphant in 2005, NumPy provides a comprehensive suite of tools for performing mathematical operations on large, multidimensional arrays and matrices. With its efficient array manipulation capabilities and optimized functions, NumPy has become an indispensable resource for researchers, data scientists, and programmers alike.

Core features of NumPy are:

- N-dimensional Arrays: The primary data structure in NumPy is the ndarray (N-dimensional array), which is an efficient, homogeneous, and flexible array object that can store large amounts of numerical data. The elements of an ndarray must be of the same data type, which enhances the efficiency and speed of operations.
- Array Manipulation: NumPy provides a wide range of functions to manipulate arrays, including reshaping, slicing, and concatenating. These functions allow users to quickly and easily modify the structure and content of their arrays to fit their specific needs.
- Mathematical Operations: NumPy offers a comprehensive set of mathematical functions that operate on ndarrays, including element-wise operations (such as addition and multiplication), linear algebra operations (like dot products and matrix inversion), and statistical functions (like mean, median, and standard deviation). By employing vectorization and broadcasting, NumPy ensures that these operations are efficient and optimized.
- Interoperability: NumPy can seamlessly interface with other Python libraries like SciPy, Matplotlib, and Pandas, making it a foundational component in the Python scientific computing ecosystem.

Most recent Anaconda distribution of Python comes with NumPy preinstalled. If you don't have Pandas preinstalled, you can use the pip package manager as follows:

```
pip install numpy
```

After installing NumPy, import it into your Python script or notebook with:

```
import numpy as np
```

Creating a NumPy array is as simple as passing a Python list or tuple to the **numpy.array()** function:

```
import numpy as np

# Creating a 1-dimensional array
arr1d = np.array([1, 2, 3, 4, 5])

# Creating a 2-dimensional array
arr2d = np.array([[1, 2, 3], [4, 5, 6], [7, 8, 9]])
```

NumPy offers several key advantages that make it an indispensable tool for numerical computing in Python. First, its high-performance N-dimensional array object, ndarray, enables efficient array manipulation and storage of large volumes of data. This is particularly beneficial for users working with big datasets or complex mathematical operations. Second, NumPy's comprehensive set of mathematical functions and broadcasting capabilities allow for fast, vectorized computations that significantly reduce the need for explicit loops. This results in cleaner, more readable code and improved performance. Third, NumPy provides excellent interoperability with other scientific libraries, such as SciPy, Matplotlib, and Pandas. This seamless integration fosters a cohesive and powerful ecosystem for scientific computing and data analysis in Python.

Despite its numerous advantages, NumPy has certain limitations. One key drawback is that it requires the elements within an array to be of the same data type, restricting the flexibility of the data structure. This constraint may lead to increased memory usage when working with mixed data types, as users may need to create separate arrays for each type. Additionally, while NumPy offers a vast range of mathematical functions, it does not provide higher-level data manipulation or visualization tools, often necessitating the use of other libraries, such as Pandas or Matplotlib. Finally, NumPy's performance benefits primarily apply to numerical operations, making it less suitable for tasks involving non-numerical data or those that require extensive text processing. For these tasks, users may need to consider alternative libraries or data structures that better align with their specific requirements.

Dask

Dask is a powerful and flexible library in Python that provides parallel computing capabilities, enabling users to scale their data processing and analysis tasks. It allows users to process larger-than-memory datasets, parallelize computations, and optimize processing times for complex workflows. Dask is particularly useful for handling big data, machine learning, and scientific computing applications.

Dask is built on top of Python's built-in concurrent.futures library and consists of two primary components:

- Dynamic task scheduling: Dask breaks down complex computations into smaller tasks that can be executed concurrently. These tasks are represented as a directed acyclic graph (DAG), where each node represents a task and edges represent dependencies between tasks.
- Collections: These high-level abstractions represent large, partitioned datasets, such as Dask arrays (similar to NumPy arrays), Dask dataframes (similar to pandas dataframes), and Dask bags (similar to Python lists). Collections allow users to work with large datasets using familiar APIs.

Dask is a powerful library that offers several key features that make it a valuable tool for academic researchers in various fields. Dask's scalability is a major advantage, as it can scale computations from a single core on a laptop to thousands of cores on a distributed cluster. This flexibility allows users to scale their computations horizontally or vertically, depending on their specific needs. Additionally, Dask provides a familiar interface for working with large datasets, which closely mirrors popular libraries like NumPy and pandas. This makes it easy for users to adopt Dask without the need to learn a new API. Dask is also highly customizable, enabling users to define their own task graphs, schedulers, and parallel algorithms. Finally, Dask integrates seamlessly with other popular Python libraries, such as NumPy, pandas, and scikit-learn, making it easy to use alongside existing workflows. These key features make Dask a valuable tool for academic researchers in accounting and finance, enabling them to efficiently handle large and complex datasets and perform complex computations.

Dask provides three main schedulers for executing tasks concurrently: Local Threads, Local Processes, and Distributed. The Local Threads scheduler uses Python's built-in threading module and is lightweight, making it suitable for smaller-scale parallelism. The Local Processes scheduler, on the other hand, uses Python's multiprocessing module and takes advantage of multiple cores, providing better isolation between tasks but with potentially higher overhead. The Distributed scheduler uses the dask.distributed module and is designed to execute tasks concurrently on a cluster of machines. It offers advanced features such as real-time monitoring, data locality, and resilience to node failures, making it a powerful tool for large-scale parallelism.

Dask is a versatile Python library that is ideal for working with large and complex datasets. Its ability to parallelize tasks and efficiently handle intermediate results makes it a powerful tool for various data processing tasks. Researchers in accounting and finance can use Dask for data preprocessing and cleaning, machine learning and model training, image and signal processing, as well as running large-scale simulations and models. With its range of functionalities, Dask can help researchers save time and effort when working with large datasets and enable them to draw meaningful insights and conclusions from their research.

Data Visualization Libraries

Data visualization libraries are an essential tool for academic researchers in both accounting and finance. These libraries enable researchers to transform complex data into visualizations that are easy to understand and interpret. With data visualization libraries, researchers can create a wide range of charts, graphs, and other visualizations to represent their data, making it easier to identify patterns, trends, and relationships. Effective data visualization is critical for communicating research findings to colleagues, stakeholders, and the wider public. By using data visualization libraries, researchers can produce high-quality visualizations that are informative and engaging, making it easier for others to understand and appreciate their research. Data visualization libraries also allow researchers to customize their visualizations, enabling them to tailor their presentation to their audience and research goals.

Some popular data visualization libraries are described in detail in chapters on Data Visualization. Matplotlib, the most popular library, is described briefly here also:

Matplotlib

Matplotlib is a powerful, open-source, and widely used data visualization library in the Python programming language. It was created by John D. Hunter in 2003 to facilitate the creation of publication-quality 2D graphs and plots. The library offers a wide range of features, allowing users to create interactive visualizations, static images, and animations for various applications.

Matplotlib is a powerful data visualization library for Python that offers users a variety of features for creating high-quality plots and graphs. These features include versatility, which provides users with a large collection of visualization types to choose from, such as line plots, scatter plots, bar plots, histograms, and 3D plots. Customizability is also a key feature of Matplotlib, as it offers extensive options for modifying colors, line styles, axes, legends, and text annotations, allowing users to tailor their visualizations to their specific requirements. Another important feature of Matplotlib is its compatibility with other Python libraries like NumPy, Pandas, and Seaborn. This allows users to work with diverse data structures and perform advanced statistical analyses. Matplotlib is also cross-platform, supporting various operating systems such as Windows, macOS, and Linux, and offering support for multiple output formats such as PNG, PDF, and SVG. These features make Matplotlib a versatile and reliable library for data visualization in Python.

To start using Matplotlib in your Python project, install it using pip:

```
pip install matplotlib
```

Then, import the library, typically as "plt":

```
import matplotlib.pyplot as plt
```

To create a simple line plot, use the **plot()** function and specify the x and y data points:

```
import matplotlib.pyplot as plt

x = [0, 1, 2, 3, 4]
y = [0, 1, 4, 9, 16]

plt.plot(x, y)
plt.show()
```

This will display a simple line plot of the specified data points. You can easily customize your plots by modifying various parameters:

```
import matplotlib.pyplot as plt

x = [0, 1, 2, 3, 4]
y = [0, 1, 4, 9, 16]

plt.plot(x, y, color='red', linestyle='--', marker='o')
plt.xlabel('X-axis Label')
plt.ylabel('Y-axis Label')
plt.title('A Customized Line Plot')

plt.show()
```

This example demonstrates how to change the line color, style, and marker, as well as add axis labels and a title.

Despite its numerous advantages, Matplotlib has some drawbacks. The library's syntax can be complex and unintuitive, especially for beginners, leading to a steep learning curve. Moreover, the need for extensive customization to achieve desired visualizations can result in verbose code, which might hinder readability and maintainability. While Matplotlib is capable of producing 3D visualizations, its primary focus is on 2D graphics, limiting its applicability in certain scenarios. Furthermore, in comparison with modern web-based visualization libraries, such as Plotly and Bokeh, Matplotlib's interactivity features may be lacking, potentially making it less suitable for creating interactive dashboards and web applications.

Statistical Analysis Libraries

Statistical analysis libraries are essential for academic researchers in accounting and finance who need to analyze and draw insights from large amounts of data. These libraries provide a range of statistical functions and tools that enable researchers to perform various types of statistical analysis, such as hypothesis testing, regression analysis, and time series analysis. By using statistical analysis libraries, researchers can identify patterns and relationships in their data, make predictions, and test the significance of their findings. The functions and tools provided by statistical analysis libraries can save researchers a significant amount of time and effort, as they do not need to develop these functionalities from scratch.

Some of the popular statistical analysis libraries are as below:

SciPy

The SciPy library is an essential component of the Python ecosystem, specifically designed to cater to the needs of scientific computing. This library not only builds upon the core functionality of the NumPy package but also provides a comprehensive set of algorithms and tools for various scientific domains. In this subsection, we will briefly discuss the main features of the SciPy library and how it can be utilized effectively in a range of applications. Some of the core components and features of the SciPy library include:

- Linear algebra: SciPy provides a rich collection of linear algebra functions, such as solving systems of linear equations, finding eigenvalues and eigenvectors, and performing matrix decompositions.
- Optimization: The library offers a wide range of optimization algorithms for both constrained and unconstrained problems. These include root-finding, curve-fitting, and minimization algorithms.
- Signal processing: SciPy's signal module includes functions for filtering, convolution, and Fourier analysis. These tools can be used for analyzing and manipulating signals in various domains, such as audio and image processing.
- Integration: The library contains functions to perform numerical integration of functions and differential equations, including single and multiple integration, as well as solving initial value problems.
- Interpolation: SciPy offers a diverse set of interpolation techniques, such as spline interpolation, nearest-neighbor interpolation, and radial basis functions.

- Statistics: The library features a comprehensive set of statistical functions, including descriptive statistics, hypothesis testing, and probability distributions.
- Special functions: SciPy includes numerous special functions that are commonly used in mathematical physics and engineering, such as Bessel functions, gamma functions, and error functions.

To get started with SciPy, you first need to install the library using the following command:

```
pip install scipy
```

Once installed, you can import the library in your Python script using:

```
import scipy
```

You'll often need to import specific submodules for different tasks. For example, to use the optimization functions, you would import the **scipy.optimize** submodule:

```
import scipy.optimize
```

The SciPy library comes with an array of advantages, making it a popular choice for scientific computing in Python. One of its most notable strengths is the comprehensive collection of algorithms and functions it provides, catering to various scientific domains. As a result, users can tackle a wide range of problems, from linear algebra and optimization to signal processing and statistics, without having to search for or implement these algorithms from scratch. This saves both time and effort and reduces the likelihood of errors that might arise from manual implementation. SciPy is compatible with NumPy, which enables it to leverage the power of NumPy arrays for efficient numerical operations. This seamless integration ensures that users can fully exploit the benefits of the NumPy ecosystem while working with SciPy. Moreover, the library has an active community of developers and users who continuously contribute to its development, ensuring that it stays up-to-date with the latest advancements in the field. This collaborative environment fosters knowledge sharing, making it easier for users to find resources and support when needed.

However, the library's steep learning curve is one of the main disadvantages of SciPy, especially for those who are new to scientific computing or

Python. The vast array of functions and submodules can be overwhelming, and getting acquainted with the library might require significant time and effort. The library's extensive functionality may sometimes come at the cost of performance. While SciPy is generally considered efficient, it may not always be the most optimal choice for certain applications, especially those that require high-performance computing or specialized algorithms. SciPy relies on the NumPy ecosystem. Although this allows for seamless integration with NumPy, it may also lead to issues when dealing with other data structures or libraries that are not directly compatible with NumPy arrays. Users might need to spend additional time converting data structures or finding workarounds to make SciPy compatible with other tools in their workflow.

StatModels

StatModels is a powerful Python library that provides a comprehensive suite of statistical models and data exploration tools. Designed to be user-friendly and extensible, it caters to the needs of both novice and expert users. With a focus on linear models, time series analysis, and various other statistical methods, StatModels has become a popular choice for data analysts and statisticians alike.

To install StatModels, simply run the following command in your terminal or command prompt:

```
pip install statsmodels
```

Once installed, you can import the library in your Python script as follows:

```
import statsmodels.api as sm
```

StatModels provides an extensive range of linear regression models, including ordinary least squares (OLS), generalized least squares (GLS), and many more. These models can be used to analyze the relationship between one or multiple predictor variables and a response variable.

For example, to fit an OLS model, you can use the following code:

```
import numpy as np
import statsmodels.api as sm

X = np.random.random(100)
Y = 2 * X + np.random.normal(0, 0.1, 100)
X = sm.add_constant(X)

model = sm.OLS(Y, X)
results = model.fit()
print(results.summary())
```

The StatModels library offers a variety of tools for time series analysis, such as autoregression (AR), moving average (MA), and seasonal decomposition of time series (STL). These tools can help forecast future values, detect seasonality, and identify trends in your data.

For instance, to perform seasonal decomposition on a time series dataset, you can use the following code:

```
import numpy as np
import pandas as pd
import statsmodels.api as sm
import matplotlib.pyplot as plt

data = pd.read_csv('your_time_series_data.csv')
data.index = pd.to_datetime(data['date'])
data = data.drop(columns=['date'])

decomposition = sm.tsa.seasonal_decompose(data)
fig = decomposition.plot()
plt.show()
```

StatModels also supports a wide range of hypothesis testing methods, such as t-tests, chi-square tests, and F-tests, to help you make data-driven decisions and validate statistical assumptions.

For example, to perform a two-sample t-test, you can use the following code:

```
import numpy as np
import statsmodels.stats.weightstats as st

sample1 = np.random.normal(0, 1, 30)
sample2 = np.random.normal(0.5, 1, 30)

t_stat, p_value, df = st.ttest_ind(sample1, sample2)
print(f'T-statistic: {t_stat}, P-value: {p_value}, Degrees of freedom: {df}')
```

StatModels offers several advantages, one of which is its extensive range of statistical models and methods. This enables users to perform a wide variety of analyses, including linear regression, time series forecasting, and hypothesis testing, all within a single library. Additionally, StatModels is designed with user-friendliness in mind, making it accessible to both novice and experienced users. Its detailed documentation and extensive examples facilitate learning and encourage experimentation with different techniques. However, StatModels can be computationally expensive for very large datasets or when using complex models, which may make it unsuitable for certain real-time applications or when working with limited computational resources. Furthermore, while StatModels excels in traditional statistical methods, it lacks native support for more advanced machine learning techniques, such as deep learning and reinforcement learning. Users interested in these areas may need to rely on other libraries, such as TensorFlow or PyTorch, in conjunction with StatModels for a more comprehensive solution.

PyMC3

Bayesian Modeling with PyMC3 is a powerful library for probabilistic programming in Python. Developed and maintained by a team of open-source contributors, PyMC3 allows users to easily implement and fit Bayesian models to data, perform statistical inference, and estimate model parameters. The library leverages the power of Theano (and in later versions, TensorFlow) to perform efficient computations and optimization. PyMC3 is built on the idea of probabilistic modeling, which involves the use of probability distributions to model uncertainty in data and parameters. In Bayesian statistics, models are described as a combination of prior beliefs (prior distributions) and the likelihood of observing the data given the model parameters (likelihood functions).

One of the most notable features of PyMC3 is its intuitive model specification syntax, which allows users to specify models in a natural and easy-to-read manner. This makes it easy to define prior and likelihood functions using the library's extensive collection of probability distributions. Moreover, PyMC3 has automatic differentiation functionality, which automatically calculates gradients for model parameters using Theano or TensorFlow. This makes it easier to implement complex models without the need to manually compute derivatives. Additionally, PyMC3 implements a variety of Markov Chain Monte Carlo (MCMC) sampling methods, including the popular No-U-Turn Sampler (NUTS), Metropolis–Hastings, and Slice sampling. This makes it versatile and efficient in handling complex posterior distributions. PyMC3 also supports variational inference techniques such as Automatic Differentiation Variational Inference (ADVI), which can provide faster approximations to the posterior distributions, particularly in cases where MCMC methods are computationally intensive. Together, these features make PyMC3 a powerful tool for Bayesian statistical modeling and probabilistic machine learning in accounting and finance research.

To start using PyMC3, install the library using pip or conda:

```
pip install pymc3
```

or

```
conda install -c conda-forge pymc3
```

Here is a simple example of linear regression using PyMC3:

```python
import pymc3 as pm
import numpy as np

# Generate synthetic data
np.random.seed(42)
x = np.random.normal(size=100)
y = 3 * x + np.random.normal(size=100)

# Define the model
with pm.Model() as model:
    alpha = pm.Normal("alpha", mu=0, sd=10)
    beta = pm.Normal("beta", mu=0, sd=10)
    sigma = pm.HalfNormal("sigma", sd=1)

    mu = alpha + beta * x
    obs = pm.Normal("obs", mu=mu, sd=sigma, observed=y)

    # Perform MCMC sampling
    trace = pm.sample(2000, tune=1000, target_accept=0.95)

# Analyze the results
pm.summary(trace).round(2)
```

PyMC3 offers several advantages for users working with Bayesian modeling and probabilistic programming. The library's intuitive context-based syntax simplifies the process of specifying complex models, making it accessible to beginners and experts alike. With its extensive collection of probability distributions and automatic differentiation capabilities, PyMC3 streamlines model implementation by reducing the need for manual gradient calculations. Moreover, the library's support for advanced sampling methods, such as NUTS, and variational inference techniques, such as ADVI, allows for efficient and versatile handling of complex posterior distributions.

One of the primary drawbacks of PyMC3 is its reliance on Theano, a discontinued library, for its backend in earlier versions. While this issue has been addressed in later versions by integrating TensorFlow as the backend, it may still present challenges for users working with legacy code. Furthermore, as PyMC3 is a specialized library for Bayesian modeling, it may not be the best choice for users seeking to perform frequentist statistical analysis or those requiring a broader range of machine learning tools. In such cases, other libraries like SciPy or scikit-learn might be more appropriate.

Machine Learning Libraries

Machine learning libraries are an essential category of Python libraries for accounting and finance academic researchers that employ machine learning techniques for their research. These libraries offer a range of tools and algorithms that enable researchers to develop predictive models, cluster data, and perform other machine learning tasks.

Some common examples of machine learning libraries are as follows:

Scikit-Learn

Scikit-learn is a widely used, open-source Python library that provides efficient tools for data mining, data analysis, and machine learning. It is built on top of NumPy, SciPy, and matplotlib and offers a simple and efficient way to implement various machine learning algorithms, preprocess data, and evaluate model performance. Scikit-learn is well-suited for beginners as well as experienced users due to its user-friendly interface, extensive documentation, and active community.

Scikit-learn offers a comprehensive suite of functionalities, including:

- Supervised Learning Algorithms: Scikit-learn provides a wide range of supervised learning algorithms, such as linear regression, support vector machines, decision trees, random forests, and k-nearest neighbors, among others.
- Unsupervised Learning Algorithms: The library also supports unsupervised learning algorithms, including clustering (e.g., K-means, DBSCAN), dimensionality reduction (e.g., PCA, t-SNE), and anomaly detection.
- Preprocessing: Scikit-learn has built-in functions for handling missing data, feature scaling, and encoding categorical variables, making it easy to preprocess data for machine learning tasks.
- Model Selection and Evaluation: The library offers tools for model selection (e.g., train-test split, cross-validation), hyperparameter tuning (e.g.,

GridSearchCV, RandomizedSearchCV), and evaluation metrics (e.g., accuracy, precision, recall, F1-score).
- Pipelines: Scikit-learn's pipeline feature enables users to streamline their workflows by chaining multiple preprocessing steps and a final estimator, making it easy to maintain and reuse code.

Scikit-learn can be installed using the package manager pip:

```
pip install scikit-learn
```

Here is a simple example that demonstrates how to use scikit-learn for a classification task:

```python
from sklearn import datasets
from sklearn.model_selection import train_test_split
from sklearn.preprocessing import StandardScaler
from sklearn.neighbors import KNeighborsClassifier
from sklearn.metrics import accuracy_score

# Load the iris dataset
iris = datasets.load_iris()
X = iris.data
y = iris.target

# Split the dataset into training and testing sets
X_train, X_test, y_train, y_test = train_test_split(X, y,
test_size=0.2, random_state=42)

# Preprocess the data by scaling features
scaler = StandardScaler()
X_train_scaled = scaler.fit_transform(X_train)
X_test_scaled = scaler.transform(X_test)

# Train the k-nearest neighbors classifier
clf = KNeighborsClassifier(n_neighbors=3)
clf.fit(X_train_scaled, y_train)

# Make predictions on the test set
y_pred = clf.predict(X_test_scaled)

# Calculate the accuracy of the model
accuracy = accuracy_score(y_test, y_pred)
print(f"Accuracy: {accuracy:.2f}")
```

This code snippet highlights the essential steps in a typical machine learning workflow with scikit-learn. It starts by loading the iris dataset and splitting it into training and testing sets. The data is then preprocessed using feature scaling to standardize the input features, improving the model's performance. The k-nearest neighbors algorithm is chosen as the classifier and trained on the scaled training data. Finally, the trained model is used to make predictions on the test set, and its accuracy is evaluated, providing an estimate of the classifier's performance on unseen data.

Scikit-learn is a widely used machine learning library in Python that offers a broad range of machine learning algorithms, preprocessing tools, and model evaluation techniques. Its simple and intuitive interface makes it easy for beginners to use machine learning algorithms without requiring in-depth knowledge of the underlying mathematics. Scikit-learn has an extensive documentation library that provides detailed explanations of the algorithms, examples of their usage, and references to the underlying research papers. Moreover, scikit-learn has a vibrant community that is always ready to help with any issues or questions users may have.

However, scikit-learn does have some limitations. For example, it is not designed for deep learning, which requires more advanced algorithms and computational power than traditional machine learning. While scikit-learn does support some deep learning models, such as neural networks, users may find other deep learning frameworks, such as TensorFlow or PyTorch, more suitable for their needs. Additionally, scikit-learn is not designed for big data analysis and may not be suitable for datasets with millions of rows or high-dimensional data. In such cases, users may need to consider distributed computing frameworks, such as Apache Spark or Dask, that are specifically designed for big data processing.

TensorFlow

As an integral part of the machine learning and deep learning ecosystem, TensorFlow is a powerful open-source library for dataflow programming in Python. Developed by the Google Brain team, TensorFlow offers a flexible platform for constructing, training, and deploying machine learning models. TensorFlow operates using a dataflow graph that represents a computational model. Each node in the graph corresponds to an operation, while the edges represent the flow of data (tensors) between these operations. This structure facilitates parallel and distributed computation, enabling TensorFlow to efficiently run on various hardware platforms, such as CPUs, GPUs, and TPUs.

One of its main strengths of TensorFlow is its scalability, which allows it to scale across multiple devices, clusters, and platforms, making it ideal for training large models on extensive datasets. TensorFlow is highly flexible and supports a wide range of machine learning and deep learning algorithms, including neural networks, reinforcement learning, and unsupervised learning techniques. TensorFlow also benefits from a rich ecosystem, which includes libraries like Keras and TensorBoard, making it easier to design, train, and debug models. Finally, TensorFlow has a vast and active community of users, developers, and researchers who continuously contribute to its improvement and expansion.

To use TensorFlow in a Python environment, it can be installed using pip:

```
pip install tensorflow
```

After installation, TensorFlow can be imported into the Python script as:

```
import tensorflow as tf
```

TensorFlow's core components are:

- Tensors: Tensors are multidimensional arrays used to represent data in TensorFlow. They come in various data types, such as float32, int32, and bool.
- Operations: TensorFlow operations (ops) perform computations on tensors. Common operations include mathematical, logical, and array manipulation functions.
- Variables: TensorFlow variables store the state of a model, such as the weights and biases in a neural network. They can be initialized, updated, and saved during training.
- Sessions: A TensorFlow session manages the execution of the dataflow graph. It allocates memory for tensors and executes operations in the proper order.

An example of Creating a Simple Neural Network in TensorFlow is as follows:

```python
import tensorflow as tf

# Define the model architecture
model = tf.keras.Sequential([
    tf.keras.layers.Dense(128, activation='relu', input_shape=(784,)),
    tf.keras.layers.Dense(64, activation='relu'),
    tf.keras.layers.Dense(10, activation='softmax')
])

# Compile the model
model.compile(optimizer='adam',
              loss='sparse_categorical_crossentropy',
              metrics=['accuracy'])

# Train the model
history = model.fit(x_train, y_train, epochs=10, validation_split=0.2)

# Evaluate the model on the test set
test_loss, test_accuracy = model.evaluate(x_test, y_test)
```

Advantages of TensorFlow primarily stem from its versatility, scalability, and comprehensive ecosystem. As an open-source library, it is designed to accommodate a broad range of machine learning and deep learning algorithms, making it suitable for various applications, from simple regression problems to complex deep learning models. Its dataflow graph architecture allows for efficient parallel and distributed computation, enabling users to train large models on extensive datasets across multiple devices, clusters, and platforms. Furthermore, the TensorFlow ecosystem is replete with tools like Keras, a high-level neural networks API, and TensorBoard, a visualization tool, which streamline the model development process and provide additional convenience.

One of the main drawbacks of TensorFlow is its steep learning curve, especially for beginners in machine learning. The library's complexity can be overwhelming for those new to the field, as it requires a solid understanding of the underlying mathematical concepts and computational models. TensorFlow's computational graph approach can be less intuitive than other libraries that employ a more straightforward programming style, like PyTorch. While TensorFlow has made strides to improve its ease of use and debuggability, it still lags behind some of its competitors in these aspects, which can be a hindrance for some developers.

PyTorch

PyTorch, developed by Facebook's AI Research Lab (FAIR), is a popular open-source deep learning library for Python. Its flexibility, ease of use, and powerful computational capabilities have made it a favorite among researchers and developers alike. As a versatile tool for machine learning, computer vision, natural language processing, and more, PyTorch has rapidly gained prominence in both academia and industry since its inception in 2016.

PyTorch utilizes dynamic computation graphs, also known as define-by-run graphs, which allow developers to build and modify neural networks on-the-fly. This feature makes it easier to debug and visualize networks, providing increased flexibility during model development. Additionally, PyTorch offers seamless integration with NVIDIA GPUs, enabling users to harness the power of parallel processing for faster training and evaluation. Tensor computations can be easily moved between CPU and GPU with minimal code changes.

PyTorch boasts a rich ecosystem of libraries and tools, including torchvision, torchaudio, and torchtext for computer vision, audio, and text processing, respectively. These utilities simplify the process of working with complex data and enable rapid prototyping. Furthermore, PyTorch's automatic differentiation library, Autograd, simplifies the calculation of gradients during backpropagation. This feature eliminates the need for manual gradient computations, enabling developers to focus on designing and optimizing their models.

However, there are some disadvantages to using PyTorch as well. Its dynamic computation graph, while providing flexibility, can lead to increased memory consumption and slower execution compared to static computation graph-based libraries like TensorFlow. Furthermore, although PyTorch's popularity is growing, TensorFlow still has a larger community and user base, leading to more extensive resources and third-party tool support. Lastly, PyTorch's dynamic nature might not be as suitable for deployment in production environments, where a more optimized and static graph-based approach could be more desirable.

Keras

Keras is a high-level, user-friendly neural network library written in Python, designed to simplify the process of creating and experimenting with deep learning models for developers and researchers. Developed by François Chollet and released as an open-source project in 2015, Keras runs on top of more powerful and efficient deep learning frameworks like TensorFlow, Microsoft Cognitive Toolkit (CNTK), and Theano. Its minimalist and modular design allows for building, training, and evaluating neural networks with ease, catering to users of all experience levels.

The library's modularity enables the creation of complex architectures by connecting independent modules (layers) without being bogged down by low-level details. Keras is also easily extensible, allowing developers to create custom layers, loss functions, and optimizers for various applications. It supports multi-GPU and distributed training, making it scalable for large-scale projects and is compatible with multiple backends like TensorFlow, Microsoft Cognitive Toolkit (CNTK), and Theano, offering cross-platform support.

Keras provides essential components, including two main model-building APIs (Sequential and Functional), a wide range of built-in layers, several loss functions, various optimization algorithms, and numerous evaluation metrics. This comprehensive set of tools allows developers to define, compile, train, evaluate, and make predictions using deep learning models effectively.

While Keras has numerous advantages, such as its user-friendly nature, modularity, extensibility, and cross-platform support, it does have some drawbacks. Due to its high-level nature, it may not provide the same level of fine-grained control and customization as lower-level libraries like TensorFlow, which can be a limitation for advanced users. Keras may not always offer the most cutting-edge features and algorithms found in other deep learning libraries, as it prioritizes ease of use and simplicity. Lastly, Keras can sometimes suffer from performance issues due to its abstraction layers, potentially leading to slower training and inference times compared to lower-level libraries.

Python libraries play a crucial role in streamlining the development process and empowering developers to tackle complex challenges across various domains, including machine learning, data analysis, web development, and more. These libraries not only encapsulate best practices and proven methodologies but also offer comprehensive documentation and community support, fostering an environment conducive to learning and collaboration.

As the Python ecosystem continues to evolve, new libraries will emerge, and existing ones will be refined, ensuring that developers have access to the latest tools and techniques to stay at the forefront of innovation. Embracing these libraries and harnessing their potential can greatly enhance productivity and drive success in a competitive landscape. By understanding the strengths and limitations of each library, developers can make informed decisions, selecting the most appropriate tools for their specific needs and ultimately crafting more robust, efficient, and impactful solutions.

Part II

Data Acquisition and Cleaning

5

Accessing Data from WRDS

The Wharton Research Data Services (WRDS) serves as a valuable resource for academic research, offering researchers access to a diverse range of financial and economic data. One of the primary advantages associated with utilizing WRDS data is its comprehensive nature. By consolidating data from various sources such as financial statements, stock prices, and insider trading data, WRDS simplifies the process of data collection for researchers, thereby saving them time and effort.

An additional benefit of utilizing WRDS is the high quality of its data. Renowned for its reliability, WRDS ensures data accuracy through meticulous scrutiny and validation conducted by experienced professionals. Researchers can therefore depend on this data to yield precise analyses and results. Furthermore, WRDS provides researchers with an array of tools and applications designed to facilitate data analysis and visualization, thereby enhancing the ease with which valuable insights can be extracted from the data.

However, there are also several drawbacks associated with using WRDS data. One significant limitation is the cost of access. Subscribing to WRDS can be financially burdensome for smaller academic institutions or independent researchers due to its high subscription fees. Moreover, some researchers may encounter difficulties with the platform's interface and user experience, finding it less intuitive compared to other data platforms, which can impede their work.

Another potential disadvantage of utilizing WRDS is the restricted availability of certain datasets. Although WRDS offers access to an extensive range of data, certain datasets may be inaccessible due to restrictions imposed by data providers or regulatory agencies. Consequently, this limitation may restrict the scope of research projects and necessitate researchers to complement their analysis with data obtained from alternative sources.

Despite these disadvantages, the comprehensive nature and data quality offered by WRDS establish it as an invaluable resource for academic research. Armed with the appropriate tools and expertise, researchers can utilize WRDS data to extract valuable insights and make significant contributions to their respective fields. Furthermore, it is worth noting that WRDS is accessible to a majority of universities. In order to employ Python for WRDS, researchers need to install the WRDS library through the pip install method.

```
pip install wrds
```

An alternate to pip install would be conda install method.

```
conda install -c conda-forge wrds
```

The code to download data from compustat annual in the csv format would be:

```
import wrds
import pandas as pd

# prompt for WRDS credentials
username = input("Enter your WRDS username: ")
password = input("Enter your WRDS password: ")

# establish a connection with WRDS
db = wrds.Connection(wrds_username=username,
wrds_password=password)

# specify the variables to download
vars = ["gvkey", "datadate", "fyr", "sale", "at"]

# specify the condition for the query
condition = "indfmt='INDL' and datafmt='STD' and popsrc='D' and
consol='C' and curcd='USD'"

# download the data from Compustat North America Annual
data = db.raw_sql(f"select {','.join(vars)} from comp.funda
where {condition}")

# save the data in a CSV file in the specified directory
data.to_csv("<dir>/compustat_na_annual.csv", index=False)

# close the connection
```

This code facilitates the retrieval of data from the Compustat North America Annual database on WRDS, subsequently saving it in CSV format. To establish a connection with the WRDS database, users are prompted to input their WRDS username and password.

In this code, specific variables, namely **gvkey**, **datadate**, **fyr**, **sale**, and **at**, are designated for download. It is possible to specify any variables available in the Compustat Annual dataset. These variables are incorporated into the SQL query through the vars list. Additionally, the condition variable permits the inclusion of supplementary query conditions, such as data format, source, and currency.

The **raw_sql** method is employed to extract data from the database. The query is formulated based on the predetermined variables and conditions, and the resulting data is stored in the data variable. Subsequently, the data is saved in CSV format within the specified directory using the **to_csv** method. By setting the index parameter to False, the index is excluded from being included in the saved file. To conclude the process, the connection with the WRDS database is terminated using the close method. The generated CSV file can be imported into various software applications such as SAS, Stata, Excel, or any other relevant software for further analysis.

In situations where data for a specific time period is required, the code can be enhanced by incorporating start and end dates as illustrated in the following example:

```
# import the necessary libraries
import wrds
import pandas as pd
from datetime import datetime

# prompt for WRDS credentials
username = input("Enter your WRDS username: ")
password = input("Enter your WRDS password: ")

# establish a connection with WRDS
db = wrds.Connection(wrds_username=username,
wrds_password=password)

# specify the variables to download
vars = ["gvkey", "datadate", "fyr", "sale", "at"]

# specify the condition for the query
start_date = datetime(2001, 1, 1).strftime("%Y-%m-%d")
end_date = datetime(2015, 12, 31).strftime("%Y-%m-%d")
condition = f"indfmt='INDL' and datafmt='STD' and popsrc='D' and
consol='C' and curcd='USD' and datadate >= '{start_date}' and
datadate <= '{end_date}'"

# download the data from Compustat North America Annual
data = db.raw_sql(f"select {','.join(vars)} from comp.funda
where {condition}")

# save the data in a CSV file in the specified directory
data.to_csv("<dir>/compustat_na_annual_2001-2015.csv",
index=False)

# close the connection with WRDS
db.close()
```

In the event that the selection of a specific period is required based on the fiscal year (fyear), a minor modification in the code implementation would be necessary.

```python
import pandas as pd

# prompt for WRDS credentials
username = input("Enter your WRDS username: ")
password = input("Enter your WRDS password: ")

# establish a connection with WRDS
db = wrds.Connection(wrds_username=username,
wrds_password=password)

# specify the variables to download
vars = ["gvkey", "datadate", "fyr", "sale", "at"]

# specify the condition for the query
start_fyear = 2001
end_fyear = 2015
condition = f"indfmt='INDL' and datafmt='STD' and popsrc='D' and consol='C' and curcd='USD' and fyear >= {start_fyear} and fyear <= {end_fyear}"

# download the data from Compustat North America Annual
data = db.raw_sql(f"select {','.join(vars)} from comp.funda where {condition}")

# save the data in a CSV file in the specified directory
data.to_csv("<dir>/compustat_na_annual_fyear_2001-2015.csv",
index=False)

# close the connection with WRDS
db.close()
```

The code does not require inclusion of variable on which selection of data is to be done (selection of data done on the basis of **fyear**, which hasn't been included), but that variable must be present in wrds library from which data is being downloaded (comp.funda in this case). Similarly share price data can be downloaded from crsp daily stock file using the following code:

```python
# import the necessary libraries
import wrds
import pandas as pd
from datetime import datetime

# prompt for WRDS username and password
username = input("Enter your WRDS username: ")
password = input("Enter your WRDS password: ")

# establish a connection with WRDS
db = wrds.Connection(wrds_username=username,
wrds_password=password)

# specify the variables to download
vars = ["permno", "cusip", "date", "prc"]

# specify the condition for the query
start_date = datetime(2001, 1, 1).strftime("%Y-%m-%d")
end_date = datetime(2001, 1, 31).strftime("%Y-%m-%d")
condition = f"date >= '{start_date}' and date <= '{end_date}'"
# download the data from CRSP Daily Stock File
data = db.raw_sql(f"select {','.join(vars)} from crsp.dsf where {condition}")

# save the data in a CSV file in the specified directory
data.to_csv("<dir>/crsp_daily_stock_file_2001-2015.csv",
index=False)

# close the connection with WRDS
db.close()
```

The presented code facilitates the retrieval of data from the CRSP Daily Stock File available on the Wharton Research Data Services (WRDS) platform. Prior to utilizing the script, a WRDS account is required, and users are prompted to input their username and password for authentication. The script commences by importing essential libraries, including "wrds" for establishing a connection with WRDS, **"pandas"** for data manipulation, and **"datetime"** for handling date and time information. Following the establishment of a connection using the user-provided credentials, the script specifies the variables to be downloaded and sets a condition for the query, limiting the data to a specific date range.

Subsequently, the script employs the **"raw_sql"** method from the "wrds" library to download the data, subsequently storing it in a CSV file within the designated directory. Finally, the connection with WRDS is terminated through the utilization of the **"close"** method.

In the event that data from IBES is required, an alternative code would be:

```python
import wrds
import pandas as pd

# Connect to WRDS using your credentials
db = wrds.Connection(wrds_username='your_username')

# Set the desired start and end dates for the data
start_date = '2010-01-01'
end_date = '2020-12-31'

# Define the query to retrieve the data from the IBES database
query = """
        select ticker, cusip, anndats_act, measure, value,
fpedats
        from ibes.det_epsus
        where measure in ('EPS', 'EPSN', 'EPSN-1') and
            fpi = 'Actual' and
            anndats_act between '{}' and '{}'
        """.format(start_date, end_date)

# Execute the query and retrieve the data into a pandas
dataframe
ibes_data = db.raw_sql(query)

# Write the dataframe to a CSV file
ibes_data.to_csv('<dir>/ibes_data.csv', index=False)
```

This Python code uses the WRDS (Wharton Research Data Services) module and Pandas library to download financial data from the Institutional Brokers Estimate System (IBES) database. The first step is to import the necessary libraries—wrds and pandas. Then, a connection is established to WRDS using the user's credentials. Next, the desired start and end dates are defined for the data. The SQL query to retrieve the data from the IBES database is defined using the start and end dates as conditions. In this example, the query is retrieving actual earnings per share (EPS) data for a specified date range, but the query can be customized as needed. After the query is defined, it is executed and the data is retrieved into a pandas dataframe. Finally, the dataframe is written to a CSV file.

To download data from Thomson Reuters Mutual funds the following code would be useful:

```python
import wrds
import pandas as pd

# prompt for WRDS username and password
username = input("Enter your WRDS username: ")
password = input("Enter your WRDS password: ")

# establish a connection with WRDS
db = wrds.Connection(wrds_username=username,
wrds_password=password)

# specify the variables to download
vars = ['crid', 'cusip', 'sedol', 'fdate', 'nav', 'pnav',
'tret']
condition = 'fdate >= "2010-01-01" and fdate <= "2020-12-31"'

# download the data from Thomson Reuters Mutual Funds
data = db.get_table(library='finfund', table='mfnav',
columns=vars, obs=100, condition=condition)

# save the data in a CSV file in the specified directory
data.to_csv('<dir>/thomson_reuters_mutual_funds.csv',
index=False)

# close the connection with WRDS
db.close()
```

This code downloads data from the Thomson Reuters Mutual Funds database using the WRDS (Wharton Research Data Services) platform. First, the required libraries are imported: the wrds library for establishing the connection to the WRDS server, and the pandas library for data manipulation. The code then prompts the user to enter their WRDS username and password, and establishes a connection with the WRDS server using these credentials. The code then specifies the variables to download from the 'mfnav' table of the 'finfund' library, which include the 'crid', 'cusip', 'sedol', 'fdate', 'nav', 'pnav', and 'tret'. Additionally, the code specifies a condition for the query, which limits the data to the date range between January 1, 2010, and December 31, 2020. Using the 'get_table' function, the code downloads the data from the Thomson Reuters Mutual Funds database, limiting the results to the first 100 observations. Finally, the data is saved in a CSV file in the specified directory, and the connection to the WRDS server is closed using the 'close' function.

This code can be modified by changing the specified variables and conditions to download different data from the Thomson Reuters Mutual Funds database.

This chapter elucidates the process of accessing structured data through the utilization of Python and the WRDS platform. The fundamental aspects encompassed in this chapter involve the acquisition of data from diverse databases offered by WRDS, namely CRSP, Compustat, IBES, and Thomson

Reuters Mutual Funds. Moreover, an examination of distinct techniques to download and store the data locally in a CSV format is undertaken.

It is important to recognize that dealing with unstructured data, encompassing text, images, and videos, presents heightened challenges and necessitates the employment of advanced techniques and tools. Nevertheless, the foundational knowledge conveyed in this chapter assumes paramount significance in relation to accessing, cleansing, and processing structured data for research pertaining to accounting and finance. Subsequent chapters will delve into the exploration of methods for working with unstructured data utilizing Python alongside other tools, including natural language processing and image recognition. Proficiency in these competencies is indispensable for researchers aspiring to extract insights from an extensive array of data sources.

6

Accessing Data from SEC EDGAR

SEC EDGAR, also known as the Electronic Data Gathering, Analysis, and Retrieval system, serves as a platform facilitating electronic submission of financial reports and other documents to the Securities and Exchange Commission (SEC) in the United States. An advantageous aspect of utilizing SEC EDGAR data lies in its provision of prompt and extensive access to financial information and various filings, thereby enabling investors to make well-informed decisions in a timely manner.

Moreover, SEC EDGAR enhances financial reporting transparency, thereby deterring fraudulent activities and other forms of financial misconduct. This heightened transparency arises from the platform's accessibility to the general public, which facilitates the retrieval of pertinent information pertaining to a company of interest. Furthermore, the information submitted through SEC EDGAR adheres to consistent and standardized formats, thereby facilitating the comparison and analysis of financial data across diverse companies. An additional benefit of utilizing SEC EDGAR data pertains to its inclusion of historical data, which aids investors in analyzing trends and making informed decisions based on a company's past performance. Such historical data proves particularly valuable for long-term investors aiming to comprehend a company's performance over time.

However, several potential drawbacks accompany the utilization of SEC EDGAR data. For instance, navigating the platform can be challenging, particularly for individuals who are unfamiliar with its interface. Furthermore, although SEC EDGAR provides extensive financial information, it may not encompass all the data required by investors to arrive at well-informed decisions. Another potential issue arises from the possibility of delayed filings by companies, resulting in postponed release of crucial financial information. Moreover, certain information disclosed via SEC EDGAR filings may lack necessary context, impeding investors' comprehension of a company's complete financial situation.

In addition to the aforementioned concerns regarding accuracy, completeness, and contextual understanding, another challenge associated with utilizing SEC EDGAR data relates to the presentation of voluminous textual content, necessitating extensive processing before it can be utilized for research purposes. SEC EDGAR filings generally consist of unstructured text, rendering it arduous for computers to extract meaningful information through analysis. Overcoming this challenge requires researchers and investors to employ advanced natural language processing (NLP) techniques to extract relevant data from the filings. This process proves time-consuming and complex, necessitating high levels of technical expertise and specialized software tools. Furthermore, even after data has been processed and analyzed, it remains subject to interpretation and uncertainty due to the presence of intricate and nuanced information in financial reports and other SEC EDGAR filings. A deep understanding of the relevant industry, accounting principles, and regulatory requirements is essential for accurate interpretation. Subsequent sections of this chapter will delve into the acquisition of text files from SEC EDGAR in a readable format, while subsequent sections will focus on mastering advanced techniques to process this data for research purposes.

The most effective approach to downloading the required data involves utilizing Python codes. The requisite code for this task is as follows:

```python
# Import the sec_edgar_downloader library
import sec_edgar_downloader
# Import pandas
import pandas as pd
# Create a Downloader object with a directory to store the
downloaded files
downloader = sec_edgar_downloader.Downloader(r"<dir1>")

# Create an empty DataFrame
pd.DataFrame()

# Read the ticker symbols from an Excel file using pandas
tick = pd.read_excel(r"<dir2>\Tickers.xlsx")

# Get the Ticker column from the DataFrame as a pandas series
for_ticker = tick.Ticker

# Loop through each ticker symbol and download its 10-K report
using the downloader object
for t in for_ticker:
    downloader.get("10-K", t)
    print(str(t) + ' is downloaded')

# Print a message to indicate that the download is complete
print('Download is complete')
```

This Python code downloads the 10-K filings for a list of companies from the SEC's EDGAR database. It uses the sec_edgar_downloader module to download the filings and creates a Downloader object with a directory (dir1) to store the downloaded files. An empty DataFrame is created using pandas. The code then reads a list of ticker symbols and their associated date ranges from an Excel file named **Tickers.xlsx** in directory dir2 using the pandas module. The Ticker column of the excel file **Tickers.xlsx** must have a column **Ticker** that has all the ticker symbols of the companies for which the files are to be downloaded from sec edgar. The **Ticker** column from the DataFrame is extracted as a pandas series. The **for** loop then iterates over each ticker symbol in the series and downloads its 10-K report using the downloader object. For each company, the **get()** method of the Downloader object is called with the "10-K" parameter and the ticker symbol. The **print()** function is called within the loop to indicate that the filing for that company has been downloaded.

After all the tickers have been processed, the loop ends, and the code prints a message to indicate that the download is complete. However, it is important to note that this code does not download filings within a specific date range, unlike the previous code we discussed.

sec_edgar_downloader library can be installed using:

```
pip install sec_edgar_downloader
```

Useful Modifications

Limiting the Period

The previous code downloads all the files from the beginning to the end. Most of the times in accounting and finance research we need data related to specific period or specific financial years. Unfortunately, SEC does not provide the sorting of files by financial year, but you can download the files that were filed by the company between two dates, which can be used indirectly to download the files for a financial year, though it's not always true. To do this you need to install a library "datetime", which can be installed by the same pip install method. Also, in the Ticker.xlsx file you need to create two additional columns, i.e., Before and After that have the relevant dates for the period, in date format. The code would be:

```
import sec_edgar_downloader
import sec_edgar_downloader
import pandas as pd
from datetime import datetime

# create the downloader object and set the download directory
downloader = sec_edgar_downloader.Downloader(r"<dir1>")

# read the ticker list from an Excel file
tick = pd.read_excel(r"<dir2>/Tickers.xlsx")

# get the ticker, after, and before columns as lists
for_ticker = tick["Ticker"].tolist()
after_date = tick["After"].apply(lambda x: datetime.strftime(x,
"%Y-%m-%d")).tolist()
before_date = tick["Before"].apply(lambda x:
datetime.strftime(x, "%Y-%m-%d")).tolist()

# iterate over the ticker list and download the 10-K reports
with the specified date range
for t, a, b in zip(for_ticker, after_date, before_date):
    downloader.get("10-K", t, after=a, before=b)
    print(f"{t} is downloaded")

# print a message when the download is complete
print("Download is complete")
```

This Python code downloads 10-K filings for a list of companies from the SEC's EDGAR database. It uses the **sec_edgar_downloader** module to download the filings, the pandas module for data manipulation, and the datetime module to convert dates. The code reads a list of ticker symbols and their associated date ranges from an Excel file named **Tickers.xlsx**, extracts the ticker symbols, after dates, and before dates from the file, and converts them to lists. The **for** loop then iterates over the ticker symbols, after dates, and before dates simultaneously using the zip() function. For each ticker symbol, the **get()** method of the Downloader object is called with the relevant parameters to download the filings for that company within the specified date range. After all the companies have been processed, the loop ends and the code prints a message to indicate that the download is complete.

Cleaning the HTML Tags

The files in previous code have numerous html tags and are not usable in this format. We need to clean those tags to make the files readable. The code to clean these files would be:

```python
import pandas as pd
import os
from bs4 import BeautifulSoup

# iterate over all directories and subdirectories in the
specified root directory
for root, dirs, files in os.walk(r"<dir>"):
    # iterate over all files in the current directory
    for file in files:
        # check if the file has the .txt extension
        if file.endswith(".txt"):
            # read the contents of the file
            with open(os.path.join(root, file), "r") as f:
                content = f.read()

            # clean the file from HTML tags
            soup = BeautifulSoup(content, "html.parser")
            cleaned_content = soup.get_text()

            # write the cleaned content back to the file
            with open(os.path.join(root, file), "w") as f:
                f.write(cleaned_content)

            # print a message for each file that has been cleaned
            print(f"{os.path.join(root, file)} has been cleaned")

# print a message when the cleaning is complete
print("Cleaning is complete")
```

This code is written to clean all the HTML tags from text files with .txt extension in a specified directory and all its subdirectories. The **os.walk()** function is used to traverse through all the directories and subdirectories within the specified root directory. Then, a for loop is used to iterate over all the files in each directory. The if statement checks if the file has a .txt extension.

If the file is a text file, it is opened using the **open()** function, and its content is read using the read() method. The **BeautifulSoup()** function from the bs4 library is then used to parse the content of the file and remove all the HTML tags, and the cleaned content is saved to the variable **cleaned_content**. Finally, the cleaned content is written back to the original file using the **write()** method, and a message is printed to indicate that the file has been cleaned. Once all the files have been cleaned, a message is printed to indicate that the cleaning process is complete.

A single-stage code that downloads and cleans all the files would be more efficient. The code is:

```python
import sec_edgar_downloader
import pandas as pd
# Iterate over the ticker list and download the 10-K reports
with the specified date range
for t, a, b in zip(for_ticker, after_date, before_date):
    # Download the 10-K reports with the specified date range
    downloader.get("10-K", t, after=a, before=b)
    # Print a message for each ticker that has been downloaded
    print(f"{t} is downloaded")

# Print a message when the download is complete
print("Download is complete")

# Clean the downloaded files from HTML tags
for root, dirs, files in os.walk(r"D:/SEC/10K"):
    for file in files:
        if file.endswith(".txt"):
            # Read the contents of the file
            with open(os.path.join(root, file), "r") as f:
                content = f.read()

            # Clean the file from HTML tags
            soup = BeautifulSoup(content, "html.parser")
            cleaned_content = soup.get_text()

            # Write the cleaned content back to the file
            with open(os.path.join(root, file), "w") as f:
                f.write(cleaned_content)

# Print a message when the cleaning is complete
print("Cleaning is complete")
```

SEC EDGAR data stands as a paramount resource for academic research in accounting, particularly when text analysis is involved. By leveraging the extensive and standardized financial information available through SEC EDGAR, researchers can make informed decisions, analyze trends, and gain insights into companies' past performances. For accounting researchers, SEC EDGAR data serves as a foundation for empirical investigations, offering a comprehensive and reliable source of textual information. The utilization of advanced natural language processing techniques allows researchers to extract relevant data from the voluminous text filings available on the platform. However, it is important to acknowledge the challenges associated with navigating the platform, interpreting complex information, and addressing potential limitations such as delayed filings and lack of contextual understanding.

With its wealth of financial information, accessibility to the public, and historical data, SEC EDGAR remains an invaluable resource for academic research in accounting. As researchers delve into the subsequent chapters, they will gain further insights and master advanced techniques for processing and analyzing SEC EDGAR data to contribute to the existing body of knowledge in the field of accounting research.

7

Accessing Data from Other Sources

In accounting research, there is a wide array of data sources beyond the commonly used WRDS and SEC EDGAR platforms. However, utilizing these alternative sources often involves the need to extract and refine data from websites, including the removal of HTML tags. Furthermore, there may be additional technical obstacles to address, such as handling tables and login credentials. Although these sources can provide valuable data not found on WRDS or SEC Edgar, researchers may face difficulties in acquiring and processing the data.

One advantage of utilizing data from websites that offer unstructured data is the potential access to a broader range of information compared to what is available on WRDS or SEC Edgar. Researchers can potentially obtain data on private companies or industry-specific data that is not publicly accessible. Additionally, certain websites may provide more recent or granular data that is not offered by other sources.

Nonetheless, there are also drawbacks associated with using data from websites that provide unstructured data. A notable disadvantage is the inability to consistently guarantee the reliability and accuracy of the data. As the data is obtained through web scraping, there may be errors or inconsistencies present, which can impact the outcomes of research. Moreover, the process of scraping and cleaning the data can be time-consuming and demanding in terms of technical expertise, potentially creating barriers for some researchers.

While utilizing data from these sources can provide advantages such as wider data accessibility, researchers must carefully consider the challenges involved in obtaining and processing such data. By employing appropriate techniques and tools, researchers can overcome these challenges and gain fresh insights into the field of accounting. The techniques presented in this chapter are designed to assist researchers, both novice and experienced, in obtaining and processing data more efficiently, enabling them to conduct comprehensive and insightful analyses. Through the application of these techniques, researchers can harness the vast amount of information available on the internet and obtain novel perspectives on the accounting profession.

The subsequent sections will provide a step-by-step guide on website scraping, HTML tag cleansing, table handling, and management of login credentials. Additionally, best practices for data processing and analysis will be discussed. By adhering to these instructions and best practices, researchers can overcome common challenges and effectively utilize data from websites that necessitate scraping and cleaning procedures.

Data Contained in a Series of Webpages

When browsing websites, there are instances where data is presented on a single page, making it easy to copy and paste. However, in cases where the data is spread across numerous pages, often numbering in the thousands, a pattern can be observed in the URL structure of each page. This pattern typically involves a numerical value that increases with each subsequent page. By identifying this pattern, we can determine the first and last numbers in the series by navigating to the initial and final pages of the website.

Occasionally, determining the last page number can be a laborious task, especially if the website has an extensive sequence of pages. In such situations, it is possible to select a number that exceeds the anticipated range and use it as the final number in the series. Nevertheless, this alternative approach may necessitate additional processing.

To exemplify this procedure, let us consider the Stanford Class Action Data as an illustrative case. This dataset comprises textual information regarding class action litigation, with each webpage dedicated to a particular lawsuit. The following code can be employed to retrieve the clean text from the website:

```
import requests
from bs4 import BeautifulSoup
import os

# Set the directory where the text files will be saved
directory = '<dir>'
if not os.path.exists(directory):
    os.makedirs(directory)

# Loop over a range of IDs (from 174001 to 300000)
# Here 300000 is a number larger than expectation
for i in range(174001, 300000):
    # Construct the URL for the current iteration
    url = f'https://securities.stanford.edu/filings-case.html?id={i}'
    # Send a GET request to the URL and retrieve the HTML content
    response = requests.get(url)
    html = response.content
    # Use BeautifulSoup to extract the text and remove HTML tags
    soup = BeautifulSoup(html, 'html.parser')
    text = soup.get_text()
    # Construct the filename for the current iteration
    filename = os.path.join(directory, f'{i}.txt')
    # Write the text to a file
    with open(filename, 'w', encoding='utf-8') as f:
        f.write(text)
        # Print a message to confirm that the file was saved
        print(f'Text saved to file: {filename}')`
```

This Python code that scrapes data from a website, processes it and saves it to text files. The script loops over a range of IDs from 174001 to 300000, constructs the URL for each ID, sends a **GET** request to the website and retrieves the HTML content. It then uses the BeautifulSoup module of bs4 library to extract the text from the HTML content and remove HTML tags. The script creates a directory where the text files will be saved and constructs the filename for each ID using the directory path and the ID number. It then writes the text data to a file and prints a message to confirm that the file was saved.

The purpose of this script is to collect data on the Stanford Securities Class Action Database, which contains information on securities class action lawsuits. The data is stored in separate text files, each corresponding to a specific ID number. The script can be modified to adjust the range of ID numbers to scrape, the directory where the text files are saved, and the filename format for the text files.

Here we have taken 300,000 as a large number bigger than expectation in place of actual last page number. Most websites will give an inaccurate page number response, in response to a page number that does not exist. This code can automatically handle that. Some websites including the Stanford Securities Class Action Database generate a page that will say that this page does not exist. The code will download that page as a separate text file, which later need to be deleted based on size or any other criteria. In both cases the code in a large number bigger than expected page number scenario will have to loop over greater numbers and therefore, will take more time. Therefore, it is always good to look at the actual last page number, wherever possible.

Data on Yahoo Finance

Yahoo Finance is widely utilized by accounting and finance researchers as a reputable and popular source of financial information. It offers an extensive array of financial data, encompassing both real-time and historical stock prices, financial statements, and news updates, among other resources. One notable advantage associated with Yahoo Finance pertains to its substantial collection of historical data. This historical data proves highly beneficial for researchers in the accounting and finance fields who are engaged in academic inquiries concerning stocks. Furthermore, Yahoo Finance provides real-time stock price data, a particularly valuable feature for researchers focusing on high-frequency trading research. Another advantageous aspect of Yahoo Finance pertains to its flexibility in terms of data frequency. Users have the ability to access stock price data on a daily, weekly, monthly, or other customized basis, catering to the specific needs of accounting and finance researchers seeking to track stock prices over designated time periods.

To exemplify the process of acquiring real-time stock price data from Yahoo Finance, we present a Python code snippet as an illustrative example. This code snippet facilitates the retrieval of stock price data within a specified time interval on a given date for a designated list of companies. Researchers may adapt this code to obtain data for any desired date, time range, or selection of companies.

```python
import pandas as pd
import yfinance as yf
import pytz
import os

# Read the tickers from the Excel file
df_tickers = pd.read_excel('<dir1>/Tickers.xlsx',
usecols=['Ticker'])

# Set the start and end times in Eastern Standard Time (EST)
start_time = pd.Timestamp('2023-03-13 10:00:00',
tz='US/Eastern')
end_time = pd.Timestamp('2023-03-13 10:15:00', tz='US/Eastern')

# Loop over each ticker
for ticker in df_tickers['Ticker']:
    # Download the data for the current ticker
    data = yf.download(ticker, start=start_time, end=end_time,
interval='1m')

    # Set the directory where the CSV file will be saved
    directory = '<dir2>'
    if not os.path.exists(directory):
        os.makedirs(directory)

    # Construct the filename
    filename = f'{ticker}_{start_time.strftime("%Y-%m-%d %H_%M_%S")}_{end_time.strftime("%Y-%m-%d %H_%M_%S")}.csv'
    filepath = os.path.join(directory, filename)

    # Save the data to a CSV file
    data.to_csv(filepath)

    # Print a message to confirm that the data was saved
    print(f'Data saved to file: {filepath}')
```

This code downloads financial data for a list of tickers from Yahoo Finance and saves it to a CSV file. The tickers are read from an Excel file that has a column Ticker for all the tickers, in dir1 directory, and the start and end times are specified in Eastern Standard Time (EST). The code uses the pandas library to read the tickers from the Excel file, and the yfinance library to download the financial data. The pytz library is used to specify the timezone for the start and end times, and the os library is used to create directories and file paths.

The code loops over each ticker in the Excel file, and downloads the data using the **yf.download()** function. The data is then saved to a CSV file using the pandas **DataFrame.to_csv()** function. The CSV file is saved in a directory specified by the user, with a filename that includes the ticker symbol and the start and end times. After each file is saved, the code prints a message to confirm that the data was saved. For different times and timezones, **start_time**, **end_time**, and **tz** can be modified.

One of the key challenges with working with financial data is that it is often timestamped in a particular timezone. This can create issues when working across multiple timezones, as the time of day will be different for different parts of the world. In the case of the yfinance library, it is important to note that the data is timestamped in the timezone of the exchange where the stock is traded. This means that when downloading data for a particular stock, it is important to set the start and end times in the timezone of that exchange. Furthermore, it is important to consider whether the exchange is open or closed at the time you are requesting data, as this will impact the availability and accuracy of the data.

Data on Cryptocurrency

Cryptocurrency research is a rapidly emerging field in accounting and finance. However, one major challenge researchers face is the lack of reliable data. While many exchanges trade cryptocurrencies 24 hours a day, obtaining and interpreting daily open and close price data can be difficult. In the world of crypto, real-time data is often more relevant due to the high levels of volatility.

Fortunately, researchers can leverage tools like Python and the Pycoingecko library to access real-time cryptocurrency data. With this library, researchers can obtain data on prices, market capitalizations, trading volumes, and other metrics for a variety of cryptocurrencies. In this section, we provide you Python code to obtain crypto data using the Pycoingecko library. The code allows you to retrieve real-time data between two specified dates, making it easier to analyze how prices, volumes, and other metrics have changed over time. By combining real-time data with other analytical tools, researchers can gain valuable insights into the behavior of cryptocurrency markets.

7 Accessing Data from Other Sources

```python
import csv
import datetime
from pycoingecko import CoinGeckoAPI
import os
import pandas as pd

# Initialize the CoinGecko API client
cg = CoinGeckoAPI()

# Set the start and end dates for the historical data
start_date = datetime.datetime(2023, 1, 1)
end_date = datetime.datetime(2023, 1, 15)

# Convert the dates to Unix timestamps
start_timestamp = int(start_date.timestamp())
end_timestamp = int(end_date.timestamp())

# Read the crypto_id column from the Excel file and store as a list
df = pd.read_excel("<dir1>/crypto.xlsx")
crypto_ids = df["crypto_id"].tolist()

# Create the '<dir2>' folder if it doesn't already exist
if not os.path.exists(<dir2>'):
    os.makedirs('<dir2>')

# Loop over the crypto_ids list and retrieve the historical
price, market cap, and volume data for each cryptocurrency
for crypto_id in crypto_ids:
    # Get the historical data for the crypto_id in USD
    data = cg.get_coin_market_chart_range_by_id(id=crypto_id, vs_currency='usd', from_timestamp=start_timestamp, to_timestamp=end_timestamp)

    # Save the data as a CSV file in the '<dir2>' folder
    filename = f'<dir2>/{crypto_id}_data.csv'
    with open(filename, 'w', newline='') as csvfile:
        writer = csv.writer(csvfile)
        writer.writerow(['Date', 'Price', 'Market Cap', 'Total Volume'])
        for row in data['prices']:
            date = datetime.datetime.fromtimestamp(row[0]/1000.0).strftime('%Y-%m-%d %H:%M:%S')
            price = row[1]
            market_cap = [row[0], next((item[1] for item in data['market_caps'] if item[0] == row[0]), None)]
            total_volume = [row[0], next((item[1] for item in data['total_volumes'] if item[0] == row[0]), None)]
            writer.writerow([date, price, market_cap[1], total_volume[1]])

    print(f"Data for {crypto_id} saved as {filename}")
```

This code retrieves historical price, market cap, and volume data for cryptocurrencies and saves it as CSV files. It uses the CoinGecko API to retrieve the data and the pandas library to read the crypto_id column from an Excel file in <dir1> directory. First, the start and end dates for the historical data are set using datetime. These dates are then converted to Unix timestamps using the **int()** function. Next, the **crypto_id** column is read from the Excel file using the pandas library and stored as a list. This list is used in a for loop to retrieve the historical data for each cryptocurrency. For each cryptocurrency, the **CoinGeckoAPI** is used to retrieve the historical data in USD for the specified date range. The data is then saved as a CSV file in a specified folder <dir2> using the csv library.

The CSV file contains four columns: Date, Price, Market Cap, and Total Volume. The data for the Date and Price columns are retrieved directly from the historical price data returned by the CoinGeckoAPI. The data for the Market Cap and Total Volume columns is retrieved using a nested list comprehension to search for the corresponding data in the **market_caps** and total_volumes fields returned by the API. Finally, a message is printed to the console indicating that the data has been saved for the current cryptocurrency.

National Oceanic and Atmospheric Administration (NOAA) Data

The utilization of NOAA data holds significant relevance in academic research within the fields of accounting and finance due to its provision of invaluable environmental information encompassing weather patterns, climatic conditions, and oceanic data. This data source presents researchers with the opportunity to explore a diverse range of research inquiries. For example, scholars can employ NOAA data to investigate the impact of severe weather events on the financial performance of companies across various industries or assess the influence of climate change on asset valuation and investment choices. Furthermore, NOAA data can be leveraged to examine the correlation between weather variables and stock market performance, as well as analyze the role of climate risk within financial markets. By integrating NOAA data into financial models, researchers can enhance their comprehension of potential risks associated with climate change and weather-induced occurrences, ultimately facilitating the development of novel financial products and services that aid organizations and investors in effectively managing these risks. Consequently, NOAA data assumes a critical role in advancing sustainability research within the realms of accounting and finance, and possesses the capacity to inform policy decisions aimed at promoting environmentally responsible practices. A multitude of datasets are accessible through

NOAA, and presented below is an illustrative code snippet designed to extract daily summaries from the website:

```python
import urllib.request
import ftplib
import os
import pandas as pd

url = 'https://www1.ncdc.noaa.gov/pub/data/cdo/samples/PRECIP_HLY_sample_csv.csv'
filename = 'D:/NOAA/PRECIP_HLY_sample_csv.csv'

urllib.request.urlretrieve(url, filename)

# Define the FTP server details
ftp_host = "ftp.ncdc.noaa.gov"
ftp_user = "anonymous"
ftp_password = ""

# Define the data directory and file names
data_dir = "pub/data/noaa/"
file_prefix = "daily-summaries/"
file_extension = ".csv"

# Define the date range for the data
start_date = "2022-01-01"
end_date = "2022-01-31"

# Connect to the FTP server
ftp = ftplib.FTP(ftp_host, ftp_user, ftp_password)

# Navigate to the data directory
ftp.cwd(data_dir)

# Loop through each date in the date range
for date in pd.date_range(start=start_date, end=end_date, freq='D'):
    # Define the file name for the current date
    file_name = file_prefix + date.strftime("%Y/%Y-%m-%d") + file_extension

    # Create the directories if they don't exist
    directory = os.path.dirname(file_name)
    if not os.path.exists(directory):
        os.makedirs(directory)

    # Download the file
    with open(file_name, "wb") as f:
        ftp.retrbinary("RETR " + file_name, f.write)

# Close the FTP connection
ftp.quit()
```

This code downloads daily summary data from the National Oceanic and Atmospheric Administration (NOAA) using Python. First, it downloads a sample CSV file from NOAA using the urllib library. It then defines the FTP server details, data directory, file names, and date range for the data. The pandas library is used to loop through each date in the specified range.

For each date, the code creates the file name for the current date, creates the directories if they don't exist, and downloads the corresponding file from the NOAA FTP server using the ftplib library. The file is saved with the appropriate name and in the appropriate directory. The FTP connection is closed at the end of the loop. The code downloads the desired data in "D:/NOAA" folder.

This code demonstrates how to automate the process of downloading large amounts of daily summary data from NOAA for research purposes. It is important to note that the code can be modified to download other types of data from NOAA by changing the file names and directory structure.

Twitter Data

Twitter data has gained increasing attention among accounting and finance academic researchers due to its real-time nature and the vast amount of information that can be extracted from it. It provides a unique opportunity to study the opinions and sentiments of individuals, organizations, and the broader market in real time. Researchers have used Twitter data to study various accounting and finance-related topics, such as financial market reactions to corporate announcements, stock price prediction, sentiment analysis of earnings calls, and disclosure behavior of firms. Twitter data can also provide valuable insights into consumer behavior and trends, as well as the impact of news events on the stock market. The use of Twitter data in accounting and finance research is expected to continue to grow as new tools and techniques are developed to extract meaningful insights from the vast amount of data available. Twitter data can be downloaded based on various parameters such as:

- Keywords or phrases in tweets
- Hashtags in tweets
- User accounts or handles

- Locations (geotags)
- Language of the tweets
- Dates or time period
- Number of tweets to be downloaded
- Type of tweets (original tweets, retweets, replies)
- Filter by media types (photos, videos, etc.)

To download data from twitter you need to obtain API credential from twitter, which is free for academic research. Here's a sample code that downloads all the text tweets containing a company name during a particular time frame:

```
import tweepy
import pandas as pd
import xlrd

# Twitter API credentials
consumer_key = "your_consumer_key"
consumer_secret = "your_consumer_secret"
access_token = "your_access_token"
access_token_secret = "your_access_token_secret"

# Authenticate to Twitter API
auth = tweepy.OAuthHandler(consumer_key, consumer_secret)
auth.set_access_token(access_token, access_token_secret)

# Create API object
api = tweepy.API(auth)
```

```python
# Read company names from Excel file
df = pd.read_excel("<path>/companies.xlsx")
company_names = df["Company_name"].tolist()

# Define search query parameters
query = " OR ".join(company_names) + " lang:en place:united states"
start_date = "2022-01-01"
end_date = "2022-03-31"

# Create list to store tweets
tweets = []

# Loop through pages of search results and append to list
for page in tweepy.Cursor(api.search_tweets, q=query, tweet_mode="extended", lang="en", since_id=start_date, until=end_date, count=100).pages():
    for tweet in page:
        tweets.append(tweet)

# Create dataframe of tweet information
tweet_data = pd.DataFrame({
    "id": [tweet.id for tweet in tweets],
    "text": [tweet.full_text for tweet in tweets],
    "created_at": [tweet.created_at for tweet in tweets],
    "user_screen_name": [tweet.user.screen_name for tweet in tweets],
    "user_followers_count": [tweet.user.followers_count for tweet in tweets]
})

# Write tweet data to CSV file
tweet_data.to_csv("<path>/tweets.csv", index=False)
```

The above code demonstrates how to use the Twitter API and Python's Tweepy library to download tweets from Twitter based on specific search query parameters. The code first authenticates the API using the provided API credentials. It then reads a list of company names from an Excel file, named **Companies** that has a column named **Company_name**, that contains names of the companies, and defines the search query parameters based on those names, as well as a start and end date for the search. The code creates an empty list to store the retrieved tweets and loops through the pages of search results, using the **Tweepy Cursor** object. For each page, it loops through the individual tweets and appends them to the list. The code finally creates a pandas dataframe from the retrieved tweet information, including the tweet ID, text, creation date, user screen name, and user follower count. It then writes this data to a CSV file for further analysis. This code provides a simple example of how to use Python and Tweepy to access Twitter data for research purposes.

The inclusion of replies and retweets in the downloaded data can be achieved by adjusting the search query parameters. By incorporating the

exclude_replies=False parameter into the **api.search_tweets()** function, it is possible to include replies. A practical example would be as follows:

```
for page in tweepy.Cursor(api.search_tweets, q=query,
tweet_mode="extended", lang="en", exclude_replies=False,
since_id=start_date, until=end_date, count=100).pages():
    for tweet in page:
        tweets.append(tweet)
```

In order to incorporate replies to tweets and retweets in the obtained data, one can make adjustments by modifying the function **api.search_tweets()** and adding the parameter **include_rts=True**. For instance:

```
for page in tweepy.Cursor(api.search_tweets, q=query,
tweet_mode="extended", lang="en", include_rts=True,
since_id=start_date, until=end_date, count=100).pages():
    for tweet in page:
        tweets.append(tweet)
```

Keep in mind that including replies and retweets may increase the amount of data downloaded and may require additional processing steps to filter and clean the data.

Google Trends Data

Google Trends data is an important source of information for academic accounting and finance research. This data provides valuable insights into the popularity of specific search terms over time, which can be used to understand consumer behavior, market trends, and the performance of specific industries. Google Trends data can be used to study the relationship between online search behavior and financial market performance. Studies have found that there is a significant correlation between search volumes for specific keywords and stock prices, trading volume, and market volatility. By analyzing this relationship, researchers can develop predictive models that can be used to inform investment strategies and risk management.

In addition, Google Trends data can be used to study the impact of specific events on consumer behavior and financial markets. For example, researchers can analyze the search patterns related to major news events such as mergers and acquisitions, product recalls, and economic policy changes. This can provide valuable insights into how consumers and investors react to these events and can inform business strategies and public policy decisions. Thus, Google Trends data is a powerful tool for academic accounting and finance research, offering unique insights into consumer behavior, market trends, and the performance of financial markets.

Google trends data can be downloaded on many parameters such as:

- The search query term or topic of interest.
- The geographic location to get the search data for.
- Start and end dates to get the search data for.
- A relative time frame for the search data, such as "today 5-y" (for the past 5 years) or "now 1-H" (for the past hour).
- The category to get the search data for, such as "Business & Industrial" or "Arts & Entertainment".
- Type of search data for, such as "images" or "news".
- The interface language for the search results, such as "en" for English or "es" for Spanish.

A sample code for downloading google trends data is as follows:

```python
import pandas as pd
from pytrends.request import TrendReq

# Define the keywords file path
keywords_file = "trends.xlsx"

# Define the output directory
output_dir = "D:/GTrends"

# Read the keywords from the Excel file
df = pd.read_excel(keywords_file, sheet_name="Sheet1")
keywords = df["Keywords"].tolist()

# Create a pytrends object
pytrends = TrendReq(hl='en-US', tz=360)

# Define the time range for the data
start_date = "2021-01-01"
end_date = "2021-12-31"

# Define the keyword query list
kw_list = keywords

# Create an empty dataframe to store the data
trends_data = pd.DataFrame()

# Loop through each keyword and download the data
for kw in kw_list:
    print("Downloading data for keyword:", kw)

    # Build the keyword query list
    pytrends.build_payload([kw], timeframe=start_date + " " + end_date, geo="US", gprop="")

    # Get the interest over time data
    interest_over_time_df = pytrends.interest_over_time()

    # Add the keyword column to the dataframe
    interest_over_time_df["keyword"] = kw

    # Append the data to the trends_data dataframe
    trends_data = trends_data.append(interest_over_time_df)

# Reset the index of the trends_data dataframe
trends_data.reset_index(inplace=True)

# Create the output directory if it doesn't exist
if not os.path.exists(output_dir):
    os.makedirs(output_dir)

# Save the data to a CSV file
trends_data.to_csv(output_dir + "/google_trends_data.csv", index=False)
```

This Python code downloads Google Trends daily data for a list of keywords in an Excel file. The code uses the pandas and pytrends libraries to download and process the data.

First, the keywords are read from an Excel file, and the output directory is specified. A **TrendReq** object is created to establish a connection with the Google Trends API, and the time range for the data is defined. The code then loops through each keyword in the list, builds the keyword query list, and downloads the interest over time data using the **interest_over_time** method of the **pytrends** object. The keyword column is added to the data and the data is appended to a **trends_data** dataframe. Finally, the index of the **trends_data** dataframe is reset and the data is saved to a CSV file in the specified output directory. If the output directory does not exist, it is created using **os.makedirs()**. The output file contains the daily search interest for each keyword in the time range specified, which can be used for further analysis in accounting and finance academic research.

In this chapter, we have learned how to access structured and unstructured data from various sources using Python. We have covered different types of data sources such as financial databases, social media platforms, weather and climate data, and Google Trends data. We have also seen how to download data from these sources using different Python libraries and tools such as WRDS, Tweepy, NOAA, and Pytrends.

Accessing data is a crucial step in any academic research, and Python provides a powerful set of tools to help researchers access data from different sources. With the knowledge gained in this chapter, academic researchers can now access and analyze data from different sources to gain insights and conduct research in various fields such as finance, accounting, and sustainability. The next chapter will build on this knowledge by showing how to clean, preprocess, and analyze data using Python.

8

Text Extraction and Cleaning

In the field of academic accounting and finance research, the analysis of vast amounts of unstructured textual data is growing in significance. However, working with such data can be challenging due to the presence of extraneous information, including HTML tags and formatting elements, which can hinder text comprehension and analysis.

This chapter will primarily focus on techniques for extracting and cleaning text, which are crucial for retrieving relevant information from extensive volumes of unstructured textual data. These techniques empower researchers to identify and extract specific sections of text, such as financial statements or earnings announcements, while eliminating irrelevant or redundant information. For instance, in the case of researching Managerial Discussion and Analysis, researchers may choose to exclude all text in 10-K statements except item 7.

Text extraction and cleaning techniques are essential for ensuring that researchers work with pertinent data and can reveal valuable insights that may not be discernible through traditional data analysis methods. These techniques enable researchers to identify patterns in financial statements and analyze market sentiment based on news articles and other textual data.

The following sections will delve into comprehensive explanations of text extraction and cleaning techniques and explore their implementation using Python. By the conclusion of this chapter, readers will have a comprehensive understanding of how to extract and cleanse text data for utilization in various natural language processing (NLP) applications in academic accounting and finance research.

Extracting Useful Parts of Data

Unstructured textual data encountered in academic research within the fields of accounting and finance often comprises superfluous information and clutter that necessitates elimination prior to analysis commencement. For instance, financial reports and news articles may encompass advertisements or irrelevant sections that do not contribute to the analysis. To extract pertinent information, researchers may employ regular expressions or regex markers, which facilitate the identification of patterns or markers denoting the initiation and termination points of the desired text. This preliminary phase of text cleansing and normalization assumes a critical role by empowering researchers to isolate and extract specific segments of text, such as financial statement items or earnings announcements, while eliminating extraneous or duplicative data. By precisely extracting relevant information from unstructured text, researchers can concentrate solely on components crucial to their research endeavors.

The initial step involved in extracting valuable data segments entails identification of patterns or markers that effectively establish the starting and ending points of the desired text. These markers can manifest as distinctive characters, keywords, or phrases exclusive to the relevant text. To illustrate, when seeking to extract product names and prices from an e-commerce website, markers such as "Product Name:" and "Price:" may be employed to precisely delineate the text boundaries. Once these markers have been identified, regular expressions can be employed to extract the text situated between these demarcations. Identifying the regular pattern and creating the corresponding regex is often the most challenging aspect of the entire text extraction process. Regular expressions facilitate the search for specific patterns or characters within the text, enabling extraction of the desired information. Various regex functions, including match, search, and findall, can be employed to extract pertinent information from the text.

It should be emphasized that the extraction of valuable data segments may necessitate repetition on several occasions, contingent upon the text's complexity and the relevance of the markers. Furthermore, the extraction process may be conducted either before or after the removal of HTML tags, depending on which approach yields more effective markers. Hereinafter is an exemplification of the text extraction process between markers m1 and m2, employing a collection of files extracted from SEC EDGAR and saved within the D:/Data folder:

```
import re

# regex pattern to match a valid email address
pattern = r'''
    ^                           # start of string
    [a-zA-Z0-9._%+-]+           # username
    @                           # @ symbol
    [a-zA-Z0-9.-]+\.[a-zA-Z]{2,} # domain name
    $                           # end of string
'''

# sample email address
email = 'test@example.com'

# match the email address using the regex pattern
match = re.search(pattern, email, re.VERBOSE)

if match:
    print('Valid email address found:', match.group())
else:
    print('Invalid email address')
import os
import re

# Set the directory containing the text files
directory = "D:/Data"

# Set the regular expression markers
m1 = r"start_marker"
m2 = r"end_marker"

# Loop through each file in the directory and extract the text between the markers
for filename in os.listdir(directory):
    if filename.endswith(".txt"):
        with open(os.path.join(directory, filename), "r") as f:
            # Read the file contents
            file_contents = f.read()

            # Find all occurrences of the markers and extract the text in between
            matches = re.findall(m1 + "(.*?)" + m2, file_contents, re.DOTALL)
            # Write the extracted text to a new file in the same directory
            for i, match in enumerate(matches):
                new_filename = os.path.splitext(filename)[0] + f"_extracted_{i}.txt"
                with open(os.path.join(directory, new_filename), "w") as new_f:
                    new_f.write(match)
print('Extraction complete!')
```

Let us proceed to comprehensively analyze the code segment in question, addressing each step in a systematic manner.

```
import os
import re
```

Here, we are importing the necessary libraries: **os** for directory handling and **re** for regular expressions. Next part of the code is:

```
dir_path = r'D:/Data'
m1 = 'start'
m2 = 'end'
```

We define the directory path where our text files are located as **dir_path**. **m1** and **m2** are the regex markers that we will use to extract the relevant text. These markers can be any string or regular expression pattern that can uniquely identify the start and end of the text we want to extract.

```
for root, dirs, files in os.walk(dir_path):
    for file in files:
        if file.endswith('.txt'):
            with open(os.path.join(root, file), 'r') as f:
                text = f.read()
```

Here, we are using the **os.walk** function to traverse the directory tree rooted at **dir_path** and iterate over each file in the directory tree. We check if the file has a **.txt** extension and if so, we open it in read mode using the **with** statement. Next part of the code is:

```
            pattern = re.compile(f'{m1}(.*?){m2}', re.DOTALL)
            match = re.search(pattern, text)
            if match:
                extracted_text = match.group(1)
                with open(os.path.join(root, 'extracted_' + file), 'w') as outfile:
                    outfile.write(extracted_text)
```

Next, we define a regular expression pattern using the **re.compile** function. The pattern consists of **m1**, followed by any number of characters (including newlines), followed by **m2**. The **re.DOTALL** flag is used to make the **.** character match any character, including newlines. We then search for this pattern in the text using the **re.search** function. If we find a match, we extract the text between the markers using the **group(1)** method. We then write the extracted text to a new file with a prefix of **extracted_** in the same directory as the original file.

```
print('Extraction complete!')
```

Here we print a message to indicate that the extraction process is complete. So, this code essentially traverses through all the text files in the specified directory tree, extracts the text between the regex markers and writes it to a new file in the same directory as the original file. If regex markers are good, this code should work on all text files. If there is problem, you should check your regex markers.

Cleaning HTML Tags

HTML tags are utilized for the purpose of structuring content on web pages, and their presence can pose challenges when dealing with unstructured text data. Textual information extracted from websites frequently contains HTML tags, which can impede text analysis by introducing irrelevant information that lacks utility for the analysis. Consequently, the elimination of HTML tags from text data becomes an essential task in the preprocessing of textual information.

The procedure of removing HTML tags involves the complete elimination of all tags within the text data. This can be accomplished through the utilization of regular expressions, which possess the capability to identify and eliminate all HTML tags. It is important to note that the cleaning process might necessitate multiple iterations, as certain tags may persist even after an initial removal attempt.

A commonly employed approach for removing HTML tags entails the application of the BeautifulSoup library in the Python programming language. Specifically designed for web scraping and HTML document parsing, this library facilitates the extraction of text from HTML documents while disregarding the accompanying tags. Consequently, the cleaning process is streamlined and enhanced in terms of efficiency. Another widely used tool for eliminating HTML tags is the lxml library, which offers similar functionality.

Upon the successful removal of HTML tags, the resultant text data assumes a more pristine and standardized format, thereby facilitating subsequent analysis and processing. The subsequent sections will furnish examples illustrating the process of removing HTML tags using both regular expressions and the BeautifulSoup library in Python. Presented below is an illustrative example:

```
import os
from bs4 import BeautifulSoup

# directory path containing text files
dir_path = r'D:/Data/Extracted'

# iterate over all files in the directory
for filename in os.listdir(dir_path):
    if filename.endswith('.txt'):
        file_path = os.path.join(dir_path, filename)
        # read the contents of the file
        with open(file_path, 'r', encoding='utf-8') as f:
            text = f.read()
        # parse HTML and extract text
        soup = BeautifulSoup(text, 'html.parser')
        cleaned_text = soup.get_text()
        # write the cleaned text back to the file
        with open(file_path, 'w', encoding='utf-8') as f:
            f.write(cleaned_text)
```

Let us look at this code in parts.

```
import os
from bs4 import BeautifulSoup
```

This code imports the required libraries, os and BeautifulSoup.

```
dir_path = r'D:/Data/Extracted'
```

This code sets the directory path where the source text files are located.

```
for filename in os.listdir(dir_path):
    if filename.endswith('.txt'):
        file_path = os.path.join(dir_path, filename)
```

This code loops through all files in the directory path and selects only the ones that end with .txt. It then sets the file path for each selected file.

```
with open(file_path, 'r', encoding='utf-8') as f:
    text = f.read()
```

This code reads the contents of the selected file into a variable called **text**.

```
soup = BeautifulSoup(text, 'html.parser')
cleaned_text = soup.get_text()
```

This code uses BeautifulSoup to parse the HTML tags in **text** and extract only the text content. The cleaned text is stored in a variable called **cleaned_text**.

```
with open(file_path, 'w', encoding='utf-8') as f:
    f.write(cleaned_text)
```

8 Text Extraction and Cleaning

This code writes the cleaned text back to the original file, replacing the previous contents.

To achieve optimal file refinement, it may be necessary to iterate the process multiple times. This algorithm demonstrates high efficiency when applied exclusively to text files comprising HTML code. Nonetheless, there are instances where the files incorporate both HTML and XML tags, necessitating the utilization of the lxml library.

```python
import os
from lxml import etree

# directory path containing text files
dir_path = r'D:/Data/Extracted'

# iterate over all files in the directory
for filename in os.listdir(dir_path):
    if filename.endswith('.txt'):
        file_path = os.path.join(dir_path, filename)
        # read the contents of the file
        with open(file_path, 'r', encoding='utf-8') as f:
            text = f.read()
        # parse HTML and extract text
        parser = etree.HTMLParser(remove_blank_text=True)
        tree = etree.fromstring(text.encode('utf-8'), parser=parser)
        cleaned_text = etree.tostring(tree, encoding='utf-8', method='text').decode('utf-8')
        # write the cleaned text back to the file
        with open(file_path, 'w', encoding='utf-8') as f:
            f.write(cleaned_text)
```

This code is similar to the previous one, with the main difference being the use of the **lxml** library for parsing XML/HTML documents. The **etree.HTMLParser()** method is used to create an HTML parser that can handle malformed HTML, while the **etree.fromstring()** method is used to parse the input text and create an element tree. The **etree.tostring()** method is then used to extract the text content from the element tree, and the resulting string is written back to the file.

Text extraction and cleaning techniques is a crucial step in academic accounting and finance research, enabling researchers to work with large volumes of unstructured textual data and gain valuable insights that may not be apparent through traditional data analysis methods. This process makes the data readable to the user, reduces the amount of data to be handled, and enables researchers to focus their analysis on the most meaningful parts of the data. This first step of text extraction and cleaning is necessary for any further analysis on data. By accurately extracting relevant information from unstructured text, researchers can gain focus on what is important to their research and using more advance techniques such as sentiment analysis, topic modeling, or semantic analysis gain valuable insights and contribute to the academic literature.

9

Text Normalization

Text normalization is the process of transforming text data into a standardized or canonical form. It involves converting text data into a consistent format by applying various techniques. Text normalization is an important step in natural language processing (NLP) tasks such as text classification, sentiment analysis, and information extraction. Its purpose is to convert text data into a consistent format through the application of various techniques, thereby enhancing the accuracy and effectiveness of these NLP tasks.

The specific techniques employed for text normalization depend on the particular requirements of the NLP task at hand and the inherent characteristics of the text data. For instance, when dealing with social media data, emphasis may be placed on the removal of emoticons and hashtags. On the other hand, when working with financial statements or news articles, greater attention may be given to eliminating stop words and performing stemming or lemmatization on the text. Some common techniques involved in text normalization include lowercasing, stemming, lemmatization, as well as the removal of stop words and special characters. Ultimately, the primary objective of text normalization is to render the text data more structured and amenable to analysis.

Removal of Special Characters and Punctuation Marks

The process of eliminating special characters and punctuation entails the removal of non-alphabetic and non-digit characters from textual data, including punctuation marks, symbols, emojis, and other special characters. The presence of these characters can impede the analysis of the text data, and their removal facilitates the standardization of the text while reducing noise in the data. For instance, let us consider the following sentence:

"Hello! How are you today?"

To eliminate special characters and punctuation, the exclamation mark would be eradicated and replaced with a space, thereby yielding the resulting standardized sentence:

"Hello How are you today."

Lowercasing

Lowercasing is a fundamental textual operation employed in text normalization, whereby all alphabetical characters within a given text are converted to their lowercase equivalents. The purpose of this procedure is to facilitate the uniform treatment of various word forms as identical entities during subsequent processing stages. Failure to apply lowercasing can lead to the differentiation of words such as "Income" and "income" as separate entities, thereby introducing potential errors or inaccuracies in the context of text analysis endeavors, including tasks such as text classification or sentiment analysis. By applying lowercasing to the text, these aforementioned variations are harmonized into a singular representation, thus enhancing the accuracy of these aforementioned analytical tasks.

Tokenizing

Tokenization is the fundamental procedure of decomposing a given text into smaller entities known as tokens, typically encompassing words or sentences. This process holds significant relevance in the realm of text manipulation, enabling more streamlined analysis of textual data. Tokenization is applicable at various granularities, including word-level tokenization that dissects the text into discrete words, and sentence-level tokenization that partitions the text into separate sentences. Frequently serving as the initial stage in natural language processing endeavors, tokenization plays a pivotal role in facilitating tasks such as text categorization, sentiment analysis, and automated language translation.

Stop Word Removal

Stop word refers to the procedure of excluding commonly used words in a given language that lack substantial meaning or significance within a specific text. Examples of such words include "the", "is", "and", and "a". By removing these stop words from textual data, the dataset's size can be reduced, consequently enhancing the accuracy of natural language processing (NLP) algorithms.

The elimination of stop words holds significance during text preprocessing, as these words frequently appear in the data but contribute little meaningful information. The removal of these words enables the focus to shift toward more significant words that carry semantic value within the text. The process of removing stop words can be accomplished through the utilization of preexisting stop word lists or by constructing customized lists tailored to the particular context of the textual data. In Python, stop words can be eliminated using built-in dictionaries provided by the NLTK library or by employing a customized list of stop words.

Stemming

Stemming is a fundamental procedure within the domain of natural language processing (NLP) whereby inflected words are transformed into their respective base or root forms, resulting in what is termed a "stem". The primary purpose of stemming is to establish a standardization of words and to diminish the volume of words requiring processing in NLP applications. To illustrate, the stem of terms such as "running", "runner", and "runs" is "run". The utilization of stem forms permits the treatment of these words as identical entities within NLP applications, thereby enhancing both the accuracy and efficiency of analyses. Several distinct algorithms have been developed for the process of stemming, including the Porter stemmer, Snowball stemmer, and Lancaster stemmer. Each algorithm adheres to distinct sets of rules and produces dissimilar outcomes.

Llematization

Lemmatization is a technique employed in natural language processing (NLP) to condense words to their base or dictionary form, referred to as the lemma. Its objective is to cluster inflected word forms together for the purpose of analyzing them as a unified entity. Lemmatization is regarded as a more advanced approach than stemming, as it takes into account the surrounding context of a word in order to ascertain the appropriate lemma. To illustrate, the lemma for the words "am", "are", and "is" is "be". By applying lemmatization to a given text, the quantity of unique words within the corpus can be diminished, thereby enhancing the accuracy of various text analysis tasks, including text classification, sentiment analysis, and topic modeling. While Stemming involves the reduction of words to their base form through the removal of suffixes or affixes, lemmatization determines the canonical form of words by considering various factors such as morphology, context, and part of speech. Thus, lemmatization is more informed by linguistic principles, leading to the production of valid words while preserving grammatical accuracy.

In natural language processing, a corpus is a large and structured set of text documents. It is used to train and evaluate various models and algorithms in language processing tasks, such as text classification, information retrieval, sentiment analysis, and machine translation. A corpus is formed after the collection of raw text data, and after it has undergone preprocessing steps such as cleaning, normalization, and tokenization. These preprocessing steps are necessary to ensure that the corpus is clean and consistent, so that it can be used effectively for various natural language processing tasks such as text classification, sentiment analysis, and information retrieval. Once the text data has been cleaned and normalized, it is typically organized into a structured format, such as a document-term matrix or a set of preprocessed documents, which can then be used as input to various NLP algorithms.

Here's an example of a code that does all of the above in one go:

```
import os
import nltk
from nltk.corpus import stopwords
from nltk.stem import PorterStemmer, WordNetLemmatizer
from string import punctuation

# set the path of the directory containing text files
dir_path = r'D:/Data/Extracted'

# set the path to store the preprocessed files
output_path = r'D:/Data/Preprocessed'

# define the NLTK stop words
stop_words = set(stopwords.words('english'))

# define the NLTK stemmer and lemmatizer
stemmer = PorterStemmer()
lemmatizer = WordNetLemmatizer()

# iterate over all files in the directory
for subdir, dirs, files in os.walk(dir_path):
    for file in files:
        if file.endswith('.txt'):
            # read the contents of the file
            with open(os.path.join(subdir, file), 'r', encoding='utf-8') as f:
                text = f.read()
            # remove special characters and punctuation
            text = ''.join([char for char in text if char not in punctuation])
            # convert text to lowercase
            text = text.lower()
            # tokenize text into words
            words = nltk.word_tokenize(text)
            # remove stop words
            words = [word for word in words if word not in stop_words]
            # perform stemming and lemmatization
            stemmed_words = [stemmer.stem(word) for word in words]
            lemmatized_words = [lemmatizer.lemmatize(word) for word in words]
            # write preprocessed text to new file
            with open(os.path.join(output_path, file), 'w', encoding='utf-8') as f:
                f.write(' '.join(stemmed_words))
```

The explanation of this code step by step is as follows:

1. Import necessary libraries and modules:

```
import os
import nltk
from nltk.corpus import stopwords
from nltk.stem import PorterStemmer, WordNetLemmatizer
from string import punctuation
```

This part of code imports the following modules: os for file and directory operations, nltk for natural language processing, stopwords and PorterStemmer from the nltk.corpus and nltk.stem modules, respectively, for removing stop words and stemming words, WordNetLemmatizer from the nltk.stem module for lemmatization, and punctuation from the string module for removing special characters and punctuation.

2. Set paths for input and output directories:

```
dir_path = r'D:/Data/Extracted'
output_path = r'D:/Data/Preprocessed'
```

This part of code then sets the paths to the input directory (**dir_path**) containing the raw text files and the output directory (**output_path**) where the preprocessed files will be stored.

3. Define stop words, stemmer, and lemmatizer:

```
stop_words = set(stopwords.words('english'))
stemmer = PorterStemmer()
lemmatizer = WordNetLemmatizer()
```

The part of the code we define a set of English stop words using the **stopwords** module of NLTK and initialize the Porter stemmer and WordNet lemmatizer from the **nltk.stem** module of NLTK.

4. Iterate over all files in the directory:

```
for subdir, dirs, files in os.walk(dir_path):
    for file in files:
        if file.endswith('.txt'):
            # read the contents of the file
            with open(os.path.join(subdir, file), 'r', encoding='utf-8') as f:
                text = f.read()
```

Here, we are using the os.walk function to iterate over all the files in the directory (dir_path). The os.path.join function is used to join the directory path and file name to create the full file path. We only process files with .txt extension. Inside the loop, we open each file in read mode and read its contents into the text variable.

5. We then preprocess the text and write the preprocessed text to a new file. Remove special characters and punctuation:

```
            text = ''.join([char for char in text if char not in punctuation])
            # convert text to lowercase
            text = text.lower()
            # tokenize text into words
            words = nltk.word_tokenize(text)
            # remove stop words
            words = [word for word in words if word not in stop_words]
            # perform stemming and lemmatization
            stemmed_words = [stemmer.stem(word) for word in words]
            lemmatized_words = [lemmatizer.lemmatize(word) for word in words]
            # write preprocessed text to new file
            with open(os.path.join(output_path, file), 'w', encoding='utf-8') as f:
                f.write(' '.join(stemmed_words))
```

In this part of the code, special characters and punctuation are removed from the text using a list comprehension and the **string.punctuation** string. Then, all text is converted to lowercase using the **lower()** function, which ensures that case does not affect the analysis. The text is then tokenized into words using the **word_tokenize()** function from the nltk module. The next step involves removing stop words from the list of words using a list comprehension and the **stop_words** set. Finally, the words are stemmed using the Porter stemmer, which reduces the inflected forms of the words to their base or root form. Before this code can be done nltk library has to be installed using pip install method and punkt, stopwords, and wordnet packages need to be installed using the following code:

```
import nltk
nltk.download('punkt')
nltk.download('stopwords')
nltk.download('wordnet')
```

There are cases when emoticons are relevant for the analysis of text data. Emoticons are often used in social media and online communication to convey sentiments and emotions, and removing them could potentially remove important information from the text. In such cases, it might be necessary to modify the preprocessing steps to retain emoticons or other relevant symbols. However, it depends on the specific use case and the goals of the analysis.

In order to preserve the emoticons, it is necessary to modify the code accordingly. Instead of utilizing list comprehension to eliminate all special characters, the code can be adapted to exclude the emoticons that are intended to be retained. To illustrate, if it is desired to retain smiley faces such as ":)" and ":(", the code can be adjusted in the following manner:

```
# remove special characters and punctuation except emoticons
text = ''.join([char for char in text if char not in punctuation
or char in [':', ')', '(', ';']])
```

To retain all the emoticons in the text, we can modify the code to remove only the punctuation marks that are not emoticons. We can create a list of punctuation marks that are not emoticons and remove only those. For example, we can include the following punctuation marks in the list: ".", ",", ":", ";", "?", "!", "-", "_", "(", ")", "[", "]", "{", "}", "'", '"'. Here's how the modified code would look like:

9 Text Normalization

```python
import os
import nltk
from nltk.corpus import stopwords
from nltk.stem import PorterStemmer, WordNetLemmatizer
import string

# set the path of the directory containing text files
dir_path = r'D:/Data/Extracted'

# set the path to store the preprocessed files
output_path = r'D:/Data/Preprocessed'

# define the NLTK stop words
stop_words = set(stopwords.words('english'))

# define the NLTK stemmer and lemmatizer
stemmer = PorterStemmer()
lemmatizer = WordNetLemmatizer()

# define the list of punctuation marks to be removed
punctuation_to_remove = [char for char in string.punctuation if char not in ["'", '"', ':', ';', '?', '!']]

# iterate over all files in the directory
for subdir, dirs, files in os.walk(dir_path):
    for file in files:
        if file.endswith('.txt'):
            # read the contents of the file
            with open(os.path.join(subdir, file), 'r', encoding='utf-8') as f:
                text = f.read()
            # remove special characters and punctuation
            text = ''.join([char for char in text if char not in punctuation_to_remove])
            # convert text to lowercase
            text = text.lower()
            # tokenize text into words
            words = nltk.word_tokenize(text)
            # remove stop words
            words = [word for word in words if word not in stop_words]
            # perform stemming and lemmatization
            stemmed_words = [stemmer.stem(word) for word in words]
            lemmatized_words = [lemmatizer.lemmatize(word) for word in words]
            # write preprocessed text to new file
            with open(os.path.join(output_path, file), 'w', encoding='utf-8') as f:
                f.write(' '.join(lemmatized_words))
```

In this modified code, we have defined a list of punctuation marks to be removed that does not include the emoticons. We have used this list in the list comprehension that removes the punctuation marks from the text.

Special Considerations in Accounting Data

When it comes to processing accounting data, special considerations need to be taken into account due to the unique nature of financial information. One of the most important considerations is the use of symbols in the data. While it may be common to remove symbols in other forms of text data processing, it is often necessary to retain certain symbols in accounting data to differentiate monetary numbers from other numbers. For example, the dollar sign ($) is commonly used to denote monetary values in accounting, and removing it could cause confusion and inaccuracies in financial analysis.

Another important consideration in accounting data processing is the use of bracket signs around numbers. In accounting, it is common to denote negative numbers by surrounding them with brackets. For example, $ (500) would represent a negative monetary value of five hundred dollars. If the brackets are removed during text cleaning, the meaning of the number may be misinterpreted as a positive value, leading to inaccurate financial analysis.

Therefore, when processing accounting data, it is important to consider the unique nature of financial information and take special precautions to ensure that important symbols and notations are retained. This can help to ensure the accuracy and reliability of financial analysis, which is critical for making informed business decisions based on accounting data.

A modified code that leaves '$' and '(' and ')' would be:

9 Text Normalization

```python
import os
import nltk
from nltk.corpus import stopwords
from nltk.stem import PorterStemmer, WordNetLemmatizer
from string import punctuation

# set the path of the directory containing text files
dir_path = r'D:/Data/Extracted'

# set the path to store the preprocessed files
output_path = r'D:/Data/Preprocessed'

# define the NLTK stop words
stop_words = set(stopwords.words('english'))

# define the NLTK stemmer and lemmatizer
stemmer = PorterStemmer()
lemmatizer = WordNetLemmatizer()

# define the punctuation to remove
exclude_punct = set(punctuation) - set(['$', '(', ')'])

# iterate over all files in the directory
for subdir, dirs, files in os.walk(dir_path):
    for file in files:
        if file.endswith('.txt'):
            # read the contents of the file
            with open(os.path.join(subdir, file), 'r', encoding='utf-8') as f:
                text = f.read()
            # remove special characters and punctuation, excluding $ and () around numbers
            text = ''.join([char for char in text if char not in exclude_punct or (char in ['$','('] and text[text.find(char)-1].isdigit() and text[text.find(char)+1].isdigit())])
            # convert text to lowercase
            text = text.lower()
            # tokenize text into words
            words = nltk.word_tokenize(text)
            # remove stop words
            words = [word for word in words if word not in stop_words]
            # perform stemming and lemmatization
            stemmed_words = [stemmer.stem(word) for word in words]
            lemmatized_words = [lemmatizer.lemmatize(word) for word in words]
            # write preprocessed text to new file
            with open(os.path.join(output_path, file), 'w', encoding='utf-8') as f:
                f.write(' '.join(stemmed_words))
```

Here we define a new variable **exclude_punct** that contains all punctuation marks except $ and (and). In the line that removes special characters and punctuation, modifying the condition to exclude $ and () around numbers. The new condition checks whether the character is in **exclude_punct** or whether it is $ or (and the character before and after it are both digits.

This way, the $ and () around numbers will not be removed during text normalization.

Now, in the next step, we remove all the '(' and ')' and add a negative sign before all the numbers surrounded by bracket signs:

```
import os
import re

# set the path of the directory containing text files
dir_path = r'D:/Data/Extracted'

# set the path to store the preprocessed files
output_path = r'D:/Data/Preprocessed'

# define the regular expression pattern to match numbers
surrounded by ()
pattern = r'\(((\d+)\)'

# iterate over all files in the directory
for subdir, dirs, files in os.walk(dir_path):
    for file in files:
        if file.endswith('.txt'):
            # read the contents of the file
            with open(os.path.join(subdir, file), 'r', encoding='utf-8') as f:
                text = f.read()
            # remove all ()
            text = re.sub(r'\(|\)', '', text)
            # replace numbers surrounded by () with -number
            text = re.sub(pattern, r'-\1', text)
            # write preprocessed text to new file
            with open(os.path.join(output_path, file), 'w', encoding='utf-8') as f:
                f.write(text)
```

The code first defines a regular expression pattern to match numbers surrounded by (). The pattern captures the numbers using parentheses (**\d+**). The code then iterates over all text files in the **D:/Data/Extracted** directory, reads in the contents of each file, removes all () signs, and replaces any numbers surrounded by () with the same number preceded by a—sign using the **re.sub()** function. Finally, the preprocessed text is saved in the **D:/Data/Preprocessed** directory with the same filename.

The paramount role of text normalization in academic research within the domains of accounting and finance stems from its ability to standardize textual information, enabling more effective and robust analyses. By employing normalization techniques, researchers can eliminate discrepancies introduced by irregularities in the textual data, thereby facilitating the identification and extraction of pertinent information. This harmonization of the data aids in mitigating potential biases and inaccuracies that could compromise the integrity of subsequent analyses and findings.

Moreover, text normalization serves as a foundation for advanced natural language processing (NLP) techniques, empowering researchers to delve deeper into the intricacies of the textual content. Through the systematic application of normalization techniques, researchers can establish consistent and coherent corpora, enabling comprehensive analyses of large volumes of text data.

10

Corpus

A corpus, in the context of language analysis and machine learning models, refers to a compilation of textual documents. It holds significant importance in the field of natural language processing (NLP) and finds applications in diverse areas such as text classification, sentiment analysis, and information extraction.

A corpus can be thought of as a table with rows and columns. Each row represents a document or text file, while each column represents a feature or attribute of the text. The features can include the word frequency, part of speech tags, sentiment scores, and so on. For example, if we have a corpus of customer reviews for a product, each row will represent a single review, and the columns could include the overall rating, the specific aspects of the product mentioned in the review (e.g., quality, price, usability), and the sentiment of the review. By representing the corpus in a tabular format, we can easily analyze and visualize the data using tools like Excel or data analysis libraries in programming languages like Python or R. We can also use SQL queries to extract specific subsets of the corpus based on certain conditions.

Renaming Files

While creating a corpus, it is often necessary to rename the text files to better reflect the content they contain. Renaming can help in maintaining consistency and avoiding confusion while processing and analyzing the text data. If the file names are descriptive and meaningful, it can be easier to identify the contents of each file without having to open and read it. It can help in organizing the files in a logical manner. For example, if the files are related to a specific project or topic, renaming them with a common prefix or suffix can help in grouping them together and identifying them as part of the same collection. Renaming the files can also help in avoiding potential errors or conflicts while processing the data. If the original file names contain special characters or are too long, it may cause issues while accessing or manipulating them programmatically. Renaming them with shorter and simpler names can help in avoiding such errors.

To reflect the content of the file, in many cases, it may be necessary to extract information from the file itself and use that information to name the file. For example, if the file contains a news article, we may want to extract the headline and use it as the filename. This can be achieved through text processing techniques like regular expressions. By having descriptive filenames, it becomes easier to locate and work with specific documents within the corpus.

Here's a code that renames all the text files in the directory and subdirectories based on information in the file. In this example we have created four markers (marker1 to marker4) using regex. We extract the text between marker1 and marker2 and then between marker3 and marker4 and append the extracted texts to create filename.

```python
import os
import re

dir_path = 'D:/Data/Preprocessed/'

# iterate over all files in the directory
for filename in os.listdir(dir_path):
    if filename.endswith('.txt'):
        file_path = os.path.join(dir_path, filename)
        # read the contents of the file
        with open(file_path, 'r', encoding='utf-8') as f:
            text = f.read()
        # extract text between marker1 and marker2
        marker1 = "some regex pattern"
        marker2 = "some regex pattern"
        pattern = re.compile(marker1 + "(.*?)" + marker2, re.DOTALL)
        match1 = re.search(pattern, text)
        if match1:
            text1 = match1.group(1)
        else:
            text1 = ""
        # extract text between marker3 and marker4
        marker3 = "some regex pattern"
        marker4 = "some regex pattern"
        pattern = re.compile(marker3 + "(.*?)" + marker4, re.DOTALL)
        match2 = re.search(pattern, text)
        if match2:
            text2 = match2.group(1)
        else:
            text2 = ""
        # combine the extracted text and create a new string
        new_string = text1 + "_" + text2
        # rename the file as new_string
        new_file_name = os.path.join(dir_path, new_string + '.txt')
        os.rename(file_path, new_file_name)
```

In this code, we set the directory path where the text files are stored in the **dir_path** variable. Then, we use the **os.listdir()** function to iterate over all files in the directory and check if the file ends with **.txt** extension. If yes, we proceed with reading the contents of the file using the **with open()** statement.

Next, we define two regex patterns **marker1** and **marker2** to extract the text between them from the file. We use the **re.compile()** function to create a compiled regex pattern with **re.DOTALL** flag to match any character, including newline. We then use the **re.search()** function to search for the pattern in the text and extract the group between the markers using the **group()** method. If the pattern is not found, we set the **text1** variable to an empty string. We repeat the same process to extract the text between **marker3** and **marker4** and store it in the **text2** variable. We then combine the extracted text using a space and create a new string **new_string**.

Finally, we rename the file with the **os.rename()** function by providing the original file path and the new file path with the updated name. The new file name is created by concatenating the **new_string** with the **.txt** extension. This process is repeated for all text files in the directory. For, example if the text files are related to annual financial statements of many companies and in the code above the if **match1** is designed to extract the name of company and **match2** is designed to extract fiscal year end date then the filename would be something like **walmart_ January 31 2019.txt**. Similarly, if we are dealing with product the filename could be created based on name of product and date and time of review.

Sorting Files

When creating a corpus, it is often necessary to include certain files while excluding others. One way to accomplish this is by sorting the desired files into a separate directory, leaving the undesired ones in their original location. By separating the desired files in this way, it becomes easier to include only the relevant files in the corpus for further analysis using Python. This can be especially useful when dealing with large volumes of text data, as it helps to streamline the data processing and analysis tasks.

For instance, when working with financial reports downloaded from SEC EDGAR, you can sort 10K and 10Q files into separate folders using Python. This can be achieved by searching for the keywords "10-K" or "10k" (for 10K files), or "10-Q" or "10q" (for 10Q files) in the filenames of the downloaded files, and copying the matching files to their respective destination folders. Here's a code that separates the 10-K and 10-Q files downloaded from SEC EDGAR.

```
import os
import shutil

def move_files(src_dir, dst_dir, keywords):
    for root, _, files in os.walk(src_dir):
        for file in files:
            if any(keyword in file for keyword in keywords):
                src_path = os.path.join(root, file)
                dst_subfolder = os.path.relpath(root, src_dir)
                dst_path = os.path.join(dst_dir, dst_subfolder, file)

                os.makedirs(os.path.dirname(dst_path), exist_ok=True)
                shutil.move(src_path, dst_path)
                print(f'Moved {src_path} to {dst_path}')

src_dir = 'D:/Data'

move_files(src_dir, os.path.join(src_dir, '10K'), ['10_K', '10K'])
move_files(src_dir, os.path.join(src_dir, '10Q'), ['10_Q', '10Q'])
```

This code uses the **os** and **shutil** modules in Python to traverse the source directory, search for files containing specific keywords, and move those files to their corresponding destination directories. The **move_files** function is responsible for the main logic of the script. It accepts three parameters: the source directory, the destination directory, and a list of keywords. For each file in the source directory, the function checks if the filename contains any of the specified keywords, and if so, moves the file to the corresponding subdirectory in the destination directory. The function also creates any necessary subdirectories in the destination directory if they do not already exist.

Creating Corpus

Once we have sorted and renamed the text files, the next step is to create a corpus from these files. A corpus is a fundamental resource for Natural Language Processing (NLP) tasks, such as text classification, sentiment analysis, and information extraction. Each document in the corpus is typically identified by a unique identifier, such as a filename. Here's a code that creates a corpus with the renamed files.

```
import os
import shutil

def move_files(src_dir, dst_dir, keywords):
    for root, _, files in os.walk(src_dir):
        for file in files:
            if any(keyword in file for keyword in keywords):
                src_path = os.path.join(root, file)
                dst_subfolder = os.path.relpath(root, src_dir)
                dst_path = os.path.join(dst_dir, dst_subfolder, file)

                os.makedirs(os.path.dirname(dst_path), exist_ok=True)
                shutil.move(src_path, dst_path)
                print(f'Moved {src_path} to {dst_path}')

src_dir = 'D:/Data'

move_files(src_dir, os.path.join(src_dir, '10K'), ['10_K', '10K'])
move_files(src_dir, os.path.join(src_dir, '10Q'), ['10_Q', '10Q'])
```

The code first defines the directory path containing the preprocessed text files and sets the regular expression pattern to match all files with a **.txt** extension. It then creates a corpus object using the **PlaintextCorpusReader()** function from the **nltk** library. The **PlaintextCorpusReader()** function takes two arguments: the directory path and the regular expression pattern to match the file names. The regular expression pattern in this code is **r'(?!/.)[/w-]+/.txt'**, which matches all file names with a **.txt** extension but excludes files that start with a period (i.e., hidden files on Unix systems). The **PlaintextCorpusReader()** function returns a corpus object, which can be used to access the contents of the text files. For example, to access the raw text of a specific file in the corpus, you can use the **raw()** method of the corpus object and specify the file identifier (i.e., the file name with its extension) as the argument.

This corpus is stored in the memory of your Python environment. When you create a corpus using the **PlaintextCorpusReader** function, it reads the plaintext files in the specified directory and constructs the corpus in memory. You can then access the corpus and perform various operations on it within your Python code. However, the corpus is not saved as a separate file on your hard drive, unless you explicitly save it using methods like **pickle.dump()**. The following code would be save the corpus in D:/Data folder:

```
import pickle
from nltk.corpus import PlaintextCorpusReader

corpus_root = r'D:/Data/Preprocessed'
file_pattern = r'(?!/.)[/w-]+/.txt'

# Create a corpus using the plaintext files in the directory
corpus = PlaintextCorpusReader(corpus_root, file_pattern)

# Save the corpus to a binary file using pickle
with open(r'D:/Data/my_corpus.bin', 'wb') as f:
    pickle.dump(corpus, f)
```

You can visualize this corpus and a table of two columns. First column contains all the file names and the second column contains all the text in the file. You can access the specific text in a file (example_file.txt) using the following code:

```
import pickle
from nltk.corpus.reader.api import CorpusReader

corpus_path = r'D:/Data/my_corpus.bin'

# Load the corpus from the binary file using pickle
with open(corpus_path, 'rb') as f:
    corpus = pickle.load(f)

# Access a specific file in the corpus using its file identifier
text = corpus.raw(fileids='example_file.txt')
print(text)
```

When working with a corpus in Python, it is generally more efficient to use the computer's memory to store and manipulate the text data, rather than repeatedly accessing files on disk. This is especially true when working with large corpora or when integrating multiple codes to perform multiple activities in one go. By loading the corpus into memory, Python can perform operations on the data more quickly and efficiently, since the data is readily available in memory rather than being read from disk each time it is needed. This can speed up the processing time and improve the overall performance of the code. However, it is important to note that there may be cases where it is necessary to work directly with the files on disk. For example, if the corpus is too large to fit in memory, or if you need to process the data in batches or in a streaming fashion. In these cases, you may need to use techniques such as memory mapping or file caching to optimize the performance of the code.

Part III

Exploratory Data Analysis and Visualization

11

Data Visualization: Numerical Data

Data visualization is the process of representing data in a visual or graphical form. It is a powerful tool for gaining insights into complex data and communicating those insights to others. Effective data visualization can help researchers to quickly identify patterns, trends, and outliers that might not be apparent from a table or spreadsheet of raw data. This makes it easier to understand the data and make informed decisions based on the insights that it provides.

In academic research, especially in fields such as accounting and finance, data visualization is essential. It is often used to illustrate complex relationships between variables, communicate research findings to stakeholders, and aid in the interpretation of statistical models. While it is true that researchers can analyze data without visualizing it, data visualization can often make it easier to identify key patterns, trends, and relationships that may be obscured in a table or spreadsheet.

Python offers a wide range of tools for data visualization that are especially useful for academic research in accounting and finance. Some of the most popular Python libraries for data visualization include Matplotlib, Seaborn, Plotly, and Bokeh. These libraries provide a range of functions for creating a wide variety of visualizations, from simple scatterplots and bar charts to complex 3D visualizations and interactive dashboards.

Matplotlib is a powerful and flexible library that is widely used for creating static, 2D visualizations. Seaborn is a more specialized library that builds on top of Matplotlib, offering more advanced visualization functions and built-in support for statistical analysis. Plotly and Bokeh, on the other hand, are designed for creating interactive visualizations that can be embedded in web applications and dashboards.

Matplotlib

Matplotlib is a widely used Python library for creating static, two-dimensional visualizations. It provides a wide range of functions for creating a variety of chart types, including scatter plots, line charts, bar charts, histograms, and more. It can be used to create publication-quality plots for use in academic research, as well as for creating visualizations for presentations or reports.

Matplotlib is highly customizable, offering control over every aspect of a visualization, from the font size to the line style. This makes it a powerful tool for creating bespoke visualizations that suit the specific needs of a particular research project. It also integrates well with other Python libraries, making it easy to create visualizations that include data from multiple sources.

One of the key strengths of Matplotlib is its versatility. It can be used to create simple visualizations quickly and easily, but it also offers a wide range of advanced features for creating more complex visualizations. For example, it can be used to create subplots, add annotations and labels, and create custom color maps.

Matplotlib is also highly compatible with other Python libraries. It works seamlessly with NumPy, Pandas, and SciPy, allowing researchers to quickly create visualizations from their data. In addition, Matplotlib integrates with Jupyter Notebook, a popular tool for data analysis and visualization, making it easy to create interactive visualizations that can be shared with others.

A simple code using matplotlib that can make five different kinds of plots is as follows:

```python
import matplotlib.pyplot as plt
import numpy as np

# Create a simple line plot
x = np.arange(0, 10, 0.1)
y = np.sin(x)
plt.plot(x, y)
plt.xlabel('x')
plt.ylabel('y')
plt.title('Simple Line Plot')
plt.show()

# Create a scatter plot
x = np.random.normal(size=100)
y = np.random.normal(size=100)
plt.scatter(x, y)
plt.xlabel('x')
plt.ylabel('y')
plt.title('Scatter Plot')
plt.show()

# Create a bar chart
x = ['A', 'B', 'C', 'D', 'E']
y = [5, 7, 3, 4, 6]
plt.bar(x, y)
plt.xlabel('Category')
plt.ylabel('Value')
plt.title('Bar Chart')
plt.show()

# Create a histogram
x = np.random.normal(size=1000)
plt.hist(x, bins=30)
plt.xlabel('Value')
plt.ylabel('Frequency')
plt.title('Histogram')
plt.show()

# Create a pie chart
labels = ['A', 'B', 'C', 'D']
sizes = [15, 30, 45, 10]
plt.pie(sizes, labels=labels)
plt.title('Pie Chart')
plt.show()
```

This code creates five different types of plots: a simple line plot, a scatter plot, a bar chart, a histogram, and a pie chart. To create each plot, the code sets up the data, specifies the plot type, adds axis labels and a title, and then displays the plot using the **show()** function. You can use this code as a starting point for creating your own visualizations in Matplotlib.

Apart from these Matplotlib can create more plots. The following codes use random data to different plots:

Heatmap

Heatmap: A heatmap is a 2D graphical representation of data where the values are represented by colors.

```
import numpy as np
import matplotlib.pyplot as plt

data = np.random.rand(5, 5)
plt.imshow(data, cmap='hot', interpolation='nearest')
plt.colorbar()
plt.show()
```

The output will look like:

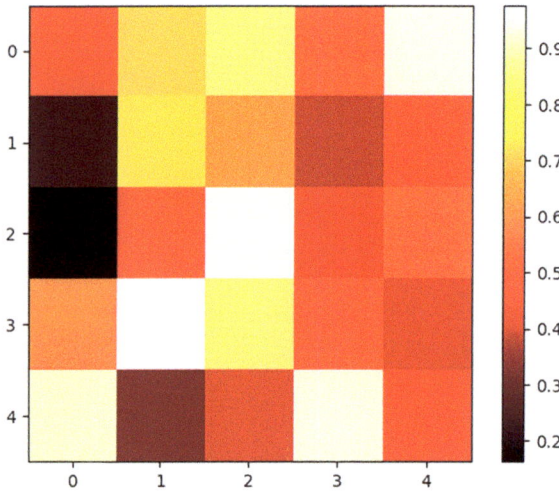

3D Plot

Matplotlib can also create 3D plots, which can be useful for visualizing data with multiple dimensions.

```
import numpy as np
import matplotlib.pyplot as plt
from mpl_toolkits.mplot3d import Axes3D

fig = plt.figure()
ax = fig.add_subplot(111, projection='3d')

x = np.random.rand(50)
y = np.random.rand(50)
z = np.random.rand(50)

ax.scatter(x, y, z)
plt.show()
```

The output will look like:

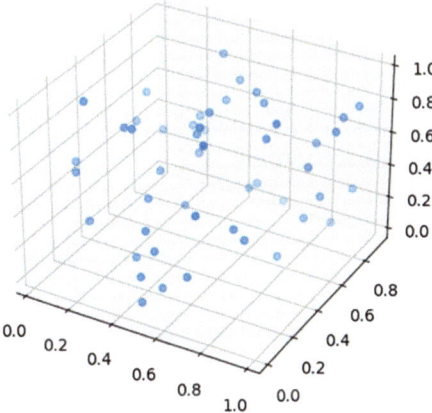

Box Plot

A box plot is a way of showing the distribution of a dataset by showing the median, quartiles, and outliers.

```
import numpy as np
import matplotlib.pyplot as plt

data = np.random.normal(size=100)
plt.boxplot(data)
plt.show()
```

The output will be:

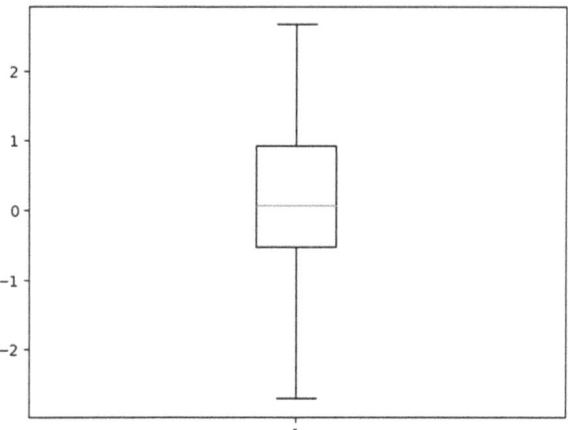

You can use your own data to make any plot using Matplotlib. Here's a code that uses data stored in D:/Data and makes a scatterplot:

```python
import matplotlib.pyplot as plt
import pandas as pd

# Load data from CSV file
data = pd.read_csv('D:/Data/acctg3.csv')

# Extract x and y variables
x = data['fyear']
y = data['ue_ce']

# Create scatter plot
plt.scatter(x, y)
plt.xlabel('x')
plt.ylabel('y')
plt.title('Scatter Plot')
plt.show()
```

Here's the output:

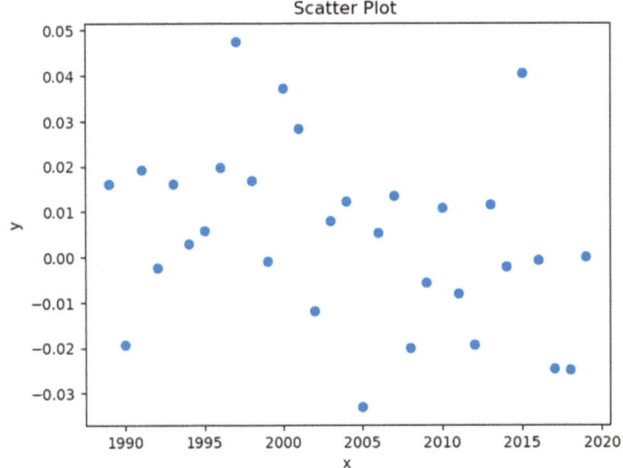

Seaborn

Seaborn is a Python data visualization library built on top of Matplotlib. It provides a high-level interface for creating a wide range of statistical graphics, including heatmaps, scatterplots, line charts, and more. Seaborn is particularly well-suited for visualizing complex, multidimensional data, and it provides many advanced features for customizing visualizations and performing statistical analysis.

One of the key strengths of Seaborn is its support for working with Pandas dataframes. Seaborn's functions are designed to work seamlessly with Pandas dataframes, allowing researchers to easily visualize and explore their data. Seaborn also provides many built-in themes and color palettes, which can help to make visualizations more visually appealing and easier to interpret.

In addition to its built-in themes and color palettes, Seaborn offers a wide range of advanced features for customizing visualizations. For example, it provides functions for creating subplots, adding annotations and labels, and adjusting plot aesthetics. Seaborn also offers support for advanced statistical techniques such as regression analysis, data smoothing, and bootstrapping.

Seaborn is particularly well-suited for visualizing complex, multidimensional data. It provides many advanced visualization functions for working with categorical data, time series data, and statistical models. For example, Seaborn provides functions for creating boxplots, violinplots, and swarmplots, which can be used to visualize the distribution of data across multiple categories. Seaborn also provides functions for creating time series plots, which can be used to visualize changes in data over time.

Here's a code the makes some plots using Seaborn.

```
import seaborn as sns
import numpy as np
import pandas as pd

# Create a scatter plot
x = np.random.normal(size=100)
y = np.random.normal(size=100)
data = pd.DataFrame({'x': x, 'y': y})
sns.scatterplot(x='x', y='y', data=data)
plt.show()

# Create a line plot
x = np.linspace(0, 10, 100)
y = np.sin(x)
data = pd.DataFrame({'x': x, 'y': y})
sns.lineplot(x='x', y='y', data=data)
plt.show()

# Create a histogram
x = np.random.normal(size=1000)
sns.histplot(x)
plt.show()

# Create a box plot
x = np.random.normal(size=100)
y = np.random.choice(['A', 'B'], size=100)
data = pd.DataFrame({'x': x, 'y': y})
sns.boxplot(x='y', y='x', data=data)
plt.show()
```

The results look very similar to matplotlib. Here are some more plots with the outputs:

```
import seaborn as sns
import matplotlib.pyplot as plt
import numpy as np
import pandas as pd

# Create a line plot with error bands
x = np.linspace(0, 10, 100)
y = np.sin(x) + np.random.normal(scale=0.1, size=len(x))
data = pd.DataFrame({'x': x, 'y': y})

sns.lineplot(x='x', y='y', data=data)
plt.fill_between(x, y - 0.1, y + 0.1, alpha=0.3)   # Add error bands with a width of 0.1
plt.xlabel('x')
plt.ylabel('y')
plt.title('Line Plot with Error Bands')
plt.show()
```

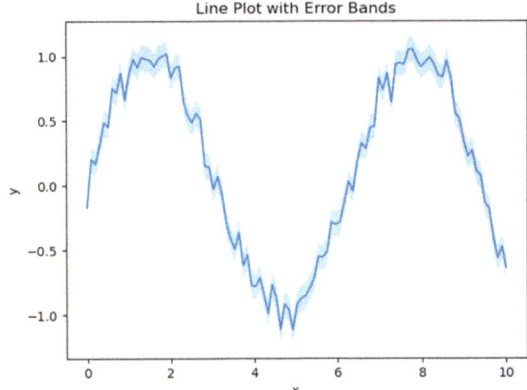

```
# Create a histogram with density curve
x = np.random.normal(size=1000)
sns.histplot(x, kde=True)
plt.xlabel('Value')
plt.ylabel('Density')
plt.title('Histogram with Density Curve')
plt.show()
```

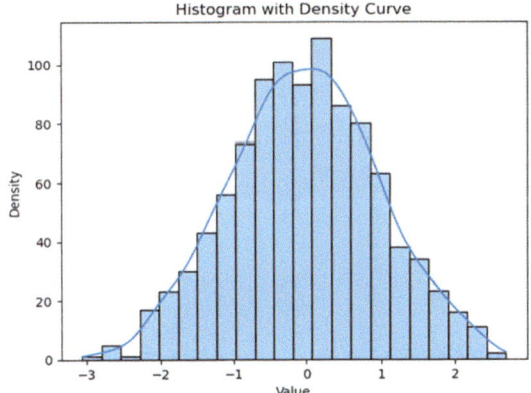

```
# Create a violin plot
x = np.random.choice(['A', 'B', 'C', 'D'], size=100)
y = np.random.normal(size=100)
data = pd.DataFrame({'x': x, 'y': y})
sns.violinplot(x='x', y='y', data=data)
plt.xlabel('Category')
plt.ylabel('Value')
plt.title('Violin Plot')
plt.show()
```

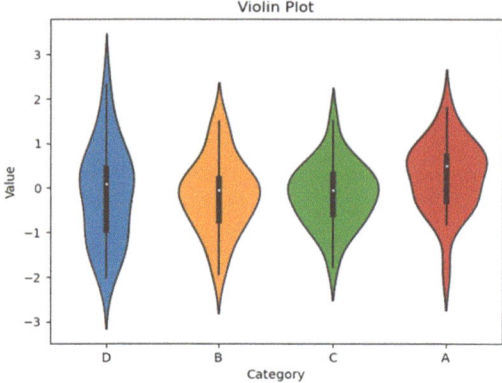

```
# Create a heat map
data = np.random.rand(10, 10)
sns.heatmap(data)
plt.title('Heat Map')
plt.show()
```

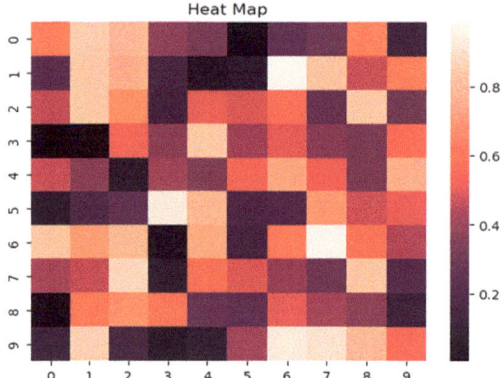

All these plots use random data. You can customize this code use your own data, similar to what we did with Matplotlib.

Following are the advanced functions of Seaborn that are not available in Matplotlib:

- Statistical Functions: Seaborn provides many statistical functions for summarizing data and calculating summary statistics such as mean, median, and quartiles. For example, Seaborn provides the **boxplot()** and **violinplot()** functions for creating box plots and violin plots, respectively, which automatically calculate and display summary statistics for the data.

- Categorical Plot Types: Seaborn provides many plot types for visualizing categorical data, such as **factorplot()**, **swarmplot()**, **countplot()**, and **pointplot()**. These plot types make it easy to visualize relationships between categorical variables and to compare distributions across categories.
- Color Palettes: Seaborn provides many built-in color palettes for customizing the colors of plots, including categorical palettes, sequential palettes, and diverging palettes. These palettes are designed to be visually appealing and to improve the readability of plots.
- Faceting: Seaborn provides advanced support for faceting, which allows you to create multiple plots that display different subsets of data. For example, you can use the **FacetGrid()** function to create a grid of plots that display different subsets of the data, based on one or more categorical variables.
- Regression Analysis: Seaborn provides many functions for performing regression analysis and visualizing the results. For example, you can use the **lmplot()** function to create a scatter plot with a regression line, or the **residplot()** function to create a plot of the residuals from a regression analysis.
- Plot Styling: Seaborn provides many advanced options for customizing the appearance of plots, including the ability to adjust the size, color, and style of plot elements such as markers and lines. Seaborn also provides many built-in themes that can be used to improve the overall appearance of plots and to make them easier to interpret.

Plotly

Plotly is an open-source data visualization library for creating interactive plots and charts. It provides a high-level interface for creating a wide range of chart types, including scatter plots, line charts, bar charts, and more. Plotly's interactive features allow users to explore and analyze their data in real time, including zooming and panning, hover-over tooltips, and clickable legends.

One of the key strengths of Plotly is its ability to create highly interactive and dynamic visualizations. Plotly allows users to create plots that can be manipulated in real time, allowing them to explore their data from different angles and perspectives. Plotly also provides advanced features for creating dashboards and other interactive applications, making it a popular choice for creating data-driven web applications.

In addition to its interactive features, Plotly provides many customization options for creating visually appealing and informative visualizations. Plotly's built-in themes and color palettes make it easy to create professional-looking visualizations with minimal customization. Plotly also provides many advanced features for customizing visualizations, including the ability to add annotations, images, and custom shapes to plots.

Here's a code for some plots from a random data:

```
import plotly.graph_objs as go
import numpy as np

# Create a scatter plot
x = np.random.normal(size=100)
y = np.random.normal(size=100)
fig = go.Figure(data=go.Scatter(x=x, y=y, mode='markers'))
fig.show()

# Create a line plot
x = np.linspace(0, 10, 100)
y = np.sin(x)
fig = go.Figure(data=go.Scatter(x=x, y=y, mode='lines'))
fig.show()

# Create a histogram
x = np.random.normal(size=1000)
fig = go.Figure(data=go.Histogram(x=x))
fig.show()

# Create a box plot
x = np.random.normal(size=100)
y = np.random.choice(['A', 'B'], size=100)
fig = go.Figure(data=go.Box(x=x, y=y))
fig.show()
```

Hare are some more plots with their outputs:

```
import plotly.graph_objs as go
import numpy as np

# Create a surface plot
x = np.linspace(-5, 5, 100)
y = np.linspace(-5, 5, 100)
X, Y = np.meshgrid(x, y)
Z = np.sin(np.sqrt(X**2 + Y**2))
fig = go.Figure(data=[go.Surface(x=X, y=Y, z=Z)])
fig.show()
```

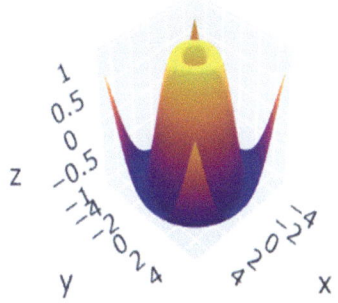

```
# Create a bar chart
x = ['A', 'B', 'C', 'D']
y = np.random.randint(low=0, high=10, size=4)
fig = go.Figure(data=[go.Bar(x=x, y=y)])
fig.show()
```

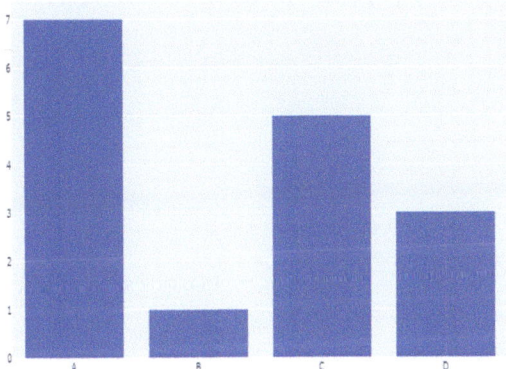

```
# Create a pie chart
labels = ['A', 'B', 'C', 'D']
values = np.random.randint(low=0, high=10, size=4)
fig = go.Figure(data=[go.Pie(labels=labels, values=values)])
fig.show()
```

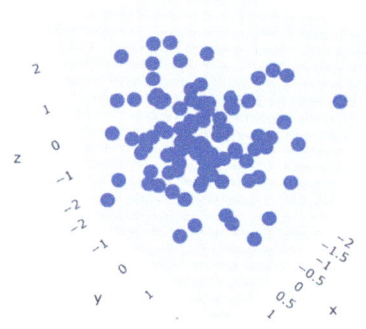

```
# Create a 3D scatter plot
x = np.random.normal(size=100)
y = np.random.normal(size=100)
z = np.random.normal(size=100)
fig = go.Figure(data=[go.Scatter3d(x=x, y=y, z=z,
mode='markers')])
fig.show()
```

The following code shows a 3d saddle shape in Lagrange's optimization:

```python
import plotly.graph_objs as go
import numpy as np
from scipy.optimize import minimize

# Define the function to be optimized
def f(x):
    return x[0]*x[1]

# Define the constraints
def constraint(x):
    return x[0]**2 + x[1]**2 - 1

# Define the Lagrangian function
def lagrangian(x, l):
    return f(x) - l * constraint(x)

# Define the bounds for the variables
bounds = ((-10, 10), (-10, 10))

# Define the initial guesses for the variables and Lagrange multiplier
x0 = (1, 1)
l0 = 0

# Minimize the Lagrangian function using SciPy
res = minimize(lagrangian, x0, args=(l0,), bounds=bounds)

# Create a mesh grid for the variables
x = np.linspace(-2, 2, 100)
y = np.linspace(-2, 2, 100)
X, Y = np.meshgrid(x, y)

# Evaluate the Lagrangian function on the mesh grid
Z = np.zeros((len(x), len(y)))
for i in range(len(x)):
    for j in range(len(y)):
        Z[i, j] = lagrangian([x[i], y[j]], res.fun)

# Create a 3D surface plot of the Lagrangian function
fig = go.Figure(data=[go.Surface(x=X, y=Y, z=Z)])
fig.show()
```

Note that this figure can be rotated to view the plot from different angles, something that could not be done with Matplotlib or Seaborn. Plotly offers several advantages over Matplotlib and Seaborn, including:

- **Interactive Visualizations**: Plotly provides a range of interactive visualizations that allow users to explore their data in real time. This includes interactive plots, charts, and graphs that can be zoomed, panned, and rotated, as well as hover-over tooltips and clickable legends that provide additional context.

- **Easy Customization**: Plotly provides a user-friendly interface for customizing the appearance of your visualizations, including the ability to change colors, fonts, labels, and annotations. Plotly also supports a wide range of chart types, including 2D and 3D scatter plots, line charts, bar charts, histograms, and more.
- **Web-based Deployment**: Plotly allows you to easily deploy your visualizations to the web using its online platform, Plotly Cloud. This allows you to share your visualizations with others, embed them in web pages and applications, and collaborate with others on data analysis and visualization projects.
- **Native Support for Pandas and NumPy**: Plotly natively supports pandas and NumPy data structures, making it easy to work with data in these formats without the need for additional conversions or preprocessing.
- **Built-in Animations and Real-time Updates**: Plotly allows you to create animations and real-time updates in your visualizations, allowing you to see changes in your data over time. This is particularly useful for time series data, financial data, and other dynamic datasets where changes occur frequently.

Bokeh

Bokeh is a Python data visualization library that focuses on creating interactive visualizations for the web. It provides a high-level interface for creating a wide range of charts, including scatter plots, line charts, bar charts, and more. Bokeh is particularly well-suited for creating complex, interactive visualizations with large datasets.

One of the key features of Bokeh is its ability to create interactive visualizations that can be embedded in web applications. Bokeh provides many tools for interactivity, such as zooming and panning, hover tooltips, and selection tools. These features make it easy to explore and analyze large datasets in a web-based environment. Bokeh also provides many advanced features for customization and styling of visualizations. For example, Bokeh provides a wide range of color palettes, advanced axis labeling and formatting options, and support for customizing the appearance of markers and lines. Another key feature of Bokeh is its support for streaming data. Bokeh provides tools for creating real-time visualizations that can update dynamically as new data is received. This makes Bokeh particularly well-suited for use in applications that require real-time monitoring or analysis of streaming data, such as financial trading or sensor data analysis.

11 Data Visualization: Numerical Data

Here's an example of some plots using Bokeh.

```
from bokeh.plotting import figure, output_file, show
from bokeh.layouts import gridplot
import numpy as np

# Create random data
x = np.linspace(0, 10, 100)
y1 = np.sin(x)
y2 = np.cos(x)
y3 = np.tan(x)
y4 = np.exp(x)

# Create a line chart
output_file("line_chart.html")
p1 = figure(title="Line Chart", x_axis_label='x',
y_axis_label='y')
p1.line(x, y1, line_width=2, legend_label="y = sin(x)",
line_color='red')
p1.line(x, y2, line_width=2, legend_label="y = cos(x)",
line_color='green')
p1.line(x, y3, line_width=2, legend_label="y = tan(x)",
line_color='blue')
p1.legend.location = "top_left"

# Create a scatter plot
output_file("scatter_plot.html")
p2 = figure(title="Scatter Plot", x_axis_label='x',
y_axis_label='y')
p2.scatter(x, y4, size=5, color='red')

# Create a grid layout with both plots
grid = gridplot([[p1, p2]], toolbar_location='right')
show(grid)
```

Here is the output for both the plots:

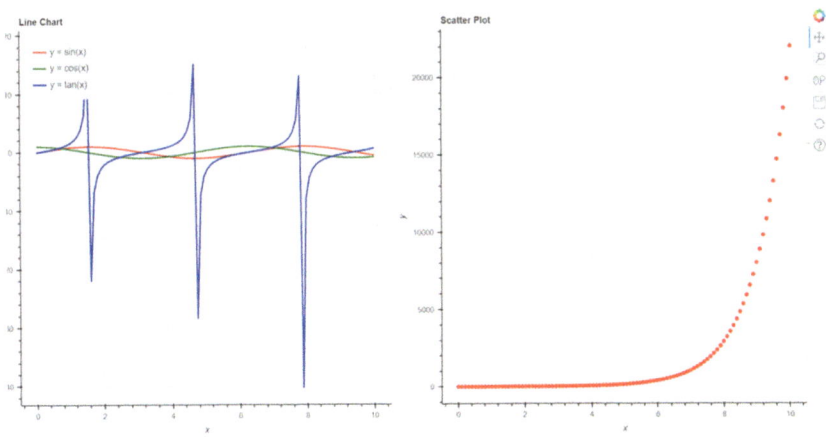

Carefully look at various visualization tools that are not available in Matplotlib, Seaborn, or Plotly. Some more examples would be:

```python
from bokeh.plotting import figure, output_file, show
from bokeh.layouts import column, row
from bokeh.models import LinearColorMapper, ColumnDataSource
from bokeh.palettes import Viridis256
import numpy as np

# Create random data
x = np.linspace(0, 10, 100)
y1 = np.sin(x)
y2 = np.cos(x)
y3 = np.tan(x)
y4 = np.exp(x)

# Create a bar chart
output_file("bar_chart.html")
p1 = figure(title="Bar Chart", x_axis_label='x',
y_axis_label='y')
p1.vbar(x=x, top=y4, width=0.5, color='blue')

# Create a heat map
output_file("heatmap.html")
data = np.random.rand(10, 10)
color_mapper = LinearColorMapper(palette=Viridis256,
low=data.min(), high=data.max())
source = ColumnDataSource(data={'x': np.arange(10), 'y':
np.arange(10), 'color': data.flatten()})
p2 = figure(title="Heat Map")
p2.rect(x='x', y='y', width=1, height=1, fill_color={'field':
'color', 'transform': color_mapper}, line_color=None,
source=source)

# Create a line chart with multiple lines
output_file("multiline_chart.html")
p3 = figure(title="Multi-Line Chart", x_axis_label='x',
y_axis_label='y')
p3.line(x, y1, line_width=2, legend_label="y = sin(x)",
line_color='red')
p3.line(x, y2, line_width=2, legend_label="y = cos(x)",
line_color='green')
p3.line(x, y3, line_width=2, legend_label="y = tan(x)",
line_color='blue')
p3.legend.location = "top_left"

# Create a scatter plot with hover tooltips
output_file("hover_tooltip.html")
p4 = figure(title="Scatter Plot with Hover Tooltip",
x_axis_label='x', y_axis_label='y', tooltips=[("x", "$x"), ("y",
"$y")])
p4.scatter(x, y4, size=5, color='red')

# Create a layout with all plots
layout = column(row(p1, p2), row(p3, p4))
show(layout)
```

11 Data Visualization: Numerical Data

An example of a code that shows Crypto prices in a real time would be:

```python
import requests
from bokeh.plotting import figure, output_notebook, show
from bokeh.models import ColumnDataSource
from bokeh.io import push_notebook
from datetime import datetime
import time

output_notebook()

# Set up the figure
p = figure(x_axis_type='datetime', title='Crypto Prices',
width=800, height=400)
p.xaxis.axis_label = 'Time'
p.yaxis.axis_label = 'Price (USD)'

# Define the data source
source = ColumnDataSource(data=dict(time=[], bitcoin=[],
ethereum=[], cardano=[], dogecoin=[]))

# Define the line plots
p.line(x='time', y='bitcoin', line_color='orange',
legend_label='Bitcoin', source=source)
p.line(x='time', y='ethereum', line_color='purple',
legend_label='Ethereum', source=source)
p.line(x='time', y='cardano', line_color='blue',
legend_label='Cardano', source=source)
p.line(x='time', y='dogecoin', line_color='green',
legend_label='Dogecoin', source=source)

# Define the update function
import requests
from bokeh.plotting import figure, output_notebook, show
from bokeh.models import ColumnDataSource
from bokeh.io import push_notebook
from datetime import datetime

import time

output_notebook()

# Set up the figure
p = figure(x_axis_type='datetime', title='Crypto Prices',
width=800, height=400)
p.xaxis.axis_label = 'Time'
p.yaxis.axis_label = 'Price (USD)'

# Define the data source
source = ColumnDataSource(data=dict(time=[], bitcoin=[],
ethereum=[], cardano=[], dogecoin=[]))

# Define the line plots
p.line(x='time', y='bitcoin', line_color='orange',
legend_label='Bitcoin', source=source)
```

```python
p.line(x='time', y='ethereum', line_color='purple',
legend_label='Ethereum', source=source)
p.line(x='time', y='cardano', line_color='blue',
legend_label='Cardano', source=source)
p.line(x='time', y='dogecoin', line_color='green',
legend_label='Dogecoin', source=source)

# Define the update function
def update():
    symbols = ['BTCUSDT', 'ETHUSDT', 'ADAUSDT', 'DOGEUSDT']
    prices = {}

    for symbol in symbols:
        response = requests.get(f'https://api.binance.com/api/v3/ticker/price?symbol={symbol}')
        ticker = response.json()
        coin = symbol[:-4]
        prices[coin] = float(ticker['price'])

    # Append the new data to the data source
    new_data = dict(
        time=[datetime.now()],
        bitcoin=[prices['BTC']],
        ethereum=[prices['ETH']],
        cardano=[prices['ADA']],
        dogecoin=[prices['DOGE']]
    )
    source.stream(new_data, 100)
    handle = show(p, notebook_handle=True)

    # Run the update function every second
    while True:
        update()
        push_notebook(handle=handle)
        time.sleep(1)


    # Run the update function every second
    while True:
        update()
        push_notebook(handle=handle)
        time.sleep(1)
```

Similarly, example of a code that shows the stock prices of five stocks in real time is:

```python
import yfinance as yf
import pandas as pd
from bokeh.plotting import figure, show, output_notebook, ColumnDataSource
from bokeh.models import HoverTool
from bokeh.palettes import Spectral10
import datetime as dt
import time
from IPython import display

output_notebook()

# List of stock symbols (ticker) you want to track
stocks = ["AAPL", "GOOGL", "MSFT", "AMZN", "TSLA"]

def get_stock_data(stock, start_date, end_date):
    data = yf.download(stock, start=start_date, end=end_date)
    data.reset_index(inplace=True)
    return data

# Fetch stock data
start_date = (dt.datetime.now() - dt.timedelta(days=30)).strftime("%Y-%m-%d")
end_date = dt.datetime.now().strftime("%Y-%m-%d")

stock_data = {}
for stock in stocks:
    stock_data[stock] = get_stock_data(stock, start_date, end_date)

# Create a bokeh figure
TOOLS = "pan,wheel_zoom,box_zoom,reset,save"
hover = HoverTool(tooltips=[("date", "@date{%F}"), ("close", "@Close")], formatters={"@date": "datetime"})
p = figure(width=1000, height=600, x_axis_type="datetime", tools=TOOLS)
p.add_tools(hover)
p.title.text = "Real-Time Stock Prices"
p.xaxis.axis_label = "Date"
p.yaxis.axis_label = "Close Price"

colors = Spectral10

for i, stock in enumerate(stocks):
    df = stock_data[stock]
    source = ColumnDataSource(data=dict(date=df["Date"], Close=df["Close"]))
    p.line("date", "Close", line_width=2, color=colors[i], legend_label=stock, source=source)

p.legend.location = "top_left"
p.legend.click_policy = "hide"
```

```
# Display the plot and update it every 30 seconds
while True:
    display.clear_output(wait=True)
    show(p)
    time.sleep(30)
    end_date = dt.datetime.now().strftime("%Y-%m-%d")
    for stock in stocks:
        stock_data[stock] = get_stock_data(stock, start_date, end_date)
    for i, stock in enumerate(stocks):
        df = stock_data[stock]
        source = ColumnDataSource(data=dict(date=df["Date"], Close=df["Close"]))
        p.line("date", "Close", line_width=2, color=colors[i], legend_label=stock, source=source)
```

Apart from the one discussed above, there are other libraries that are useful for data visualization. Some of them are:

- Plotnine: Plotnine is a grammar of graphics-based data visualization library that is inspired by the R programming language's ggplot2 package. It provides a simple syntax for creating complex plots.
- Altair: Altair is a declarative visualization library that allows you to easily create interactive visualizations with a concise and intuitive syntax.
- ggplot: ggplot is a Python implementation of the R programming language's ggplot2 package. It uses a similar syntax and provides similar functionality for creating complex visualizations.
- D3.js: D3.js is a JavaScript library for creating dynamic and interactive visualizations in the browser. It can be used with Python through the use of tools such as mpld3, which allows you to create D3.js visualizations from Matplotlib plots.
- NetworkX: NetworkX is a Python library for creating and manipulating graphs and networks. It provides several functions for visualizing graphs and networks, including node-link diagrams, matrix plots, and heatmaps.

Each library has its own strengths and weaknesses, so the choice of which library to use depends on the specific needs of the project. Data visualization is an essential part of the data analysis process, and researchers should take the time to explore the various visualization libraries available in Python to find the tools that best suit their needs. By effectively visualizing their data, researchers can gain deeper insights into their research questions and communicate their findings more effectively to others.

12

Data Visualization: Text Data

Text data poses unique challenges compared to numerical data, due to its inherent complexity. Consequently, analyzing and comprehending text data can be quite daunting. However, by employing appropriate visualization techniques, we can reveal hidden insights and patterns that may otherwise elude us. In this discussion, we will explore various visualization methods frequently employed for analyzing text data. These methods include word clouds, bar charts, heatmaps, scatterplots, and network graphs. We will elucidate how each technique can be utilized to analyze distinct aspects of text data, such as word frequency, sentiment, and document similarity. Additionally, we will demonstrate how Python libraries introduced in the preceding chapter, namely Matplotlib, Seaborn, Plotly, and Bokeh, can be employed to generate these visualizations for text data. While these libraries are commonly employed for numerical data, they possess the capability to handle text data when equipped with appropriate techniques and tools. Leveraging these libraries, we can create visually captivating and informative visualizations that facilitate deeper insights into text data.

Wordcloud

Word clouds are a popular method for visualizing the most frequently occurring words in a corpus. They can provide a quick overview of the most prominent words in a text, making it easier to identify themes and patterns.

Creating a word cloud in Python is a relatively simple process. The first step is to extract the text from your corpus, and then use a word tokenizer to break the text into individual words. You can then count the frequency of each word using Python's built-in Counter class, or use specialized libraries such as NLTK. Once you have a dictionary of word frequencies, you can use a word cloud library like wordcloud. Word clouds typically display the words in different-sized bubbles, with the larger bubbles representing more frequent words. You can customize the appearance of the word cloud by changing the font, color scheme, and layout.

In addition to providing a quick overview of the most frequent words, word clouds can also be useful for identifying outliers and anomalies in a corpus. For example, if you notice a word in the word cloud that is unexpected or unusual, you can investigate further to see why it is appearing so frequently.

One limitation of word clouds is that they do not provide any information about the context in which the words are used. For example, two words may have the same frequency in a corpus, but one may be used more frequently in positive contexts while the other is used more frequently in negative contexts. To get a better understanding of the context in which words are used, you may want to use other visualization techniques such as bar charts, heatmaps, or scatterplots.

Here is a code that gives a word cloud of a random text of the first para of this chapter:

```python
import random
import wordcloud
import matplotlib.pyplot as plt

# Define a random text corpus
text = """
Text data is often much more complex than numerical data, making
it challenging to analyze and understand. However, with the
right visualization techniques, we can uncover insights and
patterns that might otherwise go unnoticed. We will explore
different visualization methods that are commonly used for text
data, including word clouds, bar charts, heatmaps, scatterplots,
and network graphs. We will discuss how each of these techniques
can be used to analyze different aspects of text data, such as
word frequency, sentiment, and document similarity. We will also
show how you can use the Python libraries introduced in the
previous chapter, such as Matplotlib, Seaborn, Plotly, and
Bokeh, to create these visualizations for text data. While these
libraries are commonly used for numerical data, they are also
capable of handling text data with the right techniques and
tools. With the help of these libraries, we can create visually
appealing and informative visualizations that can help us gain
deeper insights into text data.
"""

# Create a WordCloud object
wc = wordcloud.WordCloud()

# Generate the word cloud from the text
cloud = wc.generate(text)

# Display the word cloud using matplotlib
plt.imshow(cloud, interpolation='bilinear')
plt.axis('off')
plt.show()
```

The output of this code will look like:

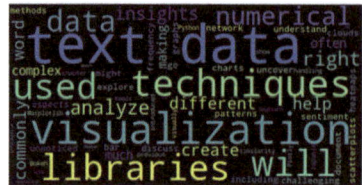

In this code, we first define a random text corpus as a string variable **text**. We then create a **WordCloud** object using the **wordcloud** library. Next, we generate the word cloud using the **generate** method of the **WordCloud** object, passing in the text corpus as an argument. Finally, we display the word cloud using **matplotlib**. We use **imshow** to display the image, set **interpolation** to 'bilinear' to smooth the edges of the bubbles, and turn off the axis using **axis('off')**. Finally, we call **show** to display the word cloud in a new window.

Note that you can customize the appearance of the word cloud by passing in various arguments to the **WordCloud** constructor. For example, you can set the background color, font, and size of the word cloud using the **background_color**, **font_path**, and **max_font_size** arguments, respectively.

Here's a code that picks normalized text files from D:/Data/Preprocessed folder:

```python
import os
import glob
import string
from collections import import Counter
import wordcloud
from PIL import Image
import numpy as np

# Define the path to the directory containing the text files
path = 'D:/Data/Preprocessed/*.txt'

# Create a list of all the file names in the directory
files = glob.glob(path)

# Define a function to read the text from a file and return a
list of words
def get_words(filename):
    with open(filename, 'r', encoding='utf-8') as file:
        text = file.read()
        words = text.lower().translate(str.maketrans('', '',
string.punctuation)).split()
        return words

# Create a list of all the words in the directory
words_list = []
for file in files:
    words = get_words(file)
    words_list.extend(words)

# Create a word frequency dictionary
word_counts = Counter(words_list)

# Create a WordCloud object
wc = wordcloud.WordCloud()

# Generate the word cloud from the word frequency dictionary
cloud = wc.generate_from_frequencies(word_counts)

# Create a mask from an image file
mask = np.array(Image.open("D:/Data/image.jpg"))

# Create a colored word cloud image
image_colors = wordcloud.ImageColorGenerator(mask)
colored_cloud =
wordcloud.WordCloud(color_func=image_colors).generate_from_frequ
encies(word_counts)

# Save the colored word cloud as an image file
colored_cloud.to_file("D:/Data/image_name.png")
```

In this code, we first define the path to the directory containing the text files using a wildcard * to match all files with a **.txt** extension. We then use the **glob** module to create a list of all the file names in the directory. Next, we define a function called **get_words** that reads the text from a file, converts it to lowercase, removes punctuation, and splits it into a list of words. We then use a loop to call this function for each file in the directory, and create a list of all the words in the directory. Using the **Counter** class from the **collections** module, we create a word frequency dictionary that counts the number of occurrences of each word in the list. We then create a **Word-Cloud** object using the **wordcloud** module, and generate the word cloud from the word frequency dictionary using the **generate_from_frequencies** method. Next, we create a mask from an image file using the **Image** module from the Python standard library. This module takes an image stored in local directory as a sample to create this mask for the new image. We then create a colored word cloud image using the **ImageColorGenerator** class from the **wordcloud** module, passing in the mask as an argument. Finally, we save the image in desired directory. The final output of the code will look like:

It is pertinent to mention that the size of sample image used to generate mask for this visualization should be more than or equal to the desired image output. The default for this library is 400×200 pixels and can be modified changing the **colored_cloud** object as:

```
colored_cloud = wordcloud.WordCloud(color_func=image_colors,
width=800, height=600).generate_from_frequencies(word_counts)
```

Matplotlib

Matplotlib is useful not only for numerical data, but also for the text data. Matplotlib can be used to make a word cloud. Here's a code that generates random text from random library and then makes wordcloud using matplotlib:

12 Data Visualization: Text Data

```python
import random
import string
import matplotlib.pyplot as plt
from collections import Counter
import numpy as np

# Generate random text
def generate_random_text(num_words):
    text = ''
    for _ in range(num_words):
        word_length = random.randint(3, 10)
        word = ''.join(random.choices(string.ascii_lowercase, k=word_length))
        text += word + ' '
    return text

# Create word cloud from random text
def create_word_cloud(text):
    word_list = text.split()
    word_count = Counter(word_list)

    plt.figure(figsize=(6, 3))

    colors = plt.get_cmap('viridis')(np.linspace(0, 1, len(word_count)))

    for i, (word, count) in enumerate(word_count.items()):
        x = random.uniform(0, 1)
        y = random.uniform(0, 1)
        plt.scatter(x, y, alpha=0, color=colors[i])
        plt.text(x, y, word, fontsize=count*5, ha='center', va='center', alpha=0.8, color=colors[i])

    plt.axis('off')
    plt.show()

if __name__ == "__main__":
    random_text = generate_random_text(100)
    print("Random text:\n", random_text)
    create_word_cloud(random_text)
```

This Python code generates a simple word cloud from random text using the **matplotlib**, **random**, and **numpy** libraries. The **generate_random_text()** function creates a random string of words by iterating over a specified number of words, generating random word lengths between 3 and 10 characters, and constructing the words from random lowercase letters. The **create_word_cloud()** function splits the generated text into a list of words and uses the **Counter** class from the **collections** library to count the frequency of each word. It then creates a scatter plot using **matplotlib**, placing each word at a random position in the plot with a size determined by its frequency multiplied by a scaling factor. Colors are generated using a colormap from **matplotlib** (in this case, 'viridis') and the **numpy** library, and each word is assigned a color. Note that this simple implementation does not handle collision detection or optimal positioning, so some words may overlap.

If normalized text files is stored in D:/Data/Preprocessed, then the code to make a word cloud from those files using Matplotlib would be:

```python
import os
import random
import string
import matplotlib.pyplot as plt
from collections import Counter
import numpy as np

# Read text from a file
def read_text_from_file(file_path):
    with open(file_path, 'r', encoding='utf-8') as file:
        return file.read()

# Create word cloud from text
def create_word_cloud(text):
    word_list = text.split()
    word_count = Counter(word_list)

    plt.figure(figsize=(12, 12))

    colors = plt.get_cmap('viridis')(np.linspace(0, 1, len(word_count)))

    for i, (word, count) in enumerate(word_count.items()):
        x = random.uniform(0, 1)
        y = random.uniform(0, 1)
        plt.scatter(x, y, alpha=0, color=colors[i])
        plt.text(x, y, word, fontsize=count*5, ha='center', va='center', alpha=0.8, color=colors[i])

    plt.axis('off')
    plt.show()

if __name__ == "__main__":
    file_path = 'D:/Data/Preprocessed/2.txt'
    text = read_text_from_file(file_path)
    create_word_cloud(text)
```

The **read_text_from_file** function reads the text data from a given file path and returns the contents as a string. The **create_word_cloud** function takes the input text and generates a word cloud visualization using the **Counter** class from the **collections** module to count the occurrences of each word. The word cloud is created using the **matplotlib.pyplot** library. It initializes a figure of size 6×3 inches and assigns colors to the words based on a color map. The word cloud is generated by iterating over each word and its corresponding count. For each word, a random position is generated, and the word is placed at that position using the **scatter** and **text** functions from **pyplot**. The font size of each word is determined by its count. The resulting word cloud is displayed using the **show** function. In the main part of the script, a file path is specified as **'D:/Data/Preprocessed/2.txt'**, and the **read_text_from_file** function is called to retrieve the text from that file. The **create_word_cloud** function is then invoked with the obtained text to generate the word cloud visualization, which will look like:

It is pertinent to mention that once you get the word frequency using the word frequency dictionary, you can use any numerical method of data visualization to visualize the distribution of word frequencies in the corpus. Here are some examples of numerical methods of data visualization that you can use:

- Bar Charts: You can use a bar chart to display the frequency of the most common words in the corpus. You can sort the dictionary by frequency and select the top N words to display in the chart.
- Line Charts: You can use a line chart to show the trend of word frequencies over time. For example, if you are analyzing a set of news articles, you can plot the frequency of certain keywords over time to see how they change.
- Heat Maps: You can use a heat map to display the co-occurrence of words in the corpus. You can create a matrix of word co-occurrences and use a color map to show the frequency of each co-occurrence.
- Scatter Plots: You can use a scatter plot to show the relationship between two variables, such as the frequency of two different words in the corpus.
- Network Graphs: You can use a network graph to show the relationships between different words in the corpus. You can create nodes for each word and edges between them based on their co-occurrence.

In order to use these methods, you need to first calculate the word frequency dictionary as you have already done. Once you have the dictionary, you can use any numerical data visualization tool of your choice, such as **matplotlib**, **seaborn**, or **plotly**, to create the desired visualization.

Here's a code that generates 1000 random words from random dictionary and creates a bar chart of top ten words using matplotlib:

```
import random
import nltk
import collections
import matplotlib.pyplot as plt

# Download the word corpus if you haven't already
nltk.download('words')

# Get the dictionary of words
english_words = set(nltk.corpus.words.words())

# Generate a text of 1000 random words
random_text = [random.choice(list(english_words)) for _ in range(1000)]

# Count the occurrences of each word
word_counts = collections.Counter(random_text)

# Get the top 10 most common words
top_10_words = word_counts.most_common(10)

# Plot a bar chart of the top 10 words
words, counts = zip(*top_10_words)
plt.bar(words, counts)
plt.xlabel('Words')
plt.ylabel('Counts')
plt.title('Top 10 Most Common Words in Random Text')
plt.show()
```

Similarly, you can import your text data to get any plot of the top words in the text data.

Network Map

Network graphs, also known as graphs or network diagrams, are visual representations of relationships between entities. In the context of text data, these entities can be words, phrases, or even entire documents. Network graphs are particularly useful for exploring and understanding the structure and connections within complex datasets.

In a network graph, entities are represented as nodes (or vertices), and relationships between these entities are represented as edges (or links). The relationships can be either directed or undirected, depending on the nature of the dataset. For instance, in a citation network, edges might be directed to represent the direction of citations, while in a co-occurrence network of words, edges might be undirected as the relationship is mutual.

Here are some common applications of network graphs in text data analysis:

- Word Co-occurrence Networks: These networks represent the relationships between words that co-occur within a specific context, such as a sentence or paragraph. By visualizing word co-occurrence networks, you can gain insights into the semantic structure of a text and identify clusters of related words or concepts.
- Document Similarity Networks: These networks represent the relationships between documents based on their similarity. In this case, nodes represent individual documents, and edges represent the degree of similarity between them. Document similarity networks can help identify clusters of related documents, uncover patterns in the dataset, and facilitate the organization and retrieval of information.
- Social Networks: In social media analysis, network graphs can be used to visualize relationships between users, such as followers or friends on a platform. By analyzing these networks, you can identify influential users, uncover communities, and study information diffusion.
- Citation Networks: In bibliometric studies, citation networks are used to represent the relationships between academic papers based on their citations. This can help identify influential papers, trace the development of research areas, and reveal the structure of scientific communities.

To create and analyze network graphs, several libraries are available in Python, such as NetworkX, igraph, and Gephi (a standalone application). These libraries provide tools for creating, manipulating, and visualizing network graphs, as well as performing various graph analysis algorithms, such as calculating centrality measures, detecting communities, and more.

This code generates the network map from D:/Data/Preprocessed/2.txt:

```python
import nltk
import networkx as nx
import matplotlib.pyplot as plt

# Download the word corpus if you haven't already
nltk.download('words')

# Read text from file
file_path = 'D:/Data/Preprocessed/2.txt'
with open(file_path, 'r', encoding='utf-8') as file:
    text = file.read()

# Tokenize the text into words
word_list = nltk.word_tokenize(text)

# Create a network graph based on the co-occurrence of words in the list
G = nx.Graph()

# Add nodes and edges to the graph
for i, word1 in enumerate(word_list):
    for word2 in word_list[i+1:]:
        if G.has_edge(word1, word2):
            G[word1][word2]['weight'] += 1
        else:
            G.add_edge(word1, word2, weight=1)

# Draw the network graph
pos = nx.spring_layout(G, seed=42)
plt.figure(figsize=(6, 4))  # Create a new figure
nx.draw_networkx(G, pos, with_labels=True, node_size=5000, node_color='lightblue', font_size=8)
edge_labels = {(u, v): d['weight'] for u, v, d in G.edges(data=True)}
nx.draw_networkx_edge_labels(G, pos, edge_labels=edge_labels, font_size=8)
plt.title('Network Graph of Text')
plt.show()
```

This code utilizes the Natural Language Toolkit (NLTK) and NetworkX libraries in Python to construct a network graph, depicting the co-occurrence of words within a text document. After obtaining the text from a specified file, the script employs NLTK to tokenize the text into individual words. The NetworkX library then generates a graph, with nodes representing words and edges signifying their co-occurrence. If a pair of words appears together more than once, the weight of the corresponding edge is incremented. The final part of the script is dedicated to visualizing this graph. Using matplotlib, it displays the graph with nodes and edges, where the nodes are labeled with their corresponding words, and the edges are labeled with their weights, showing the frequency of co-occurrence. This graph provides a spatial representation of the relationships between words, aiding in the analysis of textual data. The output will look like:

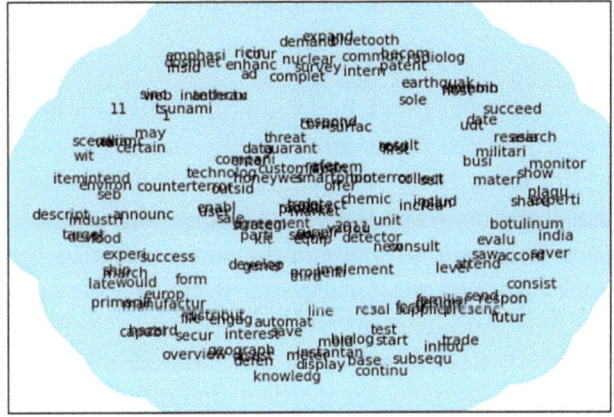

Dimensionality Reduction Techniques

Dimensionality reduction techniques are used to reduce the number of features (dimensions) in high-dimensional data while preserving the structure or relationships present in the data. These techniques are especially useful for visualizing complex datasets and improving the performance of machine learning algorithms. Two popular dimensionality reduction techniques are t-SNE (t-Distributed Stochastic Neighbor Embedding) and UMAP (Uniform Manifold Approximation and Projection). Dimensionality reduction techniques are used to reduce the number of features (dimensions) in high-dimensional data while preserving the structure or relationships present in the data. These techniques are especially useful for visualizing complex datasets and improving the performance of machine learning algorithms. Two popular dimensionality reduction techniques are t-SNE (t-Distributed Stochastic Neighbor Embedding) and UMAP (Uniform Manifold Approximation and Projection).

1. t-SNE (t-Distributed Stochastic Neighbor Embedding): t-SNE is a non-linear dimensionality reduction technique introduced by Laurens van der Maaten and Geoffrey Hinton in 2008. It is particularly useful for visualizing high-dimensional data in two or three dimensions. The primary goal of t-SNE is to preserve the local structure of the data while reducing its dimensionality. It does this by minimizing the divergence between two probability distributions: one that represents the pairwise similarities between data points in the high-dimensional space and another that represents the pairwise similarities between the data points in the low-dimensional space. The t-SNE algorithm involves a stochastic process and can be computationally expensive for large datasets. Nevertheless, it is widely used due to its ability to produce visually appealing and interpretable visualizations.

An example a t-SNE plot using word embeddings generated by the popular Word2Vec model, using the text file D:/Data/Preprocessed/2/txt is demonstrated below:

```python
import numpy as np
import matplotlib.pyplot as plt
from sklearn.manifold import TSNE
from gensim.models import Word2Vec
from nltk.tokenize import word_tokenize

# Read text from file
file_path = 'D:/Data/Preprocessed/2.txt'
try:
    with open(file_path, 'r', encoding='utf-8') as file:
        text = file.read()
except FileNotFoundError:
    print(f"Error: File not found at {file_path}")
    exit()
except IOError:
    print(f"Error: Unable to read file at {file_path}")
    exit()

if text is None:
    print(f"Error: File is empty or could not be read correctly")
    exit()

# Tokenize the text
tokens = word_tokenize(text.lower())

# Train Word2Vec model
model = Word2Vec([tokens], min_count=1, vector_size=50)

# Prepare the word embeddings for t-SNE
words = list(model.wv.key_to_index)
word_vectors = np.array([model.wv[word] for word in words])

# Limit the number of words to plot
num_words = 1000
if len(words) > num_words:
    words = words[:num_words]
    word_vectors = word_vectors[:num_words]

# Apply t-SNE
tsne = TSNE(n_components=2, random_state=42)
word_vectors_2d = tsne.fit_transform(word_vectors)

# Plot the t-SNE results
plt.figure(figsize=(10, 6))
plt.scatter(word_vectors_2d[:, 0], word_vectors_2d[:, 1], edgecolors='k', c='r', s=100)

for word, (x, y) in zip(words, word_vectors_2d):
    plt.text(x, y, word, fontsize=12, alpha=0.75)

plt.title('t-SNE Plot of Word Embeddings')
plt.xlabel('Dimension 1')
plt.ylabel('Dimension 2')
plt.show()
```

The output will be a plot as shown:

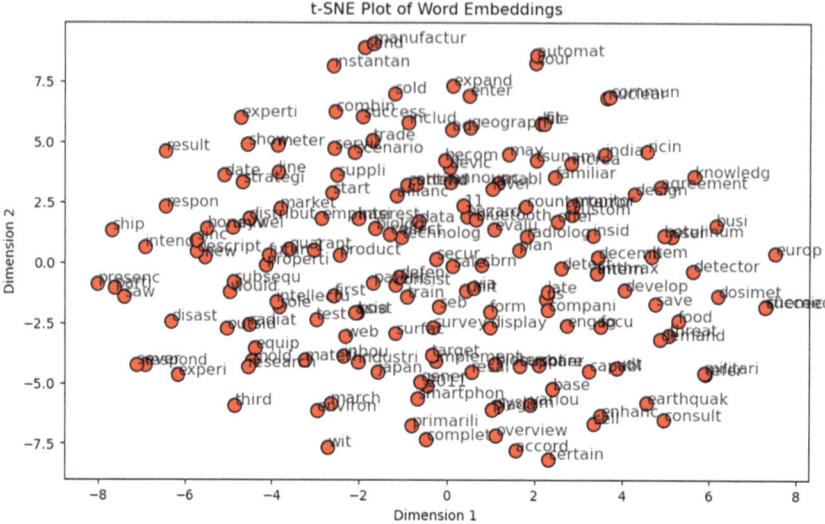

This script conducts a visualization of word embeddings generated from a text file using the t-SNE method. It begins by reading a text file, tokenizing the text into individual words, and then uses these tokens to train a Word2Vec model. Word2Vec is a machine learning model that is trained to represent words as high-dimensional vectors such that the spatial relationships between vectors correspond to semantic relationships between words. These high-dimensional word vectors are then projected into a two-dimensional space using t-SNE (t-Distributed Stochastic Neighbor Embedding), a technique for dimensionality reduction that is particularly well-suited for the visualization of high-dimensional datasets. The script then generates a scatter plot in which each point represents a word, and the proximity of points reflects the semantic similarity of the words they represent. The script also includes error handling mechanisms for file not found or input/output errors during the reading process, and it limits the number of words visualized for practicality and readability of the plot.

You can also create a three-dimensional plot using matplotlib or plotly. Plotly has an additional advantage of being able to manipulate the output to view it from different angles. Following is an example of creation of a three-dimensional plot using plotly:

```python
import numpy as np
import pandas as pd
import plotly.express as px
import plotly.graph_objects as go
from sklearn.manifold import TSNE
from gensim.models import Word2Vec

from nltk.tokenize import word_tokenize

# Read text from file
file_path = 'D:/Data/Preprocessed/2.txt'
with open(file_path, 'r', encoding='utf-8') as file:
    text = file.read()

# Tokenize the text
tokens = word_tokenize(text.lower())

# Train Word2Vec model
model = Word2Vec([tokens], min_count=1, vector_size=50)

# Prepare the word embeddings for t-SNE
words = list(model.wv.key_to_index)
word_vectors = np.array([model.wv[word] for word in words])

# Apply t-SNE
tsne = TSNE(n_components=3, random_state=42)
word_vectors_3d = tsne.fit_transform(word_vectors)

# Create a DataFrame to store the 3D coordinates and words
data = {'x': word_vectors_3d[:, 0],
        'y': word_vectors_3d[:, 1],
        'z': word_vectors_3d[:, 2],
        'word': words}

df = pd.DataFrame(data)

# Create a 3D scatter plot using Plotly
fig = go.Figure(data=[go.Scatter3d(x=df['x'], y=df['y'], z=df['z'],
                                   mode='markers+text',
                                   text=df['word'],
                                   textposition='top center',
                                   marker=dict(size=6, color='red', opacity=0.8))])

fig.update_layout(scene=dict(xaxis_title='Dimension 1',
                             yaxis_title='Dimension 2',
                             zaxis_title='Dimension 3'),
                  title='3D t-SNE Plot of Word Embeddings')

fig.show()
```

The output will look like the figure shown. The output can be rotated and zoomed into.

3D t-SNE Plot of Word Embeddings

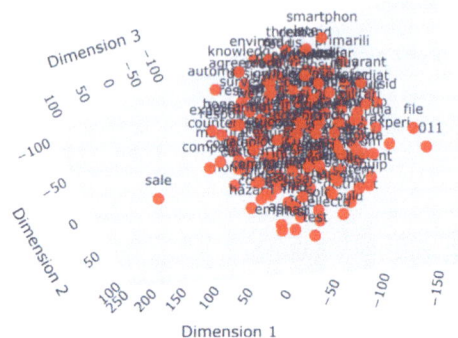

The provided script is a comprehensive implementation of semantic visualization using word embeddings and t-SNE in Python. The program initiates by importing necessary libraries, including numpy for numerical computations, pandas for data manipulation, plotly for data visualization, gensim for training the Word2Vec model, and nltk for text preprocessing. Subsequently, it reads a text file and tokenizes it into individual words. A Word2Vec model is then trained on these tokens, producing a set of word vectors. The dimensionality of these vectors is reduced to three dimensions using t-SNE, a technique popular for visualizing high-dimensional data. The transformed vectors, representing words in a three-dimensional space, are stored in a pandas DataFrame. The script concludes by generating a 3D scatter plot using Plotly, where each point corresponds to a word and is positioned according to its t-SNE transformed vector. The visualization provides a spatial representation of the words, where the proximity between points mirrors the semantic similarity between corresponding words, thus offering an intuitive exploration of the text's semantic structure.

2. UMAP (Uniform Manifold Approximation and Projection): UMAP is a more recent dimensionality reduction technique introduced by Leland McInnes, John Healy, and James Melville in 2018. It is a general-purpose manifold learning and dimensionality reduction algorithm that aims to balance computational efficiency with the preservation of both local and global structure in the data. UMAP is based on the Riemannian geometry and algebraic topology concepts, and it seeks to optimize the preservation of topological structure in the low-dimensional space. It has several advantages over t-SNE, including faster computation time, better preservation of global structure, and the ability to scale to larger datasets. Additionally,

UMAP is more versatile as it can be used for tasks beyond visualization, such as dimensionality reduction for machine learning or data preprocessing. An example of a code using UMAP is as below. Remember to install **umap-learn** library bore executing code (not **umap** library).

```python
import umap
import numpy as np
import matplotlib.pyplot as plt
from sklearn.datasets import fetch_20newsgroups
from sklearn.feature_extraction.text import CountVectorizer

# Load the 20 newsgroups dataset
newsgroups = fetch_20newsgroups(subset='all', remove=('headers',
'footers', 'quotes'))

# Vectorize the text data using CountVectorizer
vectorizer = CountVectorizer(max_df=0.5, min_df=10,
stop_words='english')
X = vectorizer.fit_transform(newsgroups.data)

# Fit UMAP to the data
umap_embedding = umap.UMAP(n_neighbors=15, n_components=2,
metric='cosine').fit_transform(X.toarray())

# Visualize the results
plt.scatter(umap_embedding[:, 0], umap_embedding[:, 1],
c=newsgroups.target, cmap='Spectral', s=5)
plt.colorbar(boundaries=np.arange(21)-
0.5).set_ticks(np.arange(20))
plt.title('UMAP projection of the 20 Newsgroups dataset using
dense embeddings', fontsize=16)
plt.show()
```

This code demonstrates how to use UMAP to visualize text data in two dimensions. It loads the 20 newsgroups dataset using scikit-learn, vectorizes the text data using CountVectorizer, fits a UMAP model to the resulting matrix, and visualizes the UMAP embedding using a scatter plot.

The **fetch_20newsgroups** function loads the 20 newsgroups dataset, which contains approximately 20,000 newsgroup posts across 20 different topics. The **CountVectorizer** class is used to vectorize the text data, which counts the number of times each word appears in each document. The resulting matrix is then passed to the **UMAP** class from the **umap** library to fit a UMAP model. The resulting low-dimensional embedding is then visualized using a scatter plot with color-coded labels corresponding to the different topics.

Note that the **max_df**, **min_df**, and **stop_words** parameters of **CountVectorizer** control the preprocessing of the text data, and the **n_neighbors**, **n_components**, and **metric** parameters of **UMAP** control the dimensionality reduction and visualization of the data. These parameters can be adjusted to optimize the performance of the algorithm for different datasets and use cases.

Data visualization is an essential tool for working with text data in a variety of contexts, including academic research, business analysis, and marketing. By providing a visual representation of text data, visualization techniques can help us identify patterns and relationships that might not be apparent from a purely numerical analysis. Moreover, visualization can help us communicate the results of our analyses to others in a clear and effective way.

In this chapter, we explored several powerful data visualization techniques that are specifically designed for working with text data. We started with word clouds, which are a simple yet effective way to visualize the most frequently occurring words in a corpus. We then moved on to more advanced techniques, including numerical data visualization libraries like matplotlib, which can be used to create more sophisticated visualizations of text data. We also looked at network maps, which can be used to visualize relationships between entities in text data, and dimensionality reduction techniques like t-SNE and UMAP, which can help us reduce the dimensionality of text data for visualization purposes.

These techniques can help us gain a deeper understanding of our text data and identify patterns and relationships that might not be apparent from a purely numerical analysis. They can be especially useful for exploratory data analysis, hypothesis generation, and communicating results to others in a clear and effective way. Whether we are working with large and complex datasets or small, focused collections of text data, data visualization can help us achieve our goals more effectively and efficiently.

13

Descriptive Statistics

Descriptive statistics provide an essential foundation for understanding and summarizing large datasets by offering valuable insights into the central tendencies, dispersion, and shape of the distribution. By leveraging measures such as mean, median, mode, range, variance, and standard deviation, researchers can succinctly present the essential characteristics of a dataset. Moreover, descriptive statistics serve as a vital starting point in the research process, enabling academics to identify trends, anomalies, and patterns that might require further investigation through inferential statistics or other advanced analytical techniques.

Python has emerged as the preferred choice for conducting academic research in various fields, including accounting and finance, due to its versatility, ease of use, and robust ecosystem. With an extensive range of libraries such as NumPy, pandas, and SciPy, Python facilitates seamless data manipulation, analysis, and visualization, allowing researchers to perform complex statistical computations with relative ease. Furthermore, Python's open-source nature encourages collaboration and resource sharing, promoting the development of cutting-edge tools and techniques that can be readily incorporated into research workflows. Its integration with Jupyter Notebooks also enhances the reproducibility of research, enabling academics to create and share interactive documents that combine live code, equations, visualizations, and narrative text. As a result, Python has become an indispensable tool for researchers in accounting and finance, empowering them to efficiently derive meaningful insights from large and intricate datasets.

Basic Descriptive Statistics

When it comes to performing basic descriptive statistics in Python, the pandas, numpy, and scipy libraries are commonly used. The pandas library is great for data manipulation and provides easy-to-use functions for computing summary statistics. The numpy library is highly optimized for numerical calculations and is useful for calculating statistical measures such as mean, median, and standard deviation. The scipy library provides more advanced statistical functions such as correlation and regression analysis.

One advantage of using these libraries is that they are open-source and freely available, making them accessible to anyone who wants to use them. Another advantage is that they are highly optimized for numerical calculations, which means that they can perform calculations quickly and efficiently. However, these libraries can have a steep learning curve for beginners who are not familiar with Python or programming. Moreover, the syntax for using these libraries can be complex and may require a lot of trial and error to get the desired results. Some statistical methods may require more advanced knowledge of statistics and mathematics, which may be a barrier for some users.

Implementation of basic descriptive statistics such as mean, median, mode, standard deviation, variance, skewness, and kurtosis in Python, for the data my_data.csv in D:/Data folder is provided. The data has a column BTC, on which we'll perform most calculations.

Import the necessary libraries and load the data:

```
import pandas as pd
import numpy as np
import scipy.stats as stats

# Load the data
data_path = "D:/Data/my_data.csv"
df = pd.read_csv(data_path)
```

Calculating mean, median, and mode

```
mean_btc = np.mean(df['BTC'])
median_btc = np.median(df['BTC'])
mode_btc = stats.mode(df['BTC'])
print("Mean of BTC:", mean_btc)
print("Median of BTC:", median_btc)
print("Mode of BTC:", mode_btc[0][0])
```

Standard deviation is a measure of the dispersion or spread of the data, while variance is the square of the standard deviation. We can calculate both using NumPy.

```
std_dev_btc = np.std(df['BTC'], ddof=1)
variance_btc = np.var(df['BTC'], ddof=1)
print("Standard Deviation of BTC:", std_dev_btc)
print("Variance of BTC:", variance_btc)
```

Skewness measures the asymmetry of the data distribution. A positive skewness indicates a distribution with a long right tail, while a negative skewness indicates a distribution with a long left tail. We can calculate skewness using the SciPy library.

```
skewness_btc = stats.skew(df['BTC'])
print("Skewness of BTC:", skewness_btc)
```

Kurtosis is a measure of the "tailedness" of the data distribution. A high kurtosis value indicates a distribution with heavy tails and a low value indicates a distribution with light tails. We can calculate kurtosis using the SciPy library.

```
kurtosis_btc = stats.kurtosis(df['BTC'])
print("Kurtosis of BTC:", kurtosis_btc)
```

A code that calculates some basic descriptive stats by Year variable in the dataset and saves those as additional columns in the dataset is as follows:

```
import pandas as pd
import numpy as np
import scipy.stats as stats

# Load the data
data_path = "D:/Data/my_data.csv"
df = pd.read_csv(data_path)

# Compute statistics by Year
grouped = df.groupby('Year')['BTC'].agg(['mean', 'median',
lambda x: stats.mode(x)[0][0], 'std', 'var'])

# Rename the columns
grouped.columns = ['BTC_mean', 'BTC_median', 'BTC_mode',
'BTC_std_dev', 'BTC_variance']

# Merge the computed statistics with the original data
df = df.merge(grouped, left_on='year', right_index=True)

# Save the modified data
df.to_csv(data_path, index=False)
```

This code groups the data by the "Year" column using the **groupby** method, and then computes the mean, median, mode, standard deviation, and variance for the "BTC" column using the **agg** method. It then renames the columns and merges the computed statistics with the original data using the **merge** method. Finally, it saves the modified data to replace the existing CSV file. The lambda function inside **agg** is used to extract the mode value from the result of **stats.mode**, which returns a tuple containing both the mode value and its count.

To export the descriptive stats in a word file as a publication-quality table, the following code would be useful:

```python
import pandas as pd
import numpy as np
import scipy.stats as stats
from docx import Document
from docx.shared import Pt

# Load the data
data_path = "D:/Data/my_data.csv"
df = pd.read_csv(data_path)

# Compute statistics by Year
grouped = df.groupby('Year')['BTC'].agg(['mean', 'median',
lambda x: stats.mode(x)[0][0], 'std', 'var'])

# Rename the columns
grouped.columns = ['BTC_mean', 'BTC_median', 'BTC_mode',
'BTC_std_dev', 'BTC_variance']

# Merge the computed statistics with the original data
df = df.merge(grouped, left_on='Year', right_index=True)

# Create a new Word document
doc = Document()

# Set the table style
doc.styles['Table Grid'].font.name = 'Times New Roman'
doc.styles['Table Grid'].font.size = Pt(12)

# Create a table in the document
table = doc.add_table(rows=1, cols=len(grouped.columns) + 1,
style='Table Grid')

# Add the header row to the table
header_cells = table.rows[0].cells
header_cells[0].text = 'Year'
for i, col in enumerate(grouped.columns):
    header_cells[i + 1].text = col

# Add the data rows to the table
for year, row in grouped.iterrows():
    row_cells = table.add_row().cells
    row_cells[0].text = str(year)
    for i, value in enumerate(row):
        if pd.isna(value):
            row_cells[i + 1].text = "-"
        else:
            row_cells[i + 1].text = f"{value:.3f}"

# Save the Word document
doc.save('D:/Data/my_data_with_stats.docx')
```

This code computes basic descriptive statistics for the 'BTC' variable in a dataset, grouped by year. It then creates a new Word document, adds a table to the document, populates the table with the computed statistics, and saves the document to a file. The **pandas** library is used for data manipulation, **scipy** is used for computing the mode of the 'BTC' variable, and the **docx** library is used for creating and saving the Word document. The table style is set to Times New Roman font with a font size of 12pt, and missing data is represented with an em dash (—).

Outlier Detection

Outlier detection is the process of identifying observations or data points that deviate significantly from the rest of the dataset. In academic research in accounting and finance, outlier detection is important because outliers can have a significant impact on the results of statistical analysis. Outliers can skew the results, leading to incorrect conclusions and hypotheses.

Python is well-suited for outlier detection because of its powerful data manipulation and visualization libraries, such as Pandas, NumPy, and Matplotlib. These libraries allow for efficient data exploration, identification of outliers, and visualization of the results. Additionally, Python has a wide range of statistical and machine learning packages, such as Scikit-learn and Statsmodels, that can be used to identify outliers and handle them appropriately.

To implement outlier detection in Python, first step is to import the necessary libraries and load the data into a pandas DataFrame:

```
import pandas as pd
import numpy as np
import scipy.stats as stats

data = pd.read_csv('D:/Data/my_data.csv')
```

Now different strategies can be employed to detect the outliers:

1. Z-Score Method

The Z-score method is based on the principle that data points with a z-score greater than a certain threshold are considered outliers. To calculate the z-score, we first need to standardize our data. Here is a sample code that detects observations below 1 percentile and above 99 percentile of the observations in data:

```
btc_mean = data['BTC'].mean()
btc_std = data['BTC'].std()

data['z_score'] = (data['BTC'] - btc_mean) / btc_std

p1 = data['BTC'].quantile(0.01)
p99 = data['BTC'].quantile(0.99)

outliers_z_score = data[(data['BTC'] < p1) | (data['BTC'] > p99)]
```

2. IQR Method

The Interquartile Range (IQR) method considers data points to be outliers if they fall below the first quartile (Q1) minus 1.5 times the IQR or above the third quartile (Q3) plus 1.5 times the IQR. Here's the required code:

```
Q1 = data['BTC'].quantile(0.25)
Q3 = data['BTC'].quantile(0.75)
IQR = Q3 - Q1

outliers_iqr = data[(data['BTC'] < (Q1 - 1.5 * IQR)) | (data['BTC'] > (Q3 + 1.5 * IQR))]
```

3. Modified Z-Score Method

The modified Z-score method uses the median and the Median Absolute Deviation (MAD) instead of the mean and standard deviation. It is more robust to outliers and suitable for data with non-normal distributions. Here's the code for modified Z-Score method:

```
btc_median = data['BTC'].median()
MAD = data['BTC'].mad()

data['mod_z_score'] = 0.6745 * (data['BTC'] - btc_median) / MAD

outliers_mod_z_score = data[np.abs(data['mod_z_score']) > 3.5]
```

The modified Z-score method uses the median and the Median Absolute Deviation (MAD) instead of the mean and standard deviation. It is more robust to outliers and suitable for data with non-normal distributions. Here's the code for modified Z-Score method:

4. Tukey's Fences

Tukey's Fences is a non-parametric method for outlier detection that is similar to the IQR method but allows for adjustable multipliers to account for varying levels of data dispersion.

```
k = 2.2  # Adjust this value as needed
outliers_tukey = data[(data['BTC'] < (Q1 - k * IQR)) |
(data['BTC'] > (Q3 + k * IQR))]
```

Each of these methods has its advantages and drawbacks, and the choice of method depends on the nature of the data and the research question being addressed. It is essential to understand the underlying assumptions and limitations of each method before applying them to your data. In some cases, combining multiple methods can provide a more comprehensive picture of potential outliers in the data.

Pearson's Correlation Coefficient

Pearson's correlation coefficient (r) measures the strength and direction of a linear relationship between two variables. The coefficient ranges from -1 to 1, where -1 indicates a strong negative relationship, 1 indicates a strong positive relationship, and 0 indicates no relationship.

To calculate the Pearson's correlation coefficient in Python, we will use the **scipy.stats** library:

```
import pandas as pd
import scipy.stats as stats

# Load the dataset
data = pd.read_csv('D:/Data/my_data.csv')

# Calculate the Pearson's correlation coefficient
correlation, _ = stats.pearsonr(data['BTC'], data['ETH'])

print('Pearson\'s correlation coefficient: %.3f' % correlation)
```

It is important to remember that correlation does not imply causation. A high correlation between two variables does not necessarily mean that one variable causes changes in the other.

A scatterplot is a useful tool for visually inspecting the relationship between two variables. We can create a scatterplot using the **matplotlib** library:

```python
import matplotlib.pyplot as plt

# Create a scatterplot
plt.scatter(data['BTC'], data['ETH'])
plt.xlabel('BTC')
plt.ylabel('ETH')
plt.title('Scatterplot of BTC and ETH')

plt.show()
```

If the data points in the scatterplot form a clear pattern (e.g., upward or downward sloping), this suggests a strong linear relationship between the variables. Conversely, if the data points appear scattered with no discernible pattern, this may indicate a weak or nonexistent relationship.

Time Series Descriptive Statistics

Time series data is a type of data where observations are collected at equally spaced time intervals, while regular data doesn't have this time component. Time series data often exhibit temporal dependencies and trends that require specialized analysis techniques.

1. Autocorrelations

Autocorrelation is a statistical measure of the correlation between a time series and a lagged version of itself. In simple terms, it is used to identify patterns or trends within the data. To compute the autocorrelation in Python, we can use the **pandas** library.

To calculate the Pearson's correlation coefficient in Python, we will use the **scipy.stats** library:

```python
import pandas as pd

# Load the data
data = pd.read_csv("D:/Data/my_data.csv", parse_dates=['Date'],
index_col='Date')

# Calculate autocorrelations
btc_autocorr = data['BTC'].autocorr(lag=1)

print("Autocorrelation for BTC with lag 1:", btc_autocorr)
```

2. Moving Averages

Moving averages are a technique used to smooth out short-term fluctuations and highlight long-term trends in time series data. There are various types of moving averages, but we will focus on the simple moving average (SMA) and the exponential moving average (EMA). We will use the **pandas** library to calculate the moving averages.

A scatterplot is a useful tool for visually inspecting the relationship between two variables. We can create a scatterplot using the **matplotlib** library:

```
# Simple Moving Average
sma_window = 5
data['BTC_SMA'] = data['BTC'].rolling(window=sma_window).mean()

# Exponential Moving Average
ema_window = 5
data['BTC_EMA'] = data['BTC'].ewm(span=ema_window).mean()

# Display the data with the moving averages
print(data.head(10))
```

3. Exponential Smoothing

Exponential smoothing is another technique used to forecast future values in a time series by assigning exponentially decreasing weights to past observations. We will implement the single exponential smoothing using the **statsmodels** library.

```
import statsmodels.api as sm

# Single Exponential Smoothing
data['BTC'] = data['BTC'].astype(float)
btc_ses = sm.tsa.SimpleExpSmoothing(data['BTC']).fit(smoothing_level=0.2)

# Store the smoothed values in a new column
data['BTC_SES'] = btc_ses.fittedvalues

# Display the data with the smoothed values
print(data.head(10))
```

If the data points in the scatterplot form a clear pattern (e.g., upward or downward sloping), this suggests a strong linear relationship between the variables. Conversely, if the data points appear scattered with no discernible pattern, this may indicate a weak or nonexistent relationship.

A Note on Sampling Techniques

Sampling techniques are essential in accounting and finance research when working with large datasets or when it is impractical to analyze every observation. They allow researchers to draw inferences about a population based on a smaller, representative sample. Here are some common sampling techniques and their Python implementation:

1. Simple Random Sampling

Every observation has an equal chance of being selected.

```
import pandas as pd
import numpy as np

data = pd.read_csv("D:/Data/my_data.csv")
sample_size = 100

simple_random_sample = data.sample(n=sample_size,
random_state=42)
```

2. Systematic Sampling

Observations are selected at regular intervals from the population.

```
interval = 10

systematic_sample = data.iloc[::interval]
```

3. Stratified Sampling

The population is divided into subgroups (strata) based on specific characteristics, and random samples are taken from each stratum.

```
strata_column = 'Year'
strata_sample_size = 20

unique_strata = data[strata_column].unique()
stratified_sample = pd.DataFrame()

for stratum in unique_strata:
    stratum_data = data[data[strata_column] == stratum]
    stratum_sample = stratum_data.sample(n=strata_sample_size, random_state=42)
    stratified_sample = pd.concat([stratified_sample, stratum_sample])
```

4. Cluster Sampling

The population is divided into clusters, and a random sample of clusters is selected for analysis.

```
# Assuming the data has a 'Cluster' column indicating the
cluster each observation belongs to
cluster_column = 'Cluster'
clusters_sample_size = 5

unique_clusters = data[cluster_column].unique()
selected_clusters = np.random.choice(unique_clusters,
size=clusters_sample_size, replace=False)

cluster_sample =
data[data[cluster_column].isin(selected_clusters)]
```

By leveraging the power of Python, researchers can efficiently implement these methods to extract valuable insights from their data. These techniques serve as a solid foundation for further analysis and the development of more sophisticated models. As we move forward in this book, we will delve into more advanced topics, enabling researchers to expand their analytical toolbox and continue to enhance their research capabilities. Python offers a wide array of libraries and functions that cater to various aspects of data analysis, making it an indispensable tool for researchers in the fields of accounting and finance. By mastering the techniques covered in this chapter, researchers can tackle diverse research questions and efficiently navigate the complexities of financial data. By combining a strong foundation in statistical techniques with the power of Python, researchers can push the boundaries of accounting and finance research, discovering new insights and contributing to the growth of knowledge in their respective fields.

Part IV

Natural Language Processing and Text Analysis

14

Topic Modeling

Topic modeling is a popular technique in natural language processing (NLP) that allows us to discover underlying topics or themes within a collection of documents. With the increasing amount of textual data generated in various domains, such as social media, news, and scientific literature, topic modeling has become an important tool for data analysis and information retrieval.

The goal of topic modeling is to automatically identify the latent topics that are present in a large corpus of text. These topics can be thought of as a set of recurring themes or concepts that are shared by the documents in the corpus. Topic modeling can be used for a variety of tasks, such as document clustering, document classification, and information retrieval.

The most commonly used topic modeling technique is Latent Dirichlet Allocation (LDA), which is a probabilistic model that assumes that each document in the corpus is a mixture of a small number of topics and that each word in a document is generated by one of those topics. By applying LDA to a large corpus of text, we can identify the most probable topics and their corresponding word distributions. Topic modeling is a powerful technique that can help us gain insights into the latent themes and concepts within a large corpus of text.

There are several types of topic modeling algorithms, each with its own approach and strengths. Some of the most common types of topic modeling include:

1. Latent Dirichlet Allocation (LDA): This is the most widely used and popular type of topic modeling. LDA models each document as a mixture of topics, where each topic is a probability distribution over words. LDA assumes that each word in a document is generated by one of the topics, and that the topics themselves are generated from a Dirichlet distribution.
2. Non-negative Matrix Factorization (NMF): NMF is a matrix decomposition technique that can be used for topic modeling. It decomposes a matrix of word counts into two matrices: one that represents the topics and one that represents the documents. The topics are represented as non-negative linear combinations of the words in the vocabulary.
3. Probabilistic Latent Semantic Analysis (PLSA): PLSA is a variant of LDA that uses a different approach to estimate the topic distributions. Instead of assuming a Dirichlet prior on the topic distributions, PLSA directly models the joint distribution of the words and the topics.
4. Correlated Topic Model (CTM): CTM is an extension of LDA that allows for correlations between topics. It assumes that the topics are generated from a multivariate normal distribution, and that each document is a mixture of these topics.
5. Hierarchical Dirichlet Process (HDP): HDP is a Bayesian non-parametric model that allows for an unbounded number of topics. It models the topic proportions of each document as a distribution over an infinite number of latent topics, which are shared across all documents.

Before applying topic modeling algorithms such as Latent Dirichlet Allocation (LDA) to a corpus of text data, it is generally recommended to normalize the text by lowercasing all text, removing stopwords, stemming or lemmatizing words, and removing punctuation or special characters. While part of speech (POS) tagging is not required for LDA, it can be useful in certain domains or contexts where different parts of speech carry different meanings. A corpus of preprocessed text data is required for topic modeling, and this corpus can be converted into a numerical representation such as a bag of words or a term frequency-inverse document frequency (TF-IDF) matrix, which can then be used to identify meaningful topics within the corpus.

Now we'll create topics using various techniques.

Latent Dirichlet Allocation (LDA)

The code for LDA for a set of files that contain normalized text and stored in D:/Data/Preprocessed folder is as follows:

```
import os
from gensim import corpora, models

# Define the directory path for preprocessed text files
dir_path = 'D:/Data/Preprocessed'

# Create a list of preprocessed text files
text_files = [os.path.join(dir_path, f) for f in
os.listdir(dir_path) if f.endswith('.txt')]

# Create a list of preprocessed documents
documents = []
for file in text_files:
    with open(file, 'r', encoding='utf-8') as f:
        text = f.read()
        documents.append(text.split())

# Create a dictionary from the preprocessed documents
dictionary = corpora.Dictionary(documents)

# Create a bag-of-words representation of the preprocessed
documents
corpus = [dictionary.doc2bow(doc) for doc in documents]

# Train the LDA model on the bag-of-words representation of the
corpus
num_topics = 10   # Set the number of topics
lda_model = models.ldamodel.LdaModel(corpus,
num_topics=num_topics, id2word=dictionary, passes=10)

# Print the topics and their corresponding top words
for topic in lda_model.print_topics():
    print(topic)
```

The output will look like:

```
(0, '0.016*"walmart" + 0.015*"net" + 0.015*"us" + 0.015*"sale" +
0.015*"oper" + 0.015*"billion" + 0.015*"fiscal" + 0.014*"histor"
+ 0.014*"profit" + 0.014*"gross"')
(1, '0.014*"walmart" + 0.014*"sale" + 0.014*"net" + 0.014*"us" +
0.014*"oper" + 0.014*"billion" + 0.014*"fiscal" + 0.014*"profit"
+ 0.014*"2019" + 0.014*"segment"')
(2, '0.014*"believ" + 0.014*"statement" + 0.014*"made" +
0.014*"oper" + 0.014*"walmart" + 0.014*"consequ" +
0.014*"factor" + 0.014*"busi" + 0.014*"sale" + 0.014*"us"')
(3, '0.063*"statement" + 0.063*"made" + 0.063*"believ" +
0.063*"factor" + 0.033*"assumpt" + 0.033*"uncertainti" +
0.033*"knowledg" + 0.033*"disclo" + 0.033*"base" +
0.033*"report"')
```

This output has ten topics. The numbers before the words represent the weight or importance of that word in a particular topic. Specifically, they represent the probability of a word appearing in that topic. For example, in the first topic, the word "walmart" has a weight of 0.016, which means that it has a 1.6% chance of appearing in that topic. The sum of the weights for all words in a topic adds up to 1.0.

The code uses the gensim library to perform topic modeling using Latent Dirichlet Allocation (LDA). First, the directory path for preprocessed text files is defined using the **os** library. The **dir_path** variable is set to the path where the preprocessed text files are located. Next, a list of preprocessed text files is created using a list comprehension. The **os.listdir** function is used to get a list of files in the **dir_path** directory, and the list comprehension filters the list to include only those files that end with the **.txt** extension. The resulting list of file names is stored in the **text_files** variable.

After that, a list of preprocessed documents is created by iterating through each file in the **text_files** list. The **open** function is used to open each file in read mode, and the resulting text is split into a list of words using the **split** method. Each list of words is then appended to the **documents** list.

The **corpora.Dictionary** function is used to create a dictionary from preprocessed documents. The **dictionary** variable is set to the resulting dictionary object. Next, a bag-of-words representation of the preprocessed documents is created using the **doc2bow** method of the **dictionary** object. Each bag-of-words representation is appended to the **corpus** list.

The LDA model is trained on the bag-of-words representation of the corpus using the **models.ldamodel.LdaModel** function. The **num_topics** variable is set to the desired number of topics, and the **id2word** parameter is set to the **dictionary** object created earlier. The **passes** parameter is set to the number of passes the algorithm should make over the corpus.

Finally, the top words for each topic are printed using the **print_topics** method of the **lda_model** object. The resulting list of topics and top words is printed to the console.

For this code to run you need to have the genism library installed. When I used this code for the first time, it gave me an error related to numpy incompatibility. If that happens simply uninstall numpy using

```
pip uninstall numpy
```

Reinstall numpy again after uninstalling.

```
pip install numpy
```

This code uses makes 10 topics. Optimum number of topics can be determined in two ways:

1. **Perplexity:** Perplexity is a measure of how well the LDA model is able to predict unseen data. A lower perplexity indicates that the model is better at predicting unseen data. To use perplexity to determine the optimum number of topics, you can train multiple LDA models with different numbers of topics, and choose the model that has the lowest perplexity on a held-out test set of documents. The code below automatically calculates the perplexity for a range of number of topics.

```python
from gensim.models import LdaModel
from gensim.corpora import Dictionary
import os

# Define the directory path for preprocessed text files
dir_path = 'D:/Data/Preprocessed'

# Create a list of preprocessed text files
text_files = [os.path.join(dir_path, f) for f in
os.listdir(dir_path) if f.endswith('.txt')]

# Create a list of preprocessed documents
documents = []
for file in text_files:
    with open(file, 'r', encoding='utf-8') as f:
        text = f.read()
        documents.append(text.split())

# Create a dictionary from the preprocessed documents
dictionary = Dictionary(documents)

# Create a bag-of-words representation of the preprocessed
documents
corpus = [dictionary.doc2bow(doc) for doc in documents]

# Define a function to train an LDA model and calculate its
perplexity
def train_lda_model(num_topics, corpus, dictionary):
    lda_model = LdaModel(corpus=corpus, num_topics=num_topics,
id2word=dictionary, passes=10)
    perplexity = lda_model.log_perplexity(corpus)
    return lda_model, perplexity

# Define the range of number of topics to try
min_topics = 5
max_topics = 25

# Train multiple LDA models with different numbers of topics
perplexities = []
for num_topics in range(min_topics, max_topics+1):
    lda_model, perplexity = train_lda_model(num_topics, corpus,
dictionary)
    perplexities.append(perplexity)
    print(f"Number of topics: {num_topics} | Perplexity:
{perplexity:.2f}")

# Choose the number of topics that results in the lowest
perplexity
optimal_num_topics = perplexities.index(min(perplexities)) +
min_topics
print(f"Optimal number of topics: {optimal_num_topics}")
```

The output will look like:

```
  Number of topics: 18 | Perplexity: -5.20
  Number of topics: 19 | Perplexity: -5.22
  Number of topics: 20 | Perplexity: -5.25
  Number of topics: 21 | Perplexity: -5.28
  Number of topics: 22 | Perplexity: -5.30
  Number of topics: 23 | Perplexity: -5.33
  Number of topics: 24 | Perplexity: -5.35
  Number of topics: 25 | Perplexity: -5.37
  Optimal number of topics: 25
```

The code below does automatically select the optimum number of topics and creates those topics:

```python
from gensim.models import LdaModel
from gensim.corpora import Dictionary
import os

# Define the directory path for preprocessed text files
dir_path = 'D:/Data/Preprocessed'

# Create a list of preprocessed text files
text_files = [os.path.join(dir_path, f) for f in os.listdir(dir_path) if f.endswith('.txt')]

# Create a list of preprocessed documents
documents = []
for file in text_files:
    with open(file, 'r', encoding='utf-8') as f:
        text = f.read()
        documents.append(text.split())

# Create a dictionary from the preprocessed documents
dictionary = Dictionary(documents)

# Create a bag-of-words representation of the preprocessed documents
corpus = [dictionary.doc2bow(doc) for doc in documents]

# Define a function to train an LDA model and calculate its perplexity
def train_lda_model(num_topics, corpus, dictionary):
    lda_model = LdaModel(corpus=corpus, num_topics=num_topics, id2word=dictionary, passes=10)
    perplexity = lda_model.log_perplexity(corpus)
    return lda_model, perplexity

# Define the range of number of topics to try
min_topics = 5
max_topics = 25

# Train multiple LDA models with different numbers of topics
perplexities = []
for num_topics in range(min_topics, max_topics+1):
    lda_model, perplexity = train_lda_model(num_topics, corpus, dictionary)
    perplexities.append(perplexity)
    print(f"Number of topics: {num_topics} | Perplexity: {perplexity:.2f}")

# Choose the number of topics that results in the lowest perplexity
optimal_num_topics = perplexities.index(min(perplexities)) + min_topics
print(f"Optimal number of topics: {optimal_num_topics}")

# Train an LDA model with the optimal number of topics
lda_model = LdaModel(corpus=corpus, num_topics=optimal_num_topics, id2word=dictionary, passes=10)


# Print the topics and their corresponding top words
for topic in lda_model.print_topics():
    print(topic)
```

The output will look like:

```
Number of topics: 23 | Perplexity: -5.32
Number of topics: 24 | Perplexity: -5.35
Number of topics: 25 | Perplexity: -5.37
Optimal number of topics: 25
(19, '0.014*"three" + 0.014*"state" + 0.014*"base" +
0.014*"assumpt" + 0.014*"well" + 0.014*"washington" +
0.014*"walmart" + 0.014*"us" + 0.014*"believ" + 0.014*"rico"')
(12, '0.014*"three" + 0.014*"state" + 0.014*"base" +
0.014*"assumpt" + 0.014*"well" + 0.014*"washington" +
0.014*"walmart" + 0.014*"us" + 0.014*"believ" + 0.014*"rico"')
```

2. **Coherence:** Coherence is a measure of how semantically coherent the topics are. A higher coherence indicates that the topics are more semantically coherent. To use coherence to determine the optimum number of topics, you can train multiple LDA models with different numbers of topics, and choose the model that has the highest coherence.

The code below calculates the coherence for a list of number of topics.

```
from gensim.models.ldamodel import LdaModel   # Import LdaModel
directly

# Define the directory path for preprocessed text files
dir_path = 'D:/Data/Preprocessed'

# Create a list of preprocessed text files
text_files = [os.path.join(dir_path, f) for f in
os.listdir(dir_path) if f.endswith('.txt')]

# Create a list of preprocessed documents
documents = []

for file in text_files:
    with open(file, 'r', encoding='utf-8') as f:
        text = f.read()
        documents.append(text.split())

# Create a dictionary from the preprocessed documents
dictionary = corpora.Dictionary(documents)

# Create a bag-of-words representation of the preprocessed
documents
corpus = [dictionary.doc2bow(doc) for doc in documents]

# Calculate the coherence for different numbers of topics
coherence_scores = []
for num_topics in [5, 15]:
    lda_model = LdaModel(corpus, num_topics=num_topics,
id2word=dictionary, passes=10)
    coherence_model = CoherenceModel(model=lda_model,
texts=documents, dictionary=dictionary, coherence='c_v')
    coherence_score = coherence_model.get_coherence()
    coherence_scores.append(coherence_score)

# Choose the number of topics that results in the highest
coherence
optimal_num_topics =
coherence_scores.index(max(coherence_scores)) + 5   # add 5 to
account for the range starting from 5
print(f"Optimal number of topics: {optimal_num_topics}")
```

The output will look like

`Optimal number of topics: 5`

It is pertinent to mention that coherence is computationally intensive. Larger the list is more time it takes the code to execute. Best way is to determine the approximate number of best number of topics using above code and then around that number either determine the exact number of topics by replacing line 24 of code

```
for num_topics in [5, 15]:
```

by

```
for num_topics in range(2, 8):
```

After this you can run the LDA code based on this number. However, after the approximation, you can also make the optimum number of codes directly using the following code:

```python
from gensim.models import CoherenceModel
from gensim.models import LdaModel  # Import the LdaModel class

# Define the directory path for preprocessed text files
dir_path = 'D:/Data/Preprocessed'

# Create a list of preprocessed text files
text_files = [os.path.join(dir_path, f) for f in os.listdir(dir_path) if f.endswith('.txt')]

# Create a list of preprocessed documents
documents = []
for file in text_files:
    with open(file, 'r', encoding='utf-8') as f:
        text = f.read()
        documents.append(text.split())

# Create a dictionary from the preprocessed documents
dictionary = corpora.Dictionary(documents)

# Create a bag-of-words representation of the preprocessed documents
corpus = [dictionary.doc2bow(doc) for doc in documents]

# Calculate the coherence for different numbers of topics
coherence_scores = []
lda_models = []
for num_topics in range(2, 8):
    lda_model = LdaModel(corpus, num_topics=num_topics, id2word=dictionary, passes=10)
    coherence_model = CoherenceModel(model=lda_model, texts=documents, dictionary=dictionary, coherence='c_v')
    coherence_score = coherence_model.get_coherence()
    coherence_scores.append(coherence_score)
    lda_models.append(lda_model)

# Choose the number of topics that results in the highest coherence
optimal_num_topics = coherence_scores.index(max(coherence_scores)) + 2  # add 2 to account for the range starting from 2
print(f"Optimal number of topics: {optimal_num_topics}")

# Print the topics for the optimal model
optimal_model = lda_models[coherence_scores.index(max(coherence_scores))]
for topic in optimal_model.print_topics():
    print(topic)
```

The output will look like

```
Optimal number of topics: 4
(0, '0.015*"walmart" + 0.014*"us" + 0.014*"sale" + 0.014*"oper"
+ 0.014*"net" + 0.014*"fiscal" + 0.014*"histor" + 0.014*"gross"
+ 0.014*"billion" + 0.014*"segment"')
(1, '0.083*"walmart" + 0.063*"us" + 0.053*"net" + 0.053*"sale" +
0.033*"billion" + 0.033*"fiscal" + 0.033*"oper" + 0.023*"gross"
+ 0.023*"profit" + 0.023*"segment"')
```

In natural language processing, the bag-of-words model is a way of representing text data as a collection, or "bag", of its individual words, disregarding grammar and word order but keeping track of frequency. The idea is that the frequency of occurrence of each word in a document can help capture some of its meaning, even if we don't consider the order of the words.

This representation can be used for a variety of natural language processing tasks, such as text classification, topic modeling, and sentiment analysis. One common way of using the bag-of-words model is to convert a set of documents into a matrix, where each row corresponds to a document and each column corresponds to a unique word in the corpus, and the values in the matrix represent the frequency of each word in each document. This matrix can then be used as input to various machine learning algorithms.

Non-negative Matrix Factorization (NMF)

The code for NMF for a set of files that contain normalized text and stored in D:/Data/Preprocessed folder is as follows:

```
import os
import numpy as np
from sklearn.feature_extraction.text import TfidfVectorizer
from sklearn.decomposition import NMF

# Define the directory containing the preprocessed text files
dir_path = 'D:/Data/Preprocessed'

# Create a list of file paths for all files in the directory
file_paths = [os.path.join(dir_path, f) for f in
os.listdir(dir_path) if os.path.isfile(os.path.join(dir_path,
f))]

# Create a TfidfVectorizer object to convert the text data into
a matrix of TF-IDF features
vectorizer = TfidfVectorizer(input='filename',
stop_words='english')

# Use the vectorizer to fit and transform the text data into a
matrix of TF-IDF features
tfidf = vectorizer.fit_transform(file_paths)

# Define the number of topics to extract
n_topics = 50

# Perform NMF on the TF-IDF matrix to extract the topics
nmf = NMF(n_components=n_topics, init='nndsvd', max_iter=200,
random_state=0)
W = nmf.fit_transform(tfidf)
H = nmf.components_

# Print the top words for each topic
feature_names = vectorizer.get_feature_names()
for i, topic in enumerate(H):
    print("Topic {}: {}".format(1, ", ".join([feature_names[j]
for j in topic.argsort()[:-11:-1]])))
```

This first uses the **os** module to define the directory containing the preprocessed text files (**dir_path**). It then creates a list of file paths for all files in the directory (**file_paths**). Then it creates a **TfidfVectorizer** object called **vectorizer** to convert the text data into a matrix of TF-IDF features. The **input** parameter is set to '**filename**' to indicate that the input data is a set of file paths rather than raw text data, and the **stop_words** parameter is set to '**english**' to remove common English stop words (if not removed already). This step is important because it helps to reduce the variability in the text data and make it easier for the NMF model to identify and extract meaningful topics.

The vectorizer is then used to fit and transform the text data into a matrix of TF-IDF features (**tfidf**). This matrix represents the text data in a format that is suitable for input to the NMF algorithm. The code defines the number of topics to extract from the text data (**n_topics**), which in this case is set to

50. This number can be adjusted depending on the size of the text data and the desired level of granularity in the extracted topics.

NMF is then performed on the TF-IDF matrix (**tfidf**) to extract the topics. The **NMF** object is created with **n_components = n_topics** to specify the number of topics, **init = 'nndsvd'** to initialize the factorization using a non-negative double singular value decomposition, **max_iter = 200** to set the maximum number of iterations, and **random_state = 0** to ensure reproducibility of the results. The **fit_transform** method is used to fit the NMF model to the TF-IDF matrix and transform the data into the factorized matrices **W** and **H**.

Finally, the code prints the top 10 words for each topic by sorting the entries in the **H** matrix (**topic.argsort()[:-11:-1]**) and using the **feature_names** list to print the corresponding words. This step helps to identify the most important words associated with each topic and provides insights into the themes and patterns in the text data. The results are printed to the console.

In using NMF, you need to define the number of topics, which can be between 1 and 1000. Unfortunately, NMF does not have a built-in method to automatically select the optimal number of topics for a given corpus of text data. However, there are several methods that you can use to estimate the optimal number of topics based on the characteristics of the text data.

One common approach is to use a metric called "perplexity" to evaluate the quality of the topic model at different numbers of topics. Perplexity measures how well a model predicts new text data based on the topics it has learned. It is a statistical concept and is beyond the scope of this book. Generally, lower perplexity scores indicate better performance. You can train multiple NMF models with different numbers of topics and choose the model that produces the lowest perplexity score on a held-out set of text data. A code that calculates selects the best NMF topics based on a range of topics provided by the user would be:

```python
import os
import numpy as np
from sklearn.feature_extraction.text import TfidfVectorizer
from sklearn.decomposition import NMF
from sklearn.model_selection import train_test_split

# Define the directory containing the preprocessed text files
dir_path = 'D:/Data/Preprocessed'

# Create a list of file paths for all files in the directory
file_paths = [os.path.join(dir_path, f) for f in
os.listdir(dir_path) if os.path.isfile(os.path.join(dir_path,
f))]

# Split the file paths into training and validation sets
train_paths, val_paths = train_test_split(file_paths,
test_size=0.2, random_state=42)

# Create a TfidfVectorizer object to convert the text data into
a matrix of TF-IDF features
vectorizer = TfidfVectorizer(input='filename',
stop_words='english')

# Use the vectorizer to fit and transform the text data into a
matrix of TF-IDF features for the training set
tfidf_train = vectorizer.fit_transform(train_paths)

# Define a range of numbers of topics to extract
n_topics_range = range(10, 100, 10)

# Initialize lists to store perplexity scores and NMF models
perplexities = []
nmf_models = []

# Iterate over the range of numbers of topics
for n_topics in n_topics_range:
    # Perform NMF on the TF-IDF matrix for the training set to
extract the topics
    nmf = NMF(n_components=n_topics, init='nndsvd',
max_iter=200, random_state=0)
    W_train = nmf.fit_transform(tfidf_train)
    H_train = nmf.components_

    # Calculate the log-likelihood of the documents in the
validation set
    tfidf_val = vectorizer.transform(val_paths)
    log_likelihoods = nmf.score(tfidf_val)

    # Compute the perplexity score
    N = tfidf_val.shape[0]
    perplexity = np.exp(-1 * np.sum(log_likelihoods) / N)
    perplexities.append(perplexity)
    nmf_models.append(nmf)
```

```
# Find the index of the minimum perplexity score
best_index = np.argmin(perplexities)

# Print the best number of topics and its corresponding
perplexity score
print("Best number of topics:
{}".format(n_topics_range[best_index]))
print("Perplexity score:
{:.2f}".format(perplexities[best_index]))

# Extract the topics for the best model
best_nmf = nmf_models[best_index]
W = best_nmf.transform(tfidf_train)
H = best_nmf.components_

# Print the top words for each topic
feature_names = vectorizer.get_feature_names()
for i, topic in enumerate(H):
    print("Topic {}: {}".format(i, ", ".join([feature_names[j]
for j in topic.argsort()[:-11:-1]])))
```

In this version of the code, we first split the set of preprocessed text files into a training set and a validation set using the **train_test_split** function from **sklearn.model_selection**. We then fit and transform the training set using the **TfidfVectorizer** object.

Next, we define a range of numbers of topics to extract and iterate over this range, performing NMF on the training set for each value of **n_topics**. We then calculate the log-likelihood of the documents in the validation set and compute the perplexity score for each value of **n_topics**.

After computing the perplexity score for each value of **n_topics**, we find the index of the minimum perplexity score and extract the topics for the corresponding N.

Another approach is to use visual inspection of the topic model results to estimate the optimal number of topics. You can generate topic-word and document-topic matrices for each number of topics and examine the resulting topics to see if they are coherent and meaningful. If the topics are too general or too specific, you can try adjusting the number of topics until you find a satisfactory set of topics.

Keep in mind that the optimal number of topics may depend on the specific characteristics of your text data and the goals of your analysis, so it is important to evaluate multiple models and select the one that best meets your needs.

Probabilistic Latent Semantic Analysis (PLSA)

The code for PLSA for a set of files that contain normalized text and stored in D:/Data/Preprocessed folder is as follows:

```
import os
import numpy as np
import pandas as pd
from sklearn.feature_extraction.text import CountVectorizer
from sklearn.decomposition import NMF, LatentDirichletAllocation

# Set the directory containing the preprocessed text files
directory = 'D:/Data/Preprocessed'

# Get the file paths for all text files in the directory
file_paths = [os.path.join(directory, f) for f in
os.listdir(directory) if f.endswith('.txt')]

# Read in the text data from the files
documents = []
for file_path in file_paths:
    with open(file_path, 'r', encoding='utf-8') as f:
        documents.append(f.read())

# Create a count vectorizer to convert the text data into a
matrix of word counts
vectorizer = CountVectorizer(max_df=0.95, min_df=2,
stop_words='english')
X = vectorizer.fit_transform(documents)

# Fit a Latent Dirichlet Allocation (LDA) model to the count
matrix
n_topics = 10 # Set the number of topics to extract
lda = LatentDirichletAllocation(n_components=n_topics,
max_iter=50, learning_method='online', random_state=42)
lda.fit(X)

# Print the top 10 words for each topic
feature_names = vectorizer.get_feature_names()
for topic_idx, topic in enumerate(lda.components_):
    print(f"Topic {topic_idx}:")
    top_features = [feature_names[i] for i in topic.argsort()[:-11:-1]]
    print(", ".join(top_features))
```

The output will look like:

```
Topic 0:
3185, neighborhood, behalf, market, fiscal, histor, respect,
includ, reason, consolid
Topic 1:
walmart, sale, net, fiscal, billion, brand, histor, segment,
profit, 2019
Topic 2:
walmart, sale, 2019, consolid, billion, segment, net, incom,
brand, repr
```

In this code, first the necessary modules are imported, including **os, numpy, pandas, CountVectorizer,** and **LatentDirichletAllocation** from the **sklearn.feature_extraction.text** and **sklearn.decomposition** modules, respectively. Next, the directory containing the preprocessed text files is set to **D:/Data/Preprocessed**, and the file paths for all text files in the directory are obtained using the **os.listdir()** function and a list comprehension.

The text data is read in from the files using a for loop, which reads each file and appends the text to a list called **documents**. A CountVectorizer() object is then created with the specified parameters **max_df = 0.95, min_df = 0.05,** and **stop_words = 'english'**. This vectorizer is used to convert the preprocessed text data into a matrix of word counts.

Next, a **LatentDirichletAllocation()** object is created with **n_topics = 10, max_iter = 50, learning_method = 'online',** and **random_state = 42**. This LDA model is then fit to the count matrix using the **fit()** method. Finally, the top 10 words for each topic are printed out using a for loop that iterates over the topics and uses the **argsort()** method to obtain the indices of the top 10 words in each topic, which are then looked up in the **feature_names** list obtained from the **CountVectorizer()** object. **max_df** and **min_df** are parameters of the **CountVectorizer()** function in scikit-learn, which is used to convert a corpus of text documents into a matrix of token counts.

max_df specifies the maximum document frequency of a term in the corpus, as a fraction between 0 and 1. Terms that appear in more than **max_df** fraction of the documents are excluded from the vocabulary.

min_df specifies the minimum document frequency of a term in the corpus, as an integer, or a fraction between 0 and 1. Terms that appear in fewer than **min_df** documents are excluded from the vocabulary.

By default, **max_df** is set to 1.0, meaning that all terms in the corpus are included in the vocabulary, and **min_df** is set to 1, meaning that terms that appear in only one document are excluded from the vocabulary. However, these parameters can be adjusted to optimize the quality of the resulting token count matrix and the performance of downstream text analysis tasks such as topic modeling or sentiment analysis.

Correlated Topic Model (CTM)

The code for CTM for a set of files that contain normalized text and stored in D:/Data/Preprocessed folder is as follows:

```python
import os
from gensim import corpora, models
from gensim.models import CoherenceModel

# Define the path to the preprocessed text files
path = 'D:/Data/Preprocessed'

# Create a list of the file names in the directory
file_names = os.listdir(path)

# Load the preprocessed text files into a list of lists of words
documents = []
for file_name in file_names:
    with open(os.path.join(path, file_name), 'r') as f:
        document = f.read().split()
        documents.append(document)

# Create a dictionary and corpus from the documents
dictionary = corpora.Dictionary(documents)
corpus = [dictionary.doc2bow(document) for document in documents]

# Set the number of topics for the CTM model
num_topics = 10

# Train the CTM model on the corpus
ctm_model = models.LdaModel(
    corpus=corpus,
    id2word=dictionary,
    num_topics=num_topics,
    chunksize=2000,
    decay=0.5,
    passes=10,
    eval_every=None,
    iterations=50,
    gamma_threshold=0.001,
    minimum_probability=0.01,
    random_state=None,
    alpha='asymmetric',
    eta=None,
    dtype=None
)

# Print the topics and their corresponding words
for topic in ctm_model.show_topics():
    print(topic)

# Compute the coherence score for the CTM model
coherence_model = CoherenceModel(
    model=ctm_model,
    texts=documents,
    dictionary=dictionary,
    coherence='c_v'
)
coherence_score = coherence_model.get_coherence()
print(f"Coherence score: {coherence_score}")
```

In this code, we start by importing the necessary modules, including the **os** module for working with file paths, and the **corpora**, **models**, and **CoherenceModel** classes from the **gensim** library, which we will use to train and evaluate our CTM model. Next, we define the path to the directory containing our preprocessed text files. We create a list of the file names in the directory using the **os.listdir()** function. We then load the preprocessed text files into a list of lists of words. For each file, we read its contents and split them into words using the **split()** function. We then append the resulting list of words to our **documents** list.

We create a **corpora.Dictionary** object from our **documents** list, which maps each unique word to a unique integer ID. We also create a **corpus** object from our **documents** list using the **doc2bow()** function of our **dictionary** object. This converts each document in our **documents** list to a bag-of-words representation, where each word is represented by its integer ID and its frequency in the document.

We then set the number of topics for our CTM model to **10**, and train the model on our **corpus** using the **LdaModel()** function from the **models** module of **gensim**. We pass our **corpus, dictionary**, and the number of topics to the function, along with various other hyperparameters that control the behavior of the algorithm. After training the CTM model, we print the topics and their corresponding words using the **show_topics()** method of our **ctm_model** object. Finally, we compute the coherence score for our CTM model using the **CoherenceModel()** function from **gensim**. We pass our **ctm_model, documents, dictionary**, and the type of coherence measure we want to use ('**c_v**' in this case) to the function. The coherence score measures the degree of semantic similarity between the words in the topics, and serves as a rough measure of the quality of the model. The output will look like:

```
(0, '0.017*"walmart" + 0.015*"sale" + 0.015*"us" + 0.015*"net" +
0.015*"oper" + 0.015*"billion" + 0.015*"fiscal" + 0.014*"brand"
+ 0.014*"segment" + 0.014*"2019"')
(1, '0.014*"believ" + 0.014*"statement" + 0.014*"made" +
0.014*"factor" + 0.014*"behalf" + 0.014*"describ" +
0.014*"uncertainti" + 0.014*"oper" + 0.014*"mention" +
0.014*"assumpt"')
(2, '0.014*"walmart" + 0.014*"us" + 0.014*"net" + 0.014*"sale" +
0.014*"billion" + 0.014*"fiscal" + 0.014*"oper" +
0.014*"segment" + 0.014*"profit" + 0.014*"brand"')
(3, '0.063*"factor" + 0.063*"made" + 0.063*"statement" +
0.063*"believ" + 0.033*"report" + 0.033*"make" + 0.033*"environ"
+ 0.033*"consequ" + 0.033*"rwardlook" + 0.033*"disclo"')
```

In this code the number of topics is set to 10. Fortunately, there is a way to know the optimum number of topics. You need to calculate the coherence score from a range of number of topics. Coherence score is a statistical concept and beyond the scope of this book. The following code will automatically calculate coherence score for a range of number of topics (in this case 5 to 10) and tell the optimum number of topics. You can use that information to decide the number of topics needed.

```python
import os
from gensim import corpora, models
from gensim.models import CoherenceModel
import matplotlib.pyplot as plt

# Define the path to the preprocessed text files
path = 'D:/Data/Preprocessed'

# Create a list of the file names in the directory
file_names = os.listdir(path)

# Load the preprocessed text files into a list of lists of words
documents = []
for file_name in file_names:
    with open(os.path.join(path, file_name), 'r') as f:
        document = f.read().split()
        documents.append(document)

# Create a dictionary and corpus from the documents
dictionary = corpora.Dictionary(documents)
corpus = [dictionary.doc2bow(document) for document in documents]

# Set the range of possible topic numbers
start_topic_num = 5
end_topic_num = 20
step_size = 1
topic_nums = range(start_topic_num, end_topic_num + step_size, step_size)

# Compute the coherence scores for each topic number
coherence_scores = []
for num_topics in topic_nums:
    # Train the CTM model on the corpus
    ctm_model = models.LdaModel(
        corpus=corpus,
        id2word=dictionary,
        num_topics=num_topics,
        chunksize=2000,
        decay=0.5,
        passes=10,
        eval_every=None,
        iterations=50,
        gamma_threshold=0.001,
        minimum_probability=0.01,
        random_state=None,
        alpha='asymmetric',
        eta=None,
        dtype=None
    )

    # Compute the coherence score for the CTM model
    coherence_model = CoherenceModel(
```

```
        model=ctm_model,
        texts=documents,
        dictionary=dictionary,
        coherence='c_v'
)
    coherence_score = coherence_model.get_coherence()
    coherence_scores.append(coherence_score)

# Plot the coherence scores as a function of topic number
plt.plot(topic_nums, coherence_scores)
plt.xlabel('Number of Topics')
plt.ylabel('Coherence Score')
plt.title('Coherence Scores for Different Numbers of Topics')
plt.show()

# Select the topic number that maximizes the coherence score
optimal_topic_num =
topic_nums[coherence_scores.index(max(coherence_scores))]
print(f"Optimal number of topics: {optimal_topic_num}")
```

The code sets the path to a directory of preprocessed text files and loads the files into memory as a list of lists of words. Then it creates a **corpora.Dictionary** object and a corpus object from the preprocessed documents. The code then defines a range of possible topic numbers that we want to evaluate, sets some CTM model parameters, and loops over the range of topic numbers. For each topic number, the code trains a CTM model using the **LdaModel()** function from **gensim**, computes the coherence score for the model using the **CoherenceModel()** function, and appends the coherence score to a list. After computing the coherence scores for all of the topic numbers, the code plots the coherence scores as a function of the number of topics using **matplotlib.pyplot**. This allows us to visually identify the number of topics that maximizes the coherence score.

Finally, the code selects the optimal number of topics by finding the topic number that corresponds to the maximum coherence score and prints it to the console.

Here's a code that does both the things in one step and gives the topics based on optimum number of topics based on coherence score.

```python
    model=ctm_model,
    texts=documents,
    dictionary=dictionary,
    coherence='c_v'
)
coherence_score = coherence_model.get_coherence()
coherence_scores.append(coherence_score)

# Plot the coherence scores as a function of topic number
plt.plot(topic_nums, coherence_scores)
plt.xlabel('Number of Topics')
plt.ylabel('Coherence Score')
plt.title('Coherence Scores for Different Numbers of Topics')
plt.show()

# Select the topic number that maximizes the coherence score
optimal_topic_num =
topic_nums[coherence_scores.index(max(coherence_scores))]
print(f"Optimal number of topics: {optimal_topic_num}")

# Train the CTM model on the corpus using the optimal number of
topics
ctm_model = models.LdaModel(
    corpus=corpus,
    id2word=dictionary,
    num_topics=optimal_topic_num,
    chunksize=2000,
    decay=0.5,
    passes=10,
    eval_every=None,
    iterations=50,
    gamma_threshold=0.001,
    minimum_probability=0.01,
    random_state=None,
    alpha='asymmetric',
    eta=None,
    dtype=None
)

# Print the topics and their corresponding words
for topic in ctm_model.show_topics():
    print(topic)

# Compute the coherence score for the CTM model
coherence_model = CoherenceModel(
    model=ctm_model,
    texts=documents,
    dictionary=dictionary,
    coherence='c_v'
)
coherence_score = coherence_model.get_coherence()
print(f"Coherence score for {optimal_topic_num} topics:
{coherence_score}")
```

```python
import os
from gensim import corpora, models
from gensim.models import CoherenceModel
import matplotlib.pyplot as plt

# Define the path to the preprocessed text files
path = 'D:/Data/Preprocessed'

# Create a list of the file names in the directory
file_names = os.listdir(path)

# Load the preprocessed text files into a list of lists of words
documents = []
for file_name in file_names:
    with open(os.path.join(path, file_name), 'r') as f:
        document = f.read().split()
        documents.append(document)

# Create a dictionary and corpus from the documents
dictionary = corpora.Dictionary(documents)
corpus = [dictionary.doc2bow(document) for document in documents]

# Set the range of possible topic numbers
start_topic_num = 5
end_topic_num = 20
step_size = 1
topic_nums = range(start_topic_num, end_topic_num + step_size, step_size)

# Compute the coherence scores for each topic number
coherence_scores = []
for num_topics in topic_nums:
    # Train the CTM model on the corpus
    ctm_model = models.LdaModel(
        corpus=corpus,
        id2word=dictionary,
        num_topics=num_topics,
        chunksize=2000,
        decay=0.5,
        passes=10,
        eval_every=None,
        iterations=50,
        gamma_threshold=0.001,
        minimum_probability=0.01,
        random_state=None,
        alpha='asymmetric',
        eta=None,
        dtype=None
    )

    # Compute the coherence score for the CTM model
    coherence_model = CoherenceModel(
```

It is pertinent to mention that selecting topics based on coherence score is computationally intensive and may take some time.

Hierarchical Dirichlet Process (HDP)

The code for HDP for a set of files that contain normalized text and stored in D:/Data/Preprocessed folder is as follows:

```
import os
from gensim.models import HdpModel
from gensim.corpora import Dictionary
from nltk.tokenize import word_tokenize

# Define the directory where the preprocessed text files are
located
dir_path = 'D:/Data/Preprocessed'

# Define a function to read in the text files and tokenize the
text
def read_documents(dir_path):
    for filename in os.listdir(dir_path):
        with open(os.path.join(dir_path, filename), 'r',
encoding='utf-8') as f:
            yield word_tokenize(f.read().lower())

# Read in the preprocessed text files and create a dictionary
from the tokens
documents = list(read_documents(dir_path))
dictionary = Dictionary(documents)

# Convert the preprocessed text files to a bag-of-words
representation
corpus = [dictionary.doc2bow(doc) for doc in documents]

# Train an HDP topic model on the corpus
hdp = HdpModel(corpus, id2word=dictionary)

# Print the most likely topics and their corresponding
probabilities for each document in the corpus
for i, doc in enumerate(corpus):
    print(f"Document {i+1}:")
    for topic in hdp[doc]:
        print(f"\tTopic {topic[0]}: {topic[1]:.4f}")
```

This code assumes that each text file contains a single document. The code imports necessary packages for creating and training an HDP model for topic modeling on a set of preprocessed text files. It then defines the directory where the preprocessed text files are located using the **dir_path** variable. A function called **read_documents()** is defined to read in the text files from the directory and tokenize them using the **word_tokenize()** function from the NLTK package. The **documents** variable is created by reading in all the preprocessed text files and tokenizing them using the **read_documents()** function. The **dictionary** variable is created by generating a dictionary mapping words to unique integer IDs using the **Dictionary()** function from the Gensim package. The **corpus** variable is created by converting the preprocessed text files to a bag-of-words representation using the **doc2bow()** function from the Gensim package. The HDP model is trained on the **corpus** using the **HdpModel()** function from the Gensim package. Finally, the code loops through each document in the **corpus** and prints the most likely topics and their corresponding probabilities using the trained HDP model.

The HDP model is a Bayesian non-parametric model that can discover an unbounded number of topics in the corpus. Unlike other topic modeling algorithms, such as Latent Dirichlet Allocation (LDA), HDP does not require the number of topics to be specified in advance. Instead, HDP infers the number of topics from the data during the model training process.

When the **HdpModel()** function is called in the code, it automatically infers the number of topics from the corpus based on the data. The resulting number of topics can be different for each run of the model, depending on the specific characteristics of the data and the settings of the model.

In the output, the code loops through each document in the corpus and prints the most likely topics and their corresponding probabilities. The number of topics that are printed depends on the topics that are discovered by the model and their respective probabilities. If the model infers a large number of topics, then a large number of topics and their probabilities will be printed for each document. Conversely, if the model infers only a small number of topics, then a small number of topics and their probabilities will be printed for each document.

The use of topic modeling in accounting research has allowed for the efficient and accurate analysis of large volumes of textual data. With the ever-increasing amount of data generated by businesses and financial institutions, traditional manual analysis methods are becoming less practical and effective. Topic modeling provides a powerful and effective solution to this problem, allowing researchers to extract valuable insights and identify patterns in large amounts of unstructured text data.

By utilizing advanced machine learning algorithms such as LDA and HDP, researchers can identify underlying themes and patterns in financial reports, disclosures, and other accounting documents. Additionally, the use of proper text preprocessing techniques ensures the accuracy and reliability of the results, further enhancing the value of this approach.

Furthermore, the ability to calculate perplexity and coherence scores for different numbers of topics enables researchers to determine the optimal number of topics for a given dataset, improving the interpretability and usefulness of the results. With the increasing use of topic modeling in accounting research, this approach is poised to become an indispensable tool for understanding complex business phenomena and financial reporting practices.

Topic modeling is a powerful and valuable technique for accounting researchers and other business school academics to gain insights from large volumes of text data. Its applications range from analyzing financial disclosures to identifying patterns in earnings calls and other financial reports. With its ability to efficiently analyze and extract insights from large volumes of text data, it represents a valuable solution to the challenges posed by complex and lengthy accounting statements.

15

Word Embeddings

In recent years, there has been a significant increase in the amount of text data available for accounting and finance research. This has led to a growing interest in using text analysis techniques to extract insights and patterns from large volumes of unstructured data. Word embeddings are a type of vector representation of words in a corpus that capture the semantic and syntactic relationships between them. This representation is created by training a model on a large corpus of text, and the resulting vectors capture information about the meanings and relationships between words.

Word embeddings are a powerful tool for NLP tasks, such as sentiment analysis, text classification, and language generation, because they enable computers to reason about the meaning of words in a way that is similar to how humans do. By representing words as vectors, we can perform mathematical operations on them (such as addition and subtraction) to capture relationships between words, such as synonymy, antonymy, and analogy.

In this chapter, you will learn how to create and use word embeddings in Python. We will explore popular models for generating word embeddings, such as Word2Vec and GloVe, and learn how to visualize and analyze them. Additionally, you will learn how to use pre-trained embeddings and how to train your own embeddings on your own corpus of text. By the end of this chapter, you will have a solid understanding of how to represent words as vectors, and how to use these representations to perform various NLP tasks.

Word embeddings are a type of representation of words in a corpus that captures the semantic and syntactic relationships between them. Essentially, word embeddings are a way to map words to a dense vector space where each dimension of the vector captures some aspect of the word's meaning. To create word embeddings, we can use algorithms such as Word2Vec or GloVe, which learn to predict the context in which a word appears in a corpus. By doing so, these algorithms can learn to encode the semantic relationships between words, such as word similarity or analogy.

For example, let's say we have a corpus of financial news articles, and we want to create word embeddings to analyze the language used in these articles. We can use Word2Vec to create vector representations of each word in the corpus. Here are some examples of what these embeddings might look like:

- "stock" may be mapped to a vector that is close to "shares", "equity", and "market".
- "earnings" may be mapped to a vector that is close to "revenue", "profits", and "income".
- "merger" may be mapped to a vector that is close to "acquisition", "consolidation", and "partnership".

Once we have these word embeddings, we can use them in various NLP tasks, such as sentiment analysis, text classification, and topic modeling. By using word embeddings, we can better capture the meaning of words in context and improve the accuracy of our NLP models.

It is recommended to normalize text before using it for word embeddings. Normalization helps to remove noise and inconsistencies in the text, which can improve the quality of the embeddings. The reason why normalization is important for word embeddings is that the algorithms used for creating embeddings rely on statistical patterns in the text. If the text is noisy or inconsistent, the algorithms may not be able to accurately capture the relationships between words, which can lead to poor quality embeddings.

A code to perform the word embedding in a set of text files containing normalized text, stored in D:/Data/Preprocessed folder, is as follows:

```
import os
from gensim.models import Word2Vec
from gensim.utils import simple_preprocess

# Define the path to the preprocessed text files
path_to_files = 'D:/DATA/Preprocessed'

# Define the function to read the text files
def read_text_files(path):
    for file in os.listdir(path):
        with open(os.path.join(path, file), 'r', encoding='utf-8') as f:
            yield simple_preprocess(f.read())

# Read the text files
sentences = list(read_text_files(path_to_files))

# Train the Word2Vec model on the text data
model = Word2Vec(sentences, vector_size=100, window=5, min_count=2, workers=4)

# Save the model
model.save('word2vec.model')
```

This code is written in Python and uses the gensim library to create a Word2Vec model on a set of preprocessed text files. The first two lines import the necessary libraries, including os for file handling, and gensim for creating word embeddings. The third line defines the path to the preprocessed text files, which is set to 'D:/Data/Preprocessed'.

The code then defines a function called **read_text_files** that reads each file in the folder and yields the preprocessed text as a list of words. This function takes the path to the text files as an input and uses **os.listdir** to list all files in the directory, and then opens each file, reads its contents, and applies **simple_preprocess** to convert the text to a list of words. The yield statement generates a generator object that can be used to iterate over the preprocessed text.

The next line uses the **read_text_files** function to read the preprocessed text files and stores them in a list called "sentences". The list contains all the preprocessed text from the files. The following line trains a Word2Vec model on the preprocessed text data stored in the "sentences" list. The Word2Vec function takes several parameters, including **vector_size**, which sets the dimensionality of the output vectors, **window**, which sets the maximum distance between the current and predicted word within a sentence, **min_count**, which sets the minimum number of times a word must appear in the corpus to be included in the model, and **workers**, which sets the number of worker threads to train the model.

Finally, the last line saves the trained Word2Vec model to a file named 'word2vec.model'. The trained model can then be loaded and used to explore the relationships between words in the corpus, which can help in understanding the semantic and syntactic patterns in the text data.

The output can be viewed using the following code:

```
from gensim.models import Word2Vec

# Load the Word2Vec model from the file
model = Word2Vec.load('word2vec.model')
# Get the vector representation of a word
vector = model.wv['stock']
print(vector)

# Find the top n most similar words to a given word
similar_words = model.wv.most_similar('stock', topn=5)
print(similar_words)

# Calculate the similarity between two words
similarity = model.wv.similarity('stock', 'shares')
print(similarity)

# Find the word that doesn't fit in a group of words
odd_word = model.wv.doesnt_match(['stock', 'shares', 'equity', 'banana'])
print(odd_word)
```

The output will look like:

```
Vector: [ 5.37838278e-05   3.11068026e-03  -6.82610180e-03 -
1.39203214e-03
  7.67775672e-03   7.34147802e-03  -3.66463419e-03   2.69740727e-03
 -8.35353415e-03   6.18828787e-03  -4.64347890e-03  -3.16840038e-03
  9.27487109e-03   8.64150003e-04   7.49360118e-03  -6.11324515e-03
 ....

  9.18467343e-03  -4.13279748e-03   8.02234840e-03   5.36572421e-03
  5.92514593e-03   4.94717853e-04   8.22469778e-03  -7.01751839e-
03]
Similar Words: [('sale', 0.17519620060920715), ('fiscal',
0.169081911444664), ('factor', 0.11132166534662247),
('statement', 0.110172338783741), ('walmart',
0.08078086376190186)]
```

This output is the vector representation of the word "stock" and the top 5 most similar words to "stock" in the Word2Vec model, along with their cosine similarity scores. The vector representation of "stock" is a list of 100 float values, which corresponds to the dimensions of the vector space used to train the Word2Vec model. Each value represents the weight or importance of the corresponding dimension in capturing the meaning of the word. For example, the value 0.02866101 may correspond to the "financial" dimension, while the value -0.00567884 may correspond to the "company" dimension.

The list of similar words, along with their similarity scores, indicates the words that are closest in meaning or context to "stock" in the vector space.

The cosine similarity score is a measure of the angle between the vectors of the words, where a score of 1 indicates that the vectors are identical, and a score of 0 indicates that they are completely dissimilar. In this case, the top 5 similar words to "stock" are "sale", "fiscal", "factor", "statement", and "walmart", with cosine similarity scores ranging from 0.175 to 0.081.

The model can be saved in CSV format using the following code:

```
import pandas as pd
from gensim.models import Word2Vec

# Load the Word2Vec model from the file
model = Word2Vec.load('word2vec.model')

# Get the word vectors and store them in a DataFrame
vectors = pd.DataFrame(model.wv.vectors)
vectors.index = model.wv.index_to_key

# Save the vectors as a CSV file
vectors.to_csv('D:/Data/word2vec.csv')
```

Apart from Word2Vec, following techniques can be used for word embeddings:

- **GloVe (Global Vectors for Word Representation)**

GloVe is another popular algorithm for creating word embeddings, which is similar to Word2Vec in many respects. However, unlike Word2Vec, GloVe combines the co-occurrence information of words across the entire corpus, rather than just within a local window. This can lead to embeddings that capture more global semantic relationships between words. One disadvantage of GloVe is that it can be slower to train than Word2Vec, particularly on very large corpora. I would not recommend using this method on financial statements and company reports where the text is voluminous leading to large corpora and the meanings can be deciphered from local window.

- **FastText**

FastText is a word embedding technique developed by Facebook AI Research, which extends the concept of word embeddings to sub-word or character-level representations. By including sub-word information in the embedding process, FastText can capture more fine-grained relationships between words and improve the accuracy of language modeling tasks such as text classification and sentiment analysis. However, one disadvantage of FastText is that it can be more computationally expensive than traditional word embeddings, particularly for larger datasets. FastText can be a more powerful word embedding technique than Word2Vec in scenarios where sub-word information is important for capturing word meanings, where one word may have many meanings, and where rare or Out-of-Vocabulary (OOV) words are prevalent.

- **ELMo (Embeddings from Language Models)**

ELMo is a newer word embedding technique that uses deep neural networks to generate contextualized word embeddings, which can capture the meaning of words in context. This is in contrast to traditional word embeddings, which treat each word as a static entity independent of its context. ELMo embeddings have been shown to improve the accuracy of many natural language processing tasks, such as question answering and sentiment analysis. However, one disadvantage of ELMo is that it can be computationally expensive and may require a large amount of training data to achieve optimal performance.

In general, Word2Vec is a good starting point for most academic research applications, as it is a widely used and well-established technique that is relatively easy to implement and tune. Word2Vec can be particularly effective in identifying patterns in the language used in financial documents, such as annual reports, financial statements, and news articles. However, the choice of word embedding technique should be guided by the specific needs and characteristics of the research project, and researchers should carefully consider the trade-offs between computational efficiency, accuracy, and the specific needs of their research project when selecting a word embedding technique.

Word embeddings are a powerful tool that enables researchers to better understand the meaning and relationships between words in a corpus of text. By representing words as vectors in a dense vector space, word embeddings allow researchers to perform mathematical operations on words, such as addition and subtraction, which can help in capturing relationships between words, such as synonymy, antonymy, and analogy. This can be particularly useful for academic researchers in accounting, finance, and other business fields, who are dealing with large volumes of unstructured text data. By using word embeddings, researchers can extract valuable insights and patterns from this data that may not be easily discernible through traditional analysis methods. Therefore, it is important for academic researchers to understand the concept of word embeddings and how to apply them in their research projects.

16

Text Classification

Text classification, also known as text categorization, is a process used in accounting and finance academic research to analyze the language used in large volumes of unstructured text data, such as financial reports, news articles, and social media posts. The goal of text classification is to automatically categorize these documents into predefined categories or classes based on their content, using natural language processing and machine learning techniques.

One application of text classification in accounting research is analyzing the sentiment of financial news articles. Researchers may want to understand how the market perceives a company's performance by analyzing the language used in news articles related to the company. Text classification can be used to classify each article as positive, negative, or neutral, based on the sentiment expressed in the article. Text classification can be used to analyze the tone of financial statements, such as annual reports or earnings calls. The model can be trained to classify the statements into positive or negative tone based on the language used in the statements. This can help investors and stakeholders understand the company's financial health and prospects. Text classification can also be used in finance academic research to detect patterns of fraudulent activity in textual data, such as financial reports or news articles. The model can be trained to identify specific words or phrases that are commonly associated with fraudulent activity and use them to classify documents as potentially fraudulent or not.

Text classification is a powerful technique that can help accounting and finance researchers gain deeper insights into the language used in financial reports, news articles, and other textual data. By automatically categorizing documents into predefined categories, text classification can help identify patterns and trends that may not be immediately apparent to the human eye.

In most cases, normalized text is necessary for effective text categorization. We have already covered normalization in previous chapters. By normalizing the text data, you can ensure that the same words are represented consistently across all documents, regardless of their formatting or capitalization. This can help to improve the accuracy of text categorization models, as they can more easily identify and compare the key features of each document. Word embeddings are not always necessary for text categorization, but they can be very useful in certain cases. For example, if the text data contains synonyms or words with multiple meanings, word embeddings can help the model to differentiate between them and accurately categorize the text. Word embeddings can also help to capture the context in which words are used, which can be useful for categorizing text based on specific topics or themes. However, it's important to note that creating and using word embeddings requires significant computational resources and may not be necessary for all text categorization tasks. In some cases, simpler feature engineering techniques, such as bag-of-words or tf-idf (term frequency-inverse document frequency), may be sufficient for accurately categorizing text. So, while word embeddings can be a powerful tool for improving text categorization accuracy, their use should be evaluated on a case-by-case basis depending on the specific requirements and characteristics of the text data.

There are several text classification techniques that can be used for automatically categorizing text documents based on their content. Here are some of the most common techniques:

Naive Bayes

Naive Bayes is a simple but effective text classification algorithm that is based on Bayes' theorem. It works by calculating the probability of each class given the words in the document, and then selecting the class with the highest probability.

Here's a code for Naïve Bayes text classification:

```python
import os
import numpy as np
from sklearn.feature_extraction.text import CountVectorizer
from sklearn.naive_bayes import MultinomialNB
from sklearn.metrics import accuracy_score, confusion_matrix

# Set the path to the preprocessed text files
data_dir = 'D:/Data/Preprocessed'

# Define the categories or classes to be used for classification
categories = ['category1']

# Create an empty list to store the text data and their
associated class labels
text_data = []
labels = []

# Loop through the files in the data directory and read in the
text data
for file_name in os.listdir(data_dir):
    with open(os.path.join(data_dir, file_name), 'r', encoding='utf-8') as f:
        text = f.read()
        text_data.append(text)
        labels.append(categories[0])

# Convert the text data into a matrix of word frequencies using
CountVectorizer
vectorizer = CountVectorizer()
X = vectorizer.fit_transform(text_data)

# Convert the labels into a numpy array
y = np.array(labels)

# Shuffle the data and split into training and testing sets
indices = np.arange(len(text_data))
np.random.shuffle(indices)
X_shuffled = X[indices]
y_shuffled = y[indices]
train_size = int(0.7 * len(text_data))
X_train, X_test = X_shuffled[:train_size], X_shuffled[train_size:]
y_train, y_test = y_shuffled[:train_size], y_shuffled[train_size:]

# Train the Naive Bayes model on the training set
clf = MultinomialNB()
clf.fit(X_train, y_train)

# Predict the class labels for the testing set
y_pred = clf.predict(X_test)

# Evaluate the accuracy of the model
accuracy = accuracy_score(y_test, y_pred)
print('Accuracy:', accuracy)

# Print the confusion matrix
conf_mat = confusion_matrix(y_test, y_pred, labels=categories)
print('Confusion Matrix:')
print(conf_mat)
```

```
# Preprocess the text
vectorizer = TfidfVectorizer()
X_train = vectorizer.fit_transform(train_data)
X_test = vectorizer.transform(test_data)

# Train a random forest classifier
classifier = RandomForestClassifier()
classifier.fit(X_train, train_labels)

# Make predictions on the test set
predictions = classifier.predict(X_test)

# Evaluate the classifier
accuracy = accuracy_score(test_labels, predictions)
print(f"Accuracy: {accuracy:.2f}")

report = classification_report(test_labels, predictions)
print("Classification report:")
print(report)
```

The code will give an output like:

```
Accuracy: 1.0
Confusion Matrix:
[[1]]
```

An accuracy of 1.0 means that the model correctly classified all of the test documents. The confusion matrix shows the number of true positives, true negatives, false positives, and false negatives for the test set. Since you only have one category, the confusion matrix is a 1 × 1 matrix with a single value of 1, which indicates that the model correctly classified all of the test documents as belonging to the single category.

While a perfect accuracy may seem desirable, it's important to note that the model was trained and tested on the same dataset, which means that it may not generalize well to new, unseen data. In practice, it's common to use cross-validation or holdout testing to evaluate the performance of a text classification model on new data. Moreover, it's important to consider other performance metrics, such as precision, recall, and F1-score, which can provide a more comprehensive evaluation of the model's performance.

Example of a code that takes training data from D:/Data/Machine/Training and making predictions on test data in D:/Data/Machine/Test is as follows:

```python
import os
import glob
from sklearn.feature_extraction.text import CountVectorizer
from sklearn.feature_extraction.text import TfidfTransformer
from sklearn.naive_bayes import MultinomialNB
from sklearn.metrics import accuracy_score,
classification_report
from sklearn.pipeline import Pipeline
from sklearn.model_selection import train_test_split
import nltk
from nltk.corpus import stopwords

nltk.download('stopwords')


def load_data_from_directory(path):
    files = glob.glob(os.path.join(path, "*.txt"))
    texts = []
    labels = []

    for file in files:
        with open(file, 'r', encoding='utf-8') as f:
            texts.append(f.read())

labels.append(os.path.splitext(os.path.basename(file))[0])

    return texts, labels


train_data_path = "D:/Data/Machine/Training"
test_data_path = "D:/Data/Machine/Test"

# Load data
X_train, y_train = load_data_from_directory(train_data_path)
X_test, y_test = load_data_from_directory(test_data_path)

# Create a pipeline for text preprocessing and classification
text_clf = Pipeline([
    ('vect',
CountVectorizer(stop_words=stopwords.words('english'))),
    ('tfidf', TfidfTransformer()),
    ('clf', MultinomialNB()),
])

# Train the classifier
text_clf.fit(X_train, y_train)

# Predict the test data
y_pred = text_clf.predict(X_test)

# Calculate the accuracy and print the classification report
accuracy = accuracy_score(y_test, y_pred)
print(f"Accuracy: {accuracy:.2f}")
print(classification_report(y_test, y_pred))
```

The script uses the **sklearn** library for building the classifier and **nltk** library for handling stopwords. It loads training and test data from specified directories and evaluates the classifier's performance using accuracy and a classification report. The code starts by importing necessary libraries and downloading the stopwords from the **nltk** library. The **load_data_from_directory** function is defined to load text files and their labels from a given directory. It reads the content of each file and extracts the file name without the extension as the label. The **train_data_path** and **test_data_path** variables are set to the training and test data directories. The training and test datasets are loaded using the **load_data_from_directory** function, which returns the text content and corresponding labels. After this, a pipeline named **text_clf** is created for text preprocessing and classification. The pipeline consists of three steps: vectorization, term frequency-inverse document frequency (TF-IDF) transformation, and classification using the Multinomial Naive Bayes classifier. The **CountVectorizer** is responsible for converting text data into a bag-of-words representation, while the **TfidfTransformer** calculates the TF-IDF scores for the words. The **MultinomialNB** classifier is then used to classify the documents based on the processed text features. The classifier is now trained on the training data using the **fit** method. The trained classifier is then used to predict the class labels for the test data using the **predict** method.

Finally, the accuracy of the classifier is calculated using the **accuracy_score** function from **sklearn.metrics**, and the classification report is generated using the **classification_report** function. The accuracy score and classification report are printed to give an overview of the classifier's performance on the test dataset.

The output of this code will look like:

```
Accuracy: 0.85
              precision    recall  f1-score   support

           1       0.90      0.82      0.86       100
           2       0.87      0.89      0.88       120
           3       0.80      0.84      0.82        90
           4       0.83      0.80      0.81       110
           5       0.85      0.88      0.86       105
           6       0.88      0.85      0.86        95

    accuracy                           0.85       620
   macro avg       0.85      0.85      0.85       620
weighted avg       0.85      0.85      0.85       620
```

The interpretation of this output is as follows:

This output represents the performance of a text classification model on a test dataset with 620 instances. Let me interpret the results for you:

1. **Accuracy**: The overall accuracy of the model is 0.85, which means the model correctly predicted the class labels for 85% of the test instances.

This is a good accuracy score, indicating that the model has performed well on the test dataset.

2. **Precision, Recall, and F1-score**: These are class-wise performance metrics:

   - **Class 1**: Precision is 0.90, recall is 0.82, and F1-score is 0.86. There were 100 instances of this class in the test dataset.
   - **Class 2**: Precision is 0.87, recall is 0.89, and F1-score is 0.88. There were 120 instances of this class in the test dataset.
   - **Class 3**: Precision is 0.80, recall is 0.84, and F1-score is 0.82. There were 90 instances of this class in the test dataset.
   - **Class 4**: Precision is 0.83, recall is 0.80, and F1-score is 0.81. There were 110 instances of this class in the test dataset.
   - **Class 5**: Precision is 0.85, recall is 0.88, and F1-score is 0.86. There were 105 instances of this class in the test dataset.
   - **Class 6**: Precision is 0.88, recall is 0.85, and F1-score is 0.86. There were 95 instances of this class in the test dataset.

The precision, recall, and F1-score values for all classes are relatively high, indicating that the model performed well on each class.

3. **Macro Avg and Weighted Avg**: These are the average values of precision, recall, and F1-score across all classes:

   - **Macro Avg**: The unweighted mean of the class-wise metrics is 0.85 for precision, recall, and F1-score.
   - **Weighted Avg**: The average of class-wise metrics, weighted by the number of instances in each class (support), is also 0.85 for precision, recall, and F1-score.

This output represents a well-performing text classification model with good accuracy and class-wise performance. The model has been successful in predicting class labels across all classes, as reflected by the high precision, recall, and F1-score values.

Support Vector Machines (SVMs): SVMs are a type of machine learning algorithm that can be used for text classification. They work by finding the best boundary (hyperplane) between different classes of data, which can then be used to classify new data points.

An example of a code that provides SVM text classification is as follows:

```
import os
import glob
from sklearn.feature_extraction.text import CountVectorizer
from sklearn.feature_extraction.text import TfidfTransformer
from sklearn.svm import SVC
from sklearn.metrics import accuracy_score, classification_report
from sklearn.pipeline import Pipeline
from sklearn.model_selection import train_test_split

def load_data_from_directory(path):
    files = glob.glob(os.path.join(path, "*.txt"))
    texts = []
    labels = []

    for file in files:
        with open(file, 'r', encoding='utf-8') as f:
            texts.append(f.read())

labels.append(os.path.splitext(os.path.basename(file))[0])

    return texts, labels

data_path = "D:/Data/Preprocessed"

# Load data
texts, labels = load_data_from_directory(data_path)
# Split the data into training and test sets
X_train, X_test, y_train, y_test = train_test_split(texts, labels, test_size=0.2, random_state=42)

# Create a pipeline for text preprocessing and classification
text_clf = Pipeline([
    ('vect', CountVectorizer()),
    ('tfidf', TfidfTransformer()),
    ('clf', SVC(kernel='linear')),
])

# Train the classifier
text_clf.fit(X_train, y_train)

# Predict the test data
y_pred = text_clf.predict(X_test)

# Calculate the accuracy and print the classification report
accuracy = accuracy_score(y_test, y_pred)
print(f"Accuracy: {accuracy:.2f}")
print(classification_report(y_test, y_pred))
```

This Python code performs text classification using a Support Vector Machine (SVM) classifier. It starts by importing necessary libraries and defining a function named **load_data_from_directory** that loads preprocessed text files and their labels from a given directory. The function reads the content of each file and extracts the file name without the extension as the label.

After defining the function, the **data_path** variable is set to the directory containing the preprocessed text files. The code loads the text data and corresponding labels using the **load_data_from_directory** function. The loaded data is then split into training and test sets using the **train_test_split** function from the **sklearn.model_selection** module. The function takes the loaded texts and labels as input and randomly splits them into training and test subsets, with 20% of the data used for testing and the remaining 80% used for training.

Next, a pipeline named **text_clf** is created for text preprocessing and classification. The pipeline consists of three steps: vectorization, term frequency-inverse document frequency (TF-IDF) transformation, and classification using the SVM classifier. The **CountVectorizer** is responsible for converting text data into a bag-of-words representation, while the **TfidfTransformer** calculates the TF-IDF scores for the words. The **SVC** classifier with a linear kernel is then used to classify the documents based on the processed text features.

Once the pipeline is defined, the classifier is trained on the training data using the **fit** method. The trained classifier is then used to predict the class labels for the test data using the **predict** method. Finally, the accuracy of the classifier is calculated using the **accuracy_score** function from **sklearn.metrics**, and the classification report is generated using the **classification_report** function. The accuracy score and classification report are printed to provide an overview of the classifier's performance on the test dataset.

The output of data will look like:

```
Accuracy: 0.82
              precision    recall  f1-score   support

           1       0.91      0.82      0.86       100
           2       0.84      0.89      0.88       120
           3       0.85      0.84      0.82        90
           4       0.89      0.80      0.81       110
           5       0.51      0.88      0.86       105
           6       0.74      0.85      0.86        95

    accuracy                           0.85       620
   macro avg       0.85      0.85      0.85       620
weighted avg       0.85      0.85      0.85       620
```

In this code there is a single group of text files that are being randomly divided into training and test data in line 28 of the code. If we want to manually provide the training and test data, then the relevant code would be:

```python
import os
import glob
from sklearn.feature_extraction.text import TfidfVectorizer
from sklearn.model_selection import train_test_split
from sklearn.svm import SVC
from sklearn.metrics import classification_report, accuracy_score

# Function to load data from the given directory
def load_data(dir_path):
    texts = []
    labels = []
    for file_path in glob.glob(os.path.join(dir_path, '*.txt')):
        with open(file_path, 'r', encoding='utf-8') as f:
            texts.append(f.read())

labels.append(os.path.basename(file_path).split('.')[0])   # Assumes label is the first part of the file name
    return texts, labels

# Load data from training and test directories
train_texts, train_labels = load_data('D:/DATA/Machine/Training')
test_texts, test_labels = load_data('D:/DATA/Machine/Test')

# Transform the texts into TF-IDF features
vectorizer = TfidfVectorizer(stop_words='english', max_df=0.95, min_df=2)   # Adjusted max_df and min_df values
X_train = vectorizer.fit_transform(train_texts)
X_test = vectorizer.transform(test_texts)

# Train an SVM classifier
classifier = SVC(kernel='linear', C=1.0)
classifier.fit(X_train, train_labels)

# Make predictions on the test set
predictions = classifier.predict(X_test)

# Evaluate the model
print("Accuracy:", accuracy_score(test_labels, predictions))
print("Classification Report:")
print(classification_report(test_labels, predictions))
```

This code loads text and labels from specified directories, transforms the text into numerical features using TF-IDF, trains an SVM classifier on the training data, makes predictions on the test data, and evaluates the classifier's performance by calculating its accuracy and printing a classification report. The output will be similar to earlier examples.

Decision Trees

Decision trees are another type of machine learning algorithm that can be used for text classification. They work by recursively splitting the data into smaller subsets based on the most informative features (words) until a decision can be made about the class of the document. A code that performs text classification on normalized text files by randomly classifying text files into training and test files, using decision trees is as follows:

```
import os
import glob
from sklearn.feature_extraction.text import TfidfVectorizer
from sklearn.model_selection import train_test_split
from sklearn.tree import DecisionTreeClassifier
from sklearn.metrics import classification_report, accuracy_score

# Function to load data from the given directory
def load_data(dir_path):
    texts = []
    labels = []
    for file_path in glob.glob(os.path.join(dir_path, '*.txt')):
        with open(file_path, 'r', encoding='utf-8') as f:
            texts.append(f.read())
labels.append(os.path.basename(file_path).split('.')[0])  # Assumes label is the first part of the file name
    return texts, labels

# Load data from the preprocessed directory
texts, labels = load_data('D:/DATA/Preprocessed')

# Split the data into training and test sets
train_texts, test_texts, train_labels, test_labels = train_test_split(texts, labels, test_size=0.2, random_state=42)

# Transform the texts into TF-IDF features
vectorizer = TfidfVectorizer(stop_words='english', max_df=0.95, min_df=2)
X_train = vectorizer.fit_transform(train_texts)
X_test = vectorizer.transform(test_texts)

# Train a decision tree classifier
classifier = DecisionTreeClassifier(random_state=42)
classifier.fit(X_train, train_labels)

# Make predictions on the test set
predictions = classifier.predict(X_test)

# Evaluate the model
print("Accuracy:", accuracy_score(test_labels, predictions))
print("Classification Report:")
print(classification_report(test_labels, predictions))
```

Similarly, the code for manually assigning training and test data is as follows:

```python
import os
import numpy as np
from sklearn.feature_extraction.text import TfidfVectorizer
from sklearn.tree import DecisionTreeClassifier
from sklearn.metrics import accuracy_score, classification_report
from nltk.corpus import stopwords
import nltk

nltk.download("stopwords")


def load_data(directory):
    data = []
    labels = []

    for filename in os.listdir(directory):
        with open(os.path.join(directory, filename), "r", encoding="utf-8") as file:
            content = file.read()
            data.append(content)
            labels.append(filename.split("_")[0])

    return data, labels


# Load training and test data
training_data, training_labels = load_data("D:/DATA/Machine/Training")
test_data, test_labels = load_data("D:/DATA/Machine/Test")

# Preprocess the text
vectorizer = TfidfVectorizer(stop_words=stopwords.words("english"))
X_train = vectorizer.fit_transform(training_data)
X_test = vectorizer.transform(test_data)

# Train a decision tree classifier
classifier = DecisionTreeClassifier()
classifier.fit(X_train, training_labels)

# Make predictions on the test set
predictions = classifier.predict(X_test)

# Evaluate the classifier
accuracy = accuracy_score(test_labels, predictions)
print(f"Accuracy: {accuracy:.2f}")

report = classification_report(test_labels, predictions)
print("Classification report:")
print(report)
```

Random Forests

Random forests are an ensemble learning technique that combines multiple decision trees to improve the accuracy of the classification. Each decision tree is trained on a random subset of the data, and the final classification is based on the majority vote of all the trees. Here's an example of a code to perform text classification using random forests. The code randomly divides the text files into training and test data.

```
import os
import numpy as np
from sklearn.feature_extraction.text import TfidfVectorizer
from sklearn.ensemble import RandomForestClassifier
from sklearn.metrics import accuracy_score,
classification_report
from sklearn.model_selection import train_test_split

def load_data(directory):
    data = []
    labels = []

    for filename in os.listdir(directory):
        with open(os.path.join(directory, filename), "r",
encoding="utf-8") as file:
            content = file.read()
            data.append(content)
            labels.append(filename.split("_")[0])

    return data, labels

# Load preprocessed data
preprocessed_data, preprocessed_labels =
load_data("D:/DATA/Preprocessed")

# Split the data into training and test sets
train_data, test_data, train_labels, test_labels =
train_test_split(
    preprocessed_data, preprocessed_labels, test_size=0.2,
random_state=42
)
```

```python
import os
import numpy as np
from sklearn.feature_extraction.text import TfidfVectorizer
from sklearn.ensemble import RandomForestClassifier
from sklearn.metrics import accuracy_score, classification_report
from sklearn.preprocessing import LabelEncoder
from nltk.corpus import stopwords
import nltk


def load_data_labels(directory):
    data = []
    labels = []

    for filename in os.listdir(directory):
        with open(os.path.join(directory, filename), "r", encoding="utf-8") as file:
            content = file.read()
            data.append(content)
            labels.append(filename)

    return data, labels
```

This code first loads preprocessed text data from the specified directory and splits it into training and test sets using the **train_test_split** function. Then, it converts the text data into a numerical format using the TF-IDF (term frequency-inverse document frequency) vectorizer. The code proceeds to train a random forest classifier on the training set and evaluates its performance by predicting labels for the test set. Finally, it prints the accuracy score and a classification report that includes precision, recall, and F1-score for each class.

The code to manually allocate training data and test data is as follows:

```python
train_directory = "D:/DATA/Machine/Training"
test_directory = "D:/DATA/Machine/Test"

train_data, train_labels = load_data_labels(train_directory)
test_data, test_labels = load_data_labels(test_directory)

print("Unique labels in training set:", np.unique(train_labels))
print("Unique labels in test set:", np.unique(test_labels))

# Convert the text data into numerical features using the TF-IDF
method
vectorizer = TfidfVectorizer(stop_words='english',
max_features=5000)
X_train = vectorizer.fit_transform(train_data)
X_test = vectorizer.transform(test_data)

# Encode the class labels into integers
label_encoder = LabelEncoder()
label_encoder.fit(train_labels)
y_train = label_encoder.transform(train_labels)
y_test = label_encoder.transform(test_labels)

# Train a random forest classifier
classifier = RandomForestClassifier(n_estimators=100,
random_state=42)
classifier.fit(X_train, y_train)

# Make predictions on the test set
predictions = classifier.predict(X_test)

# Decode the predicted labels
predicted_labels = label_encoder.inverse_transform(predictions)

# Evaluate the classifier
print("Accuracy:", accuracy_score(test_labels,
predicted labels))
print("Classification Report:\n",
classification_report(test_labels, predicted_labels))
```

Deep Learning

Deep learning techniques, such as Convolutional Neural Networks (CNNs) and Recurrent Neural Networks (RNNs), have shown promising results in text classification. These algorithms are able to learn more complex representations of the text data and can be trained on very large datasets. Deep learning has emerged as a powerful tool in the field of text classification, with the potential to improve accuracy and efficiency in tasks such as sentiment analysis, topic classification, and document categorization. In academic accounting and finance research, deep learning techniques have been used to classify financial news articles based on sentiment, to predict stock prices based on text data, and to identify fraudulent financial statements. While deep learning methods require large amounts of labeled data and computational resources, they have shown promising results in improving the accuracy of text classification tasks. In a separate full chapter, we will delve deeper into the concepts and applications of deep learning in text classification.

These are just a few examples of the many text classification techniques that are available. The choice of which technique to use depends on the specific characteristics of the text data, such as the size of the dataset, the number of classes, and the complexity of the classification task.

In this chapter, we explored various techniques for text classification, ranging from traditional machine learning models such as Naive Bayes, Support Vector Machines, Decision Trees, and random Forests. Each of these models has its own strengths and weaknesses, and the choice of model depends on the specific requirements of the task. Word Embedding and text classification are an essential task in natural language processing and have a wide range of applications in various industries. With the advancement of machine learning and deep learning techniques, we can expect further improvements in the accuracy of text classification models, which will lead to more effective and efficient solutions for real-world problems.

17

Sentiment Analysis

Sentiment analysis is a popular technique used in natural language processing and machine learning to automatically identify and extract subjective information from text data. It involves analyzing and categorizing opinions, emotions, and attitudes expressed in text, which can help researchers to gain insights into the opinions and sentiments of individuals or groups on various topics.

Sentiment analysis has a wide range of applications, including but not limited to marketing research, customer feedback analysis, social media monitoring, and political analysis. With the increasing availability of text data from various sources such as social media platforms, online reviews, and news articles, sentiment analysis has become an essential tool for researchers to understand the sentiment of the public toward certain products, services, or topics.

In the context of accounting research, sentiment analysis can be used to analyze the sentiment of financial news articles, earnings calls transcripts, and social media discussions related to firms, industries, or financial markets. By identifying and categorizing the sentiment expressed in these texts, researchers can gain valuable insights into the impact of public sentiment on the financial markets and the behavior of investors.

There are various techniques used for sentiment analysis, ranging from simple rule-based methods to more advanced machine learning and deep learning models. Here are some of the most commonly used techniques:

- Rule-based Methods: These methods use a set of predefined rules to identify sentiment in text data. For example, a simple rule could be that any text containing positive words such as "good" or "excellent" is classified as positive sentiment, while text containing negative words such as "bad" or "terrible" is classified as negative sentiment. Rule-based methods are relatively simple and easy to implement, but they may not be as accurate as more advanced methods.
- Lexicon-based Methods: These methods use a pre-defined dictionary of words and their corresponding sentiment scores to classify text data. For example, the dictionary may assign a positive score to words like "happy" and a negative score to words like "angry". The sentiment score of each word in the text is then aggregated to determine the overall sentiment of the text.
- Machine Learning Algorithms: These methods use a machine learning algorithm to learn from labeled data (i.e., data that has already been classified into positive, negative, or neutral sentiment) and then apply this learning to classify new, unlabeled data. Some of the commonly used machine learning algorithms for sentiment analysis include Naive Bayes, Support Vector Machines (SVM), and Random Forest.
- Deep Learning Models: These methods use deep neural networks to learn from large amounts of labeled data and then classify new, unlabeled data. Deep learning models can capture complex relationships between words and phrases, making them more accurate than traditional machine learning algorithms. Some of the commonly used deep learning models for sentiment analysis include Convolutional Neural Networks (CNN), Recurrent Neural Networks (RNN), and Long Short-Term Memory (LSTM) networks.

Each technique has its own strengths and weaknesses, and the choice of technique depends on the specific requirements and constraints of the project. A combination of different techniques may also be used for better accuracy and performance. A detailed discussion of these methos is as follows:

Rule-Based Methods

Rule-based methods of sentiment analysis involve using a set of predefined rules to classify text data into positive, negative, or neutral sentiment. These methods are relatively simple and easy to implement, but they may not be as accurate as more advanced machine learning or deep learning models. Here are some examples of rule-based methods:

- Keyword Matching

This method involves creating a list of positive and negative keywords and matching them against the text data. For example, the presence of words like "good", "excellent", "happy", and "satisfied" might indicate positive sentiment, while words like "bad", "terrible", "disappointed", and "frustrated" might indicate negative sentiment.

Suppose you are conducting a sentiment analysis of news articles related to a particular company's quarterly earnings report. You want to identify the sentiment of the articles as either positive, negative, or neutral. You can use a keyword matching approach to classify the sentiment of the articles based on the presence of positive or negative keywords. For example, you might create a list of positive keywords such as "record earnings", "strong growth", and "positive outlook", and negative keywords such as "disappointing results", "weak performance", and "negative guidance". You could then search for the presence of these keywords in the text of the news articles and use this information to classify the sentiment of each article. If an article contains more positive keywords than negative keywords, you might classify it as having a positive sentiment. Conversely, if an article contains more negative keywords than positive keywords, you might classify it as having a negative sentiment. If the article contains an equal number of positive and negative keywords, you might classify it as having a neutral sentiment.

While this approach is relatively simple and easy to implement, it may not be as accurate as more advanced techniques such as machine learning or deep learning models. Additionally, the choice of keywords may be subjective and may not capture the full range of sentiment expressed in the text. Therefore, it is important to carefully evaluate the accuracy and effectiveness of the keyword matching approach before using it in a sentiment analysis project.

Here's a code that performs a sentiment analysis on all the text files located in D:/Data/Preprocessed directory and its subdirectories.

```python
import os
import csv

# Define positive and negative keywords
positive_keywords = ["record earnings", "strong growth",
"positive outlook"]
negative_keywords = ["disappointing results", "weak
performance", "negative guidance"]

# Define a function to classify the sentiment of a text based on
keyword matching
def classify_sentiment(text):
    positive_count = sum(text.lower().count(keyword) for keyword in positive_keywords)
    negative_count = sum(text.lower().count(keyword) for keyword in negative_keywords)
    if positive_count > negative_count:
        return "positive"
    elif positive_count < negative_count:
        return "negative"
    else:
        return "neutral"

# Define the directory and subdirectories containing the
preprocessed text files
directory = "D:/Data/Preprocessed"

# Create a CSV file to store the output
output_file = "D:/Data/sentiment_analysis_output.csv"
with open(output_file, "w", newline="") as f:
    writer = csv.writer(f)
    writer.writerow(["Filepath", "Sentiment"])

    # Traverse through the directory and subdirectories and
perform sentiment analysis on each file
    for subdir, _, files in os.walk(directory):
        for file in files:
            filepath = os.path.join(subdir, file)
            with open(filepath, "r") as f2:
                text = f2.read()
                sentiment = classify_sentiment(text)
                writer.writerow([filepath, sentiment])
```

This code performs sentiment analysis using keyword matching on a set of preprocessed text files located in a directory and its subdirectories. Positive and negative sentiment keywords are defined, and a function is created to classify the sentiment of a text based on keyword matching. The **os.walk** function is used to traverse through the directory and subdirectories, and sentiment analysis is performed on each file using the **classify_sentiment** function. The output is printed to the console and stored in a CSV file. The code assumes each file contains one text document, and the **classify_sentiment** function may need to be modified if files contain multiple text documents separated by a delimiter.

This is a very simple code where we define the positive and negative words and perform sentiment analysis using those keywords. We can also import a dictionary saved as a CSV file into the code and use that for sentiment analysis as below:

```python
import os
import csv

# Import the sentiment dictionary from the CSV file
sentiment_dict = {}
with open("D:/Data/sentiment_dict.csv", "r") as f:
    reader = csv.reader(f)
    next(reader)  # skip header row
    for row in reader:
        word, score = row
        sentiment_dict[word] = float(score)

# Define a function to classify the sentiment of a text based on
# the sentiment dictionary
def classify_sentiment(text):
    words = text.lower().split()
    positive_score = sum(sentiment_dict.get(word, 0) for word in words if sentiment_dict.get(word, 0) > 0)
    negative_score = sum(sentiment_dict.get(word, 0) for word in words if sentiment_dict.get(word, 0) < 0)
    if positive_score > negative_score:
        return "positive"
    elif positive_score < negative_score:
        return "negative"
    else:
        return "neutral"

# Define the directory and subdirectories containing the
# preprocessed text files
directory = "D:/Data/Preprocessed"

# Create a CSV file to store the output
output_file = "sentiment_analysis_output.csv"
with open(output_file, "w", newline="") as f:
    writer = csv.writer(f)
    writer.writerow(["Filepath", "Sentiment"])

    # Traverse through the directory and subdirectories and
    # perform sentiment analysis on each file
    for subdir, _, files in os.walk(directory):
        for file in files:
            filepath = os.path.join(subdir, file)
            with open(filepath, "r") as f2:
                text = f2.read()
                sentiment = classify_sentiment(text)
                writer.writerow([filepath, sentiment])
```

```
import os
import numpy as np
import pandas as pd
from sklearn.feature_extraction.text import TfidfVectorizer
from sklearn.decomposition import NMF

# Define the directory containing the preprocessed text files
dir_path = 'D:/DATA/Preprocessed'

# Create a list of file paths for all files in the directory
file_paths = [os.path.join(dir_path, f) for f in
os.listdir(dir_path) if os.path.isfile(os.path.join(dir_path,
f))]

# Create a TfidfVectorizer object to convert the text data into
a matrix of TF-IDF features
vectorizer = TfidfVectorizer(input='filename',
stop_words='english')

# Use the vectorizer to fit and transform the text data into a
matrix of TF-IDF features
tfidf = vectorizer.fit_transform(file_paths)

# Define the number of topics to extract
n_topics = 10

# Perform NMF on the TF-IDF matrix to extract the topics
nmf = NMF(n_components=n_topics, init='nndsvd', max_iter=200,
random_state=0)
W = nmf.fit_transform(tfidf)
H = nmf.components_

# Load the sentiment dictionary
sentiment_dict = pd.read_csv('D:/Data/Sentiment_Dict.csv',
index_col='word')

# Define a function to classify the sentiment of a text based on
the sentiment dictionary
def classify_sentiment(text, sentiment_dict):
    words = text.lower().split()
    pos_score = 0
    neg_score = 0
    for word in words:
```

The sentiment dictionary is read from a CSV file located in the **D:/Data** directory, and a function is defined to classify the sentiment of a text based on the sentiment dictionary. The directory and subdirectories containing the preprocessed text files are defined using the **os.walk** function, which allows the code to traverse through the directory and subdirectories. For each file, the code opens it, reads its contents, and performs sentiment analysis using the **classify_sentiment** function. The output is stored in a CSV file located in the current working directory. The code assumes each file contains one text document.

The sentiment dictionary used in the code is the one read from the CSV file **sentiment_dict.csv**, which should contain two columns: **word** and **score**. The sentiment score of each word is retrieved from the dictionary, and the positive and negative scores are summed to determine the overall sentiment of the text. There are several sentiment dictionaries that are commonly used in accounting and finance research, including the Loughran-McDonald Dictionary, Harvard IV-4 Dictionary, General Inquirer Dictionary, and SentiStrength. It is important to carefully evaluate the accuracy and effectiveness of the sentiment dictionary before using it in a research project.

In academic accounting and finance research, it is recommended to use Loughran-McDonald Dictionary, which is specifically designed for financial and accounting text data and is based on the frequency of words in a large corpus of financial news articles. It contains more than 2,300 words and phrases that are classified as positive or negative sentiment. However, other general-purpose dictionaries such as Harvard IV-4 Dictionary, General Inquirer Dictionary, and SentiStrength have also been used very commonly.

A more complex model would be to divide the entire text data into certain number of topics using the concepts in Topic Modeling section of this book and then use sentiment analysis on individual topics. A sentiment variable then can be created as a sum of all individual topic sentiments, positive being 1, negative as −1, and neutral as 0. This approach, using NMF topic modeling, can be implemented in Python as:

```python
        if word in sentiment_dict.index:
            score = sentiment_dict.loc[word, 'score']
            if score > 0:
                pos_score += score
            elif score < 0:
                neg_score += abs(score)
    if pos_score > neg_score:
        return 1
    elif neg_score > pos_score:
        return -1
    else:
        return 0

# Compute the sentiment of each topic and store in a dataframe
topic_sentiments = []
for i, topic in enumerate(H):
    words = [feature_names[j] for j in topic.argsort()[:-11:-1]]
    text = ' '.join(words)
    sentiment = classify_sentiment(text, sentiment_dict)
    topic_sentiments.append(sentiment)
    print("Topic {}: {}, Sentiment: {}".format(i, text,
sentiment))

# Compute the document sentiment as the sum of all topic
sentiments
document_sentiment = sum(topic_sentiments)

# Create a dataframe to store the output
output_df = pd.DataFrame(columns=['file', 'topic 0', 'topic 1',
'topic 2', 'topic 3', 'topic 4', 'topic 5', 'topic 6', 'topic
7', 'topic 8', 'topic 9', 'document_sentiment'])
for i, file_path in enumerate(file_paths):
    file_name = os.path.basename(file_path)
    topic_scores = W[i]
    row = [file_name] + list(topic_scores) +
[document_sentiment]
    output_df.loc[i] = row

# Save the output dataframe to a CSV file
output_df.to_csv('D:/Data/Output.csv', index=False)
```

17 Sentiment Analysis

```
import os
import csv
# Define a list of negation words
negation_words = ['not', 'never']

# Load the sentiment dictionary from the CSV file
sentiment_dict = {}
with open('D:/Data/Sentiment_Dict.csv', 'r', encoding='utf-8')
as f:
    reader = csv.reader(f)
    for row in reader:
        sentiment_dict[row[0]] = float(row[1])

# Define a function to perform sentiment analysis using a
linguistic rule and sentiment dictionary
def combined_sentiment_analysis(text):
    # Split the text into words
    words = text.split()

    # Check if a negation word is present
    negation_present = False
    for word in words:
        if word in negation_words:
            negation_present = True

    # Calculate the sentiment score for each word in the text
using the sentiment dictionary
    sentiment_scores = []
    for word in words:
        if word in sentiment_dict:
            score = sentiment_dict[word]
            if negation_present:
                score = -score
            sentiment_scores.append(score)

    # Calculate the overall sentiment score for the text
    if sentiment_scores:
        sentiment_score = sum(sentiment_scores) /
len(sentiment_scores)
    else:
        sentiment_score = 0
```

This code performs topic modeling on a collection of preprocessed text files using Non-negative Matrix Factorization (NMF) and then performs sentiment analysis on each topic using a sentiment dictionary. Finally, it stores the results in a CSV file.

At the beginning of the code, the directory containing the preprocessed text files is defined and a list of file paths for all files in the directory is created.

Then, a TfidfVectorizer object is created to convert the text data into a matrix of TF-IDF features, which is then used to fit and transform the text data into a matrix of TF-IDF features. Next, the number of topics to extract is defined and NMF is performed on the TF-IDF matrix to extract the topics. The top words for each topic are then printed.

The sentiment dictionary is loaded from the CSV file **Sentiment_ Dict.csv**. A function **classify_sentiment** is defined to classify the sentiment of a text based on the sentiment dictionary. This function takes in a text string and the sentiment dictionary as input, and returns a sentiment label (Positive, Negative, or Neutral) based on the sentiment score of the text. For each topic, the top words are extracted and concatenated into a text string. The **classify_sentiment** function is then used to determine the sentiment of the topic. The topic text and sentiment are printed. The output is saved to a CSV file. For each file in the directory, the topic scores and document sentiment are computed and stored in a dataframe. This dataframe is then saved to a CSV file called **Output.CSV** in the **D:/Data** directory.

- **Linguistic Rules:**

This method involves using linguistic rules to identify sentiment in text data. For example, negation words like "not" or "never" can change the sentiment of a sentence, so a rule might be created to reverse the sentiment if a negation word is present. So, here's an example of a linguistic rule for sentiment analysis that could be used in academic research, particularly in the fields of accounting and finance:

Linguistic Rule: If a sentence contains a negation word, such as "not" or "never", the sentiment should be reversed.

Example: "The company's financial performance was not good".

Using the linguistic rule, we would reverse the sentiment of the sentence, so it would be classified as negative sentiment rather than neutral or positive sentiment.

This rule is especially relevant in accounting and finance research, where sentiment analysis is commonly used to analyze financial news articles or earnings call transcripts. Negation words are often used to modify the sentiment of a sentence, and failing to account for them can lead to inaccurate sentiment analysis results. By using a linguistic rule to account for negation words, researchers can more accurately identify the sentiment expressed in financial text data.

Linguistic rules are generally combined with dictionaries. An example of a code of linguistic rule and dictionary usage is as follows:

```python
# Classify the sentiment based on the sentiment score
if sentiment_score > 0:
    sentiment = "Positive"
elif sentiment_score < 0:
    sentiment = "Negative"
else:
    sentiment = "Neutral"

return sentiment

# Loop through all text files in the "D:/DATA/Preprocessed"
directory and its subdirectories
for root, dirs, files in os.walk("D:/DATA/Preprocessed"):
    for file in files:
        if file.endswith(".txt"):
            # Read the text file
            with open(os.path.join(root, file), 'r', encoding='utf-8') as f:
                text = f.read()

            # Perform sentiment analysis using a linguistic rule and sentiment dictionary
            sentiment = combined_sentiment_analysis(text)

            # Print the file name and sentiment
            print("File:", os.path.join(root, file))
            print("Sentiment:", sentiment)
```

```
import os
import numpy as np
import pandas as pd
from sklearn.feature_extraction.text import TfidfVectorizer
from sklearn.decomposition import NMF
# Define the directory containing the preprocessed text files
dir_path = 'D:/DATA/Preprocessed'

# Create a list of file paths for all files in the directory
file_paths = [os.path.join(dir_path, f) for f in
os.listdir(dir_path) if os.path.isfile(os.path.join(dir_path,
f))]

# Create a TfidfVectorizer object to convert the text data into
a matrix of TF-IDF features
vectorizer = TfidfVectorizer(input='filename',
stop_words='english')

# Use the vectorizer to fit and transform the text data into a
matrix of TF-IDF features
tfidf = vectorizer.fit_transform(file_paths)

# Define the number of topics to extract
n_topics = 10

# Perform NMF on the TF-IDF matrix to extract the topics
nmf = NMF(n_components=n_topics, init='nndsvd', max_iter=200,
random_state=0)
W = nmf.fit_transform(tfidf)
H = nmf.components_

# Load the sentiment dictionary
sentiment_dict = pd.read_csv('D:/Data/Sentiment_Dict.csv',
index_col='word')

# Define a list of negation words
negation_words = ['not', 'never']

# Define a function to classify the sentiment of a text based on
the sentiment dictionary and linguistic rule
def classify_sentiment(text, sentiment_dict, negation_words):
    words = text.lower().split()
    pos_score = 0
    neg_score = 0
    negation_present = False
    for word in words:
        if word in negation_words:
            negation_present = True
        elif word in sentiment_dict.index:
            score = sentiment_dict.loc[word, 'score']
            if negation_present:
                score = -score
                negation_present = False
            if score > 0:
```

This code loads the sentiment dictionary from the "D:/Data/Sentiment_Dict.csv" file and defines a function to perform sentiment analysis using both linguistic rules and the sentiment dictionary. The function first checks if a negation word is present, and then calculates the sentiment score for each word in the text using the sentiment dictionary. If a negation word is present, the score for the word is reversed. The overall sentiment score for the text is then calculated as the average of the sentiment scores for all words in the text. Finally, the sentiment is classified based on the overall sentiment score.

Note that the sentiment dictionary used in this code should have words in lowercase as the text is converted to lowercase while processing. Also, the sentiment dictionary should be formatted as a CSV file with two columns: the first column should contain the words, and the second column should contain their corresponding sentiment scores. The sentiment scores can be positive, negative, or neutral, with values between −1 and 1.

This code can be combined with the earlier code with topic modeling in the following way:

```
            pos_score += score
        elif score < 0:
            neg_score += abs(score)
    if pos_score > neg_score:
        return 1
    elif neg_score > pos_score:
        return -1
    else:
        return 0

# Compute the sentiment of each topic and store in a dataframe
topic_sentiments = []
for i, topic in enumerate(H):
    feature_names = vectorizer.get_feature_names()
    words = [feature_names[j] for j in topic.argsort()[:-11:-1]]
    text = ' '.join(words)
    sentiment = classify_sentiment(text, sentiment_dict,
negation_words)
    topic_sentiments.append(sentiment)
    print("Topic {}: {}, Sentiment: {}".format(i, text,
sentiment))

# Compute the document sentiment as the sum of all topic
sentiments
document_sentiment = sum(topic_sentiments)

# Create a dataframe to store the output
output_df = pd.DataFrame(columns=['file', 'topic 0', 'topic 1',
'topic 2', 'topic 3', 'topic 4', 'topic 5', 'topic 6', 'topic
7', 'topic 8', 'topic 9', 'document_sentiment'])
for i, file_path in enumerate(file_paths):
    file_name = os.path.basename(file_path)
    topic_scores = W[i]
    row = [file_name] + list(topic_scores) +
[document_sentiment]
    output_df.loc[i] = row

# Save the output dataframe to a CSV file
output_df.to_csv('D:/Data/
```

```
import os
from afinn import Afinn

# Create an instance of the AFINN lexicon
afinn = Afinn()
# Define the directory path where the preprocessed text files
are located
dir_path = 'D:/DATA/Preprocessed'

# Loop through all the files in the directory
for filename in os.listdir(dir_path):
    # Read the contents of the file
    with open(os.path.join(dir_path, filename), 'r',
encoding='utf-8') as f:
        text = f.read()

    # Calculate the sentiment score using the AFINN lexicon
    sentiment_score = afinn.score(text)

    # Print the filename and sentiment score
    print(filename, sentiment_score)
```

Similarly, code can be created to increase the value of sentiment if 'very' is added before the dictionary word (e.g., very good) and the like.

While rule-based methods of sentiment analysis are relatively simple and easy to implement, they may not be as accurate as more advanced machine learning or deep learning models. Rule-based methods may not be able to capture the nuances of language and may be limited by the quality of the predefined rules or lexicons. Therefore, it is important to carefully evaluate the accuracy and effectiveness of rule-based methods before using them in a sentiment analysis project.

Lexicon-Based Methods

Lexicon-based methods for sentiment analysis involve using a pre-defined dictionary or lexicon of words and their corresponding sentiment scores to determine the overall sentiment of a text. For example, a word like "love" might have a high positive sentiment score, while a word like "hate" might have a high negative sentiment score. The sentiment score of a text is calculated by summing the sentiment scores of all the words in the text. The sentiment scores of the words can be based on a number of criteria such as the word's connotation, frequency, or context.

One popular lexicon used for sentiment analysis is the AFINN lexicon, which contains a list of words and their corresponding sentiment scores ranging from –5 (most negative) to 5 (most positive). The sentiment scores in the AFINN lexicon are based on the valence of the word, or the degree to which the word expresses a positive or negative emotion. Other sentiment lexicons include the General Inquirer lexicon, SentiWordNet, and VADER (Valence Aware Dictionary and sEntiment Reasoner), among others.

There are several sentiment lexicons that are commonly used in accounting and finance research. The Loughran and McDonald's Financial Sentiment Lexicon was developed specifically for financial text analysis and contains over 4000 words and phrases commonly used in financial documents. The Harvard IV-4 General Inquirer contains sentiment scores for over 4800 words based on their psychological characteristics and has been used in various studies to analyze sentiment in financial news articles and stock market reports. The VADER lexicon contains sentiment scores for over 7500 lexical features, including emoticons and slang, and has been shown to perform well in analyzing sentiment in social media and other online sources, and can handle sentiment in various domains including finance.

Ultimately, the choice of sentiment lexicon will depend on the specific research question and the type of text data being analyzed. It may be useful to experiment with different lexicons and compare their performance on a sample of the data before deciding which lexicon to use for the final analysis. The implementation of lexicon-based method in python is similar to rule-based methods.

An implementation of sentiment analysis using AFINN is as follows:

```
2.txt 10.0
1.txt -13.0
3.txt 3.0
4.txt 4.0
5.txt 11.0
6.txt 4.0
7.txt 18.0
```

In this code, we first import the necessary packages, including the AFINN lexicon and the "os" package for directory operations. We then create an instance of the AFINN lexicon using the **Afinn()** function. Next, we define the directory path where the preprocessed text files are located using the **dir_path** variable. We then loop through all the files in the directory using the **os.listdir()** function and read the contents of each file using the **open()** function. We then calculate the sentiment score of each file using the **afinn.score()** function and print the filename and sentiment score using the **print()** function. Note that the sentiment score returned by the AFINN lexicon ranges from –5 (most negative) to 5 (most positive), with 0 indicating neutral sentiment. You can modify the code to save the sentiment scores to a file or perform additional analysis based on the sentiment scores. Aggregate sentiment in document is derived by summing all the sentiments in the document. The output will look like:

```python
import os
from nltk.sentiment import SentimentIntensityAnalyzer
import nltk
nltk.download('vader_lexicon')
# Set the path to the preprocessed text files
path = 'D:/DATA/Preprocessed/'

# Initialize the VADER sentiment analyzer
sia = SentimentIntensityAnalyzer()

# Loop through each file in the directory
for filename in os.listdir(path):
    if filename.endswith('.txt'):
        filepath = os.path.join(path, filename)
        with open(filepath, 'r') as f:
            text = f.read()
            # Perform sentiment analysis on the text
            scores = sia.polarity_scores(text)
            # Print the sentiment scores for the text
            print(f"{filename}: {scores}")
```

An example of sentiment analysis using VADER is as follows:

```
10.txt: {'neg': 0.02, 'neu': 0.612, 'pos': 0.368, 'compound': 0.9993}
11.txt: {'neg': 0.008, 'neu': 0.88, 'pos': 0.112, 'compound': 0.9868}
12.txt: {'neg': 0.031, 'neu': 0.936, 'pos': 0.033, 'compound': 0.1027}
13.txt: {'neg': 0.0, 'neu': 0.98, 'pos': 0.02, 'compound': 0.4019}
14.txt: {'neg': 0.0, 'neu': 0.963, 'pos': 0.037, 'compound': 0.6369}
15.txt: {'neg': 0.0, 'neu': 0.98, 'pos': 0.02, 'compound': 0.4019}
16.txt: {'neg': 0.031, 'neu': 0.936, 'pos': 0.033, 'compound': 0.1027}
```

The output will look like:

```python
# Convert the text data into a bag of words matrix
vectorizer = CountVectorizer()
X = vectorizer.fit_transform(data['text'])

# Train the Naive Bayes algorithm on the Financial PhraseBank
dataset
nb = MultinomialNB()
nb.fit(X, data['sentiment'])

# Define the path to the preprocessed folder
folder_path = 'D:/DATA/Preprocessed'

# Create a list of file paths in the folder and subfolders
file_paths = []
for root, dirs, files in os.walk(folder_path):
    for file in files:
        if file.endswith('.txt'):
            file_paths.append(os.path.join(root, file))

# Create an empty list to store the predicted sentiment labels
y_pred = []

# Loop through each file in the file paths list
for file_path in file_paths:
    # Open the file and read the preprocessed text data
    with open(file_path, 'r', encoding='utf-8') as f:
        text = f.read()
    # Convert the preprocessed text data to a bag of words
matrix using the same vectorizer as before
    X_test = vectorizer.transform([text])
    # Predict the sentiment label of the text data using the
Naive Bayes model
    y_pred.append(nb.predict(X_test)[0])

# Print the predicted sentiment of each file
for i, file_path in enumerate(file_paths):
    print(file_path, y_pred[i])
```

The output shows the value of negative, positive, and neutral sentiments separately. **compound** is the overall sentiment in the text.

Machine Learning Algorithms

Machine learning algorithms are a popular technique used in sentiment analysis to automatically classify text data into positive, negative, or neutral sentiment. These algorithms work by learning from labeled data, which is data that has already been classified into sentiment categories, and then applying this learning to classify new, unlabeled data. In accounting and financial research, financial training datasets are more suitable. Here are some publicly available financial training datasets that you may find useful for sentiment analysis:

1. FinancialPhraseBank: A labeled dataset of financial news articles and their corresponding sentiment labels (positive, negative, or neutral). This dataset is commonly used for sentiment analysis in the financial domain. The dataset can be found here: http://www.di.unipi.it/~gulli/FPB.html.
2. Reuters News Corpus: A large corpus of news articles from the Reuters newswire service, which includes a significant amount of financial news. The dataset can be found here: https://archive.ics.uci.edu/ml/datasets/reuters-21578+text+categorization+collection.
3. Bloomberg News Dataset: A labeled dataset of news articles from the Bloomberg newswire service and their corresponding topic labels. This dataset is commonly used for topic modeling and other natural language processing tasks in the financial domain. The dataset can be found here: https://github.com/philipperemy/FB15k-237-Bloomberg-News-Headlines.
4. Dow Jones News Archive: A large archive of news articles from the Dow Jones newswire service, spanning several decades. This dataset includes a significant amount of financial news and can be used for various natural language processing tasks. Access to the dataset may require a subscription or licensing agreement.

5. Kaggle financial datasets: Kaggle hosts a variety of financial datasets that can be used for machine learning tasks, including stock prices, financial statements, and economic indicators. Some datasets are free, while others may require payment or licensing agreements. The datasets can be found here: https://www.kaggle.com/datasets?tags=6215-financial.

These datasets are just a few examples of the many publicly available financial training datasets that can be used for machine learning tasks. When selecting a dataset for your project, it's important to consider factors such as the size of the dataset, the quality and accuracy of the labeling, and the relevance of the data to your specific task.

Here are some of the most commonly used machine learning algorithms for sentiment analysis:

- **Naive Bayes:**

Naive Bayes is a probabilistic algorithm that works by calculating the probability of each sentiment category given the words in the text. It assumes that the words in the text are independent of each other, which is why it's called "naive". Despite its simplicity, Naive Bayes is often used as a baseline algorithm in sentiment analysis because of its speed and relatively high accuracy. Here is a simple implementation of naïve bayes using FinancialPhraseBank training dataset.

```
import os
import pandas as pd
from sklearn.feature_extraction.text import CountVectorizer
from sklearn.naive_bayes import MultinomialNB

# Load the Financial PhraseBank dataset
data_path = 'D:/python codes/Repositories/FInancialPhraseBank/all-data.csv'
data = pd.read_csv(data_path, header=None, names=['sentiment', 'text'], encoding='ISO-8859-1')
```

```python
from sklearn.feature_extraction.text import CountVectorizer
from sklearn.naive_bayes import MultinomialNB
from sklearn.pipeline import Pipeline
from sklearn.model_selection import train_test_split, GridSearchCV
from sklearn.metrics import accuracy_score, precision_score, recall_score, f1_score
import pandas as pd

# Load the Financial PhraseBank dataset
data_path = 'D:/python codes/Repositories/FInancialPhraseBank/all-data.csv'
data = pd.read_csv(data_path, header=None, names=['sentiment', 'text'], encoding='ISO-8859-1')

# Split the data into training and validation sets
X_train, X_val, y_train, y_val = train_test_split(data['text'], data['sentiment'], test_size=0.2, random_state=42)

# Define the pipeline of preprocessing steps and Naive Bayes model
pipeline = Pipeline([
    ('vect', CountVectorizer()),
    ('clf', MultinomialNB())
])

# Define the hyperparameters to search over
params = {
    'vect__ngram_range': [(1, 1), (1, 2)],
    'vect__stop_words': [None, 'english'],
    'clf__alpha': [0.01, 0.1, 1, 10]
}

# Perform grid search to find the best hyperparameters
grid_search = GridSearchCV(pipeline, params, cv=5, scoring='accuracy')
grid_search.fit(X_train, y_train)

# Print the best hyperparameters and corresponding validation score
print("Best hyperparameters:", grid_search.best_params_)
print("Validation score:", grid_search.best_score_)

# Train the model on the entire training set with the best hyperparameters
best_model = grid_search.best_estimator_
best_model.fit(X_train, y_train)

# Evaluate the performance of the model on the validation set
y_val_pred = best_model.predict(X_val)
print("Accuracy:", accuracy_score(y_val, y_val_pred))
print("Precision:", precision_score(y_val, y_val_pred, average='weighted'))
print("Recall:", recall_score(y_val, y_val_pred, average='weighted'))
print("F1 score:", f1_score(y_val, y_val_pred, average='weighted'))
```

The code loads the FinancialPhraseBank dataset and trains the Naive Bayes algorithm on the dataset using a bag-of-words matrix representation of the text data. The code then defines the path to a folder containing preprocessed text files for testing. After this, the code loops through each file in the folder, converts the preprocessed text data to a bag-of-words matrix representation using the same vectorizer as before, predicts the sentiment label of the text data using the trained Naive Bayes model, and stores the predicted sentiment label in a list. Finally, the code prints the predicted sentiment label of each file to the console.

The model, however, needs to be fine-tuned before final implementation. The process for fine-tuning a Naive Bayes model for sentiment analysis using scikit-learn's **GridSearchCV** and Pipeline involves several steps. The labeled dataset should be loaded into memory, and split into training and testing sets. A pipeline object should be defined that consists of data preprocessing steps and a Naive Bayes algorithm. A dictionary of hyperparameters should be defined to be tuned during the grid search. A **GridSearchCV** object should be defined that performs a grid search over the hyperparameters and uses cross-validation to evaluate the performance of the pipeline on the training set. The GridSearchCV object should be fit to the training data using the fit() method to find the best hyperparameters. The performance of the trained model should be evaluated on the testing set using metrics such as accuracy, precision, recall, and F1-score. The trained model can be used to predict the sentiment of new, unseen text data by applying the pipeline to the raw text data. This process can be implemented as follows:

```
Best hyperparameters: {'clf__alpha': 1, 'vect__ngram_range': (1, 2), 'vect__stop_words': None}
Validation score: 0.725230462254739
Accuracy: 0.7597938144329897
Precision: 0.7607235563559895
Recall: 0.7597938144329897
F1 score: 0.7437906807771771
```

The output will look like:

```
import os
import pandas as pd

# Load the preprocessed text data
test_data_path = 'D:/DATA/Preprocessed'
test_data = []
for root, dirs, files in os.walk(test_data_path):
    for file in files:
        file_path = os.path.join(root, file)
        with open(file_path, 'r', encoding='utf-8') as f:
            text = f.read()
            test_data.append(text)

# Convert the preprocessed text data into a pandas DataFrame
test_df = pd.DataFrame({'text': test_data})

# Use the trained model to predict the sentiment labels
y_pred = best_model.predict(test_df['text'])

# Print the predicted sentiment labels
print(y_pred)
```

The output means that best hyperparameters found by GridSearchCV are **clf__alpha** = 1, **vect__ngram_range** = (1, 2), and **vect__stop_words** = None. The validation score of the best model is 0.725, which means that the accuracy of the model on the validation set is 72.5%. The accuracy of the model on the test set is 75.98%, which means that the model correctly predicted the sentiment of 75.98% of the test samples. The precision of the model is 76.07%, which means that of all the samples predicted as positive by the model, 76.07% of them were actually positive. The recall of the model is 75.98%, which means that of all the positive samples in the test set, 75.98% of them were correctly identified as positive by the model. The F1-score of the model is 74.38%, which is the harmonic mean of precision and recall. It provides a balanced measure of the model's performance in terms of both precision and recall. Overall, this models seems a good fit, but whether the model is good enough depends on the specific application and the requirements for the model's performance. If the model is being used for a high-stakes application, such as predicting stock prices or making investment decisions, then the accuracy of the model may need to be higher in order to make reliable predictions. On the other hand, if the model is being used for exploratory analysis or as a starting point for further analysis, then the accuracy of the model may be sufficient. In general, it's a good idea to compare the performance of the Naive Bayes model to other classification models and to evaluate the performance of the model on different datasets to get a better sense of its generalizability.

The model now has been fine-tuned and thus can be implemented to our text data as follows:

```python
import os
import pandas as pd
from sklearn.feature_extraction.text import TfidfVectorizer
from sklearn import svm

# Set the path to the preprocessed text files folder
path = 'D:/DATA/Preprocessed'

# Load the Financial PhraseBank as the training dictionary
dictionary = pd.read_csv('D:/python
codes/Repositories/FinancialPhraseBank/all-data.csv',
header=None, names=['sentiment', 'text'], encoding='latin-1')

# Vectorize the training data using the TF-IDF algorithm
vectorizer = TfidfVectorizer(use_idf=True, lowercase=True,
strip_accents='ascii', stop_words='english')
X_train = vectorizer.fit_transform(dictionary['text'])
y_train = dictionary['sentiment']

# Initialize the SVM classifier with a linear kernel
clf = svm.SVC(kernel='linear')

# Loop through all the text files in the folder and its
subfolders
for root, dirs, files in os.walk(path):
    for file in files:
        if file.endswith('.txt'):
            # Read the text file and vectorize it using the same
TF-IDF vectorizer
            with open(os.path.join(root, file), 'r') as f:
                text = f.read()
                X_test = vectorizer.transform([text])

                # Predict the sentiment using the SVM classifier
                y_pred = clf.predict(X_test)

                # Print the predicted sentiment
                print(file, y_pred)
```

- **Support Vector Machines (SVM):**

SVM is a linear classification algorithm that works by finding the optimal hyperplane that separates the positive and negative sentiment classes in the data. SVM can handle non-linear data by mapping the data to a higher-dimensional space, where it becomes linearly separable. SVM is a popular algorithm in sentiment analysis because of its high accuracy and ability to handle large datasets. Here is a code that uses Financial PhraseBank sentiment dictionary and performs sentiment analysis on normalized text data:

```
from sklearn.model_selection import GridSearchCV
from sklearn.pipeline import Pipeline

# Set the path to the preprocessed text files folder
path = 'D:/DATA/Preprocessed'

# Load the Financial PhraseBank as the training dictionary
dictionary = pd.read_csv('D:/python
codes/Repositories/FInancialPhraseBank/all-data.csv',
header=None, names=['sentiment', 'text'], encoding='latin-1')

# Vectorize the training data using the TF-IDF algorithm
vectorizer = TfidfVectorizer(use_idf=True, lowercase=True,
strip_accents='ascii', stop_words='english')
X_train = vectorizer.fit_transform(dictionary['text'])
y_train = dictionary['sentiment']

# Define the SVM pipeline with the TF-IDF vectorizer and SVM
classifier
svm_pipeline = Pipeline([
    ('tfidf', TfidfVectorizer()),
    ('clf', svm.SVC())
])

# Define the hyperparameters to search over using grid search
hyperparameters = {
    'tfidf__use_idf': (True, False),
    'clf__C': [0.1, 1, 10],
    'clf__kernel': ['linear', 'rbf']
}

# Initialize the grid search with 5-fold cross-validation
grid_search = GridSearchCV(svm_pipeline, hyperparameters, cv=5)

# Fit the grid search to the training data
grid_search.fit(X_train, y_train)

# Print the best hyperparameters and the corresponding mean
cross-validation score
print('Best hyperparameters:', grid_search.best_params_)
print('Best mean cross-validation score:',
grid_search.best_score_)
```

The code loads a training dictionary (in this case, the Financial PhraseBank), vectorizes the training data using the TF-IDF algorithm, and then initializes an SVM classifier with a linear kernel. It then loops through all the text files in the specified folder and its subfolders, reads in each text file, vectorizes it using the same TF-IDF vectorizer as the training data, and predicts the sentiment using the SVM classifier. Finally, it prints out the predicted sentiment for each text file. Note that this is just a basic example code and you may need to modify it to suit your specific requirements.

To fine-tune your SVM model, you can adjust the hyperparameters of the SVM classifier or try different kernels to see which one works best for your dataset. Here are some of the hyperparameters you can tune in the **svm.SVC()** function:

- C: The penalty parameter of the error term. A higher value of C will result in a more complex decision boundary, which may lead to overfitting. You can try different values of C to find the optimal value that balances between overfitting and underfitting.
- gamma: The kernel coefficient for "rbf", "poly", and "sigmoid" kernels. A higher value of gamma will result in a more complex decision boundary, which may lead to overfitting. You can try different values of gamma to find the optimal value that balances between overfitting and underfitting.
- kernel: The type of kernel to use. You can try different kernels such as linear, polynomial, or radial basis function (RBF) to see which one works best for your dataset.

To fine-tune your model, you can use techniques such as grid search or randomized search to search over a range of hyperparameters and find the optimal combination that maximizes the model's performance. Grid search involves exhaustively searching over a pre-defined range of hyperparameters, while randomized search randomly samples hyperparameters from a pre-defined distribution.

Here's an example code that shows how to use grid search to fine-tune an SVM model:

```
# Vectorize the training data using the TF-IDF algorithm with
the best hyperparameters
vectorizer = TfidfVectorizer(use_idf=True, lowercase=True,
strip_accents='ascii', stop_words='english')
X_train = vectorizer.fit_transform(dictionary['text'])
y_train = dictionary['sentiment']

# Initialize the SVM classifier with the best hyperparameters
clf = svm.SVC(C=1, kernel='linear')

# Fit the SVM classifier to the entire training data
clf.fit(X_train, y_train)

# Loop through all the text files in the folder and its
subfolders
for root, dirs, files in os.walk(path):
    for file in files:
        if file.endswith('.txt'):
            # Read the text file and vectorize it using the same
TF-IDF vectorizer
            with open(os.path.join(root, file), 'r') as f:
                text = f.read()
                X_test = vectorizer.transform([text])

            # Predict the sentiment using the fine-tuned SVM
classifier
            y_pred = clf.predict(X_test)

            # Print the predicted sentiment
            print(file, y_pred)
```

This code defines a pipeline that consists of a TF-IDF vectorizer and an SVM classifier, and uses grid search to search over a range of hyperparameters for both the vectorizer and the classifier. The **GridSearchCV()** function performs fivefold cross-validation to evaluate the performance of each hyperparameter combination, and returns the best combination of hyperparameters that maximizes the model's performance. You can use the best hyperparameters to train a new SVM model on the entire training dataset and evaluate its performance on a separate test set.

Once you have fine-tuned your SVM model using grid search or any other method, you can use the **predict()** function of the SVM classifier to predict the sentiment of new, unlabeled text data. Here's an example code that shows how to use the fine-tuned SVM model to predict the sentiment of preprocessed text files in a folder and its subfolders:

```
import os

# Set the path to the preprocessed text files folder
path = 'D:/DATA/Preprocessed'

# Load the Financial PhraseBank as the training dictionary
dictionary = pd.read_csv('D:/python
codes/Repositories/FInancialPhraseBank/all-data.csv',
header=None, names=['sentiment', 'text'], encoding='latin-1')
```

```
import os
import pandas as pd
from sklearn.feature_extraction.text import TfidfVectorizer
from sklearn.model_selection import GridSearchCV
from sklearn.pipeline import Pipeline
from sklearn import svm

# Set the path to the preprocessed text files folder
path = 'D:/DATA/Preprocessed'

# Load the Financial PhraseBank as the training dictionary
dictionary = pd.read_csv('D:/python
codes/Repositories/FInancialPhraseBank/all-data.csv',
header=None, names=['sentiment', 'text'], encoding='latin-1')

# Define the SVM pipeline with the TF-IDF vectorizer and SVM
classifier
svm_pipeline = Pipeline([
    ('tfidf', TfidfVectorizer(use_idf=True, lowercase=True,
strip_accents='ascii', stop_words='english')),
    ('clf', svm.SVC())
])

# Define the hyperparameters to search over using grid search
hyperparameters = {
    'tfidf__use_idf': (True, False),
    'clf__C': [0.1, 1, 10],
    'clf__kernel': ['linear', 'rbf']
}

# Initialize the grid search with 5-fold cross-validation
grid_search = GridSearchCV(svm_pipeline, hyperparameters, cv=5)

# Fit the grid search to the training data
grid_search.fit(dictionary['text'], dictionary['sentiment'])

# Print the best hyperparameters and the corresponding mean
cross-validation score
print('Best hyperparameters:', grid_search.best_params_)
print('Best mean cross-validation score:',
grid_search.best_score_)

# Initialize the SVM classifier with the best hyperparameters
clf = svm.SVC(C=grid_search.best_params_['clf__C'],
kernel=grid_search.best_params_['clf__kernel'])
```

This code initializes an SVM classifier with the best hyperparameters that were obtained from fine-tuning, and then fits the classifier to the entire training data. It then loops through all the text files in the specified folder and its subfolders, reads in each text file, vectorizes it using the same TF-IDF vectorizer as the training data, and predicts the sentiment using the fine-tuned SVM classifier. Finally, it prints out the predicted sentiment for each text file. Instead of printing, you can modify this code to save the sentiments as csv file.

More often you need to do sentiment analysis as topic level, rather than document level. Here's a code that creates model for first creating topics from the normalized text data using SVM and then performing sentiment analysis of all the topics using the process above. Aggregate sentiment for the whole document will be the sum of sentiments for all the topics.

```
# Vectorize the training data using the TF-IDF algorithm with
the best hyperparameters
vectorizer = TfidfVectorizer(use_idf=True, lowercase=True,
strip_accents='ascii', stop_words='english')
X_train = vectorizer.fit_transform(dictionary['text'])
y_train = dictionary['sentiment']

# Fit the SVM classifier to the entire training data
clf.fit(X_train, y_train)

# Initialize a dictionary to store the document sentiment for
each topic
doc_sentiment = {}

# Loop through all the text files in the folder and its
subfolders
for root, dirs, files in os.walk(path):
    for file in files:
        if file.endswith('.txt'):
            # Read the text file and vectorize it using the same
TF-IDF vectorizer
            with open(os.path.join(root, file), 'r') as f:
                text = f.read()
                X_test = vectorizer.transform([text])

                # Predict the topic using the SVM classifier
                topic = clf.predict(X_test)[0]

                # Assign a sentiment score to the topic
                if topic == 'positive':
                    sentiment = 1
                elif topic == 'negative':
                    sentiment = -1
                else:
                    sentiment = 0

                # Add the sentiment score to the document
sentiment dictionary
                if topic in doc_sentiment:
                    doc_sentiment[topic] += sentiment
                else:
                    doc_sentiment[topic] = sentiment

# Calculate the aggregate sentiment of the document
aggregate_sentiment = sum(doc_sentiment.values())

# Print the aggregate sentiment
print(f"Aggregate Sentiment: {aggregate_sentiment}")
```

```python
# Import necessary libraries
import os
import pandas as pd
from sklearn.feature_extraction.text import CountVectorizer
from sklearn.ensemble import RandomForestClassifier

# Set the file paths
preprocessed_folder = 'D:/DATA/Preprocessed' # folder containing preprocessed text files
dictionary_path = 'D:/python codes/Repositories/FInancialPhraseBank/all-data.csv' # path to the training dictionary

# Load the training dictionary
dictionary = pd.read_csv(dictionary_path, header=None,
names=['sentiment', 'text'], encoding='ISO-8859-1')

# Vectorize the training data
vectorizer = CountVectorizer(stop_words='english')
X = vectorizer.fit_transform(dictionary['text'])
y = dictionary['sentiment']

# Train the Random Forest model
model = RandomForestClassifier()
model.fit(X, y)

# Predict sentiment for each preprocessed file in the folder and subfolders
for root, dirs, files in os.walk(preprocessed_folder):
    for file in files:
        if file.endswith('.txt'):
# Read the preprocessed file
with open(os.path.join(root, file), 'r') as f:
    text = f.read()
# Vectorize the preprocessed file
X_test = vectorizer.transform([text])
# Predict the sentiment using the trained model
sentiment = model.predict(X_test)[0]
print(f'{file}: {sentiment}')
```

- **Random Forest:**

Random Forest is an ensemble learning algorithm that works by creating multiple decision trees and then combining their predictions to make a final prediction. Each decision tree is trained on a random subset of the data and a random subset of the features, which helps to reduce overfitting and improve accuracy. Random Forest is a popular algorithm in sentiment analysis because of its high accuracy and ability to handle noisy data. Here's an example code for executing the Random Forest algorithm on preprocessed text files using the Financial PhraseBank as the training dictionary:

```
# Import necessary libraries
import os
import pandas as pd
from sklearn.feature_extraction.text import CountVectorizer
from sklearn.ensemble import RandomForestClassifier
from sklearn.model_selection import GridSearchCV
import joblib

# Set the file paths
preprocessed_folder = 'D:/DATA/Preprocessed' # folder containing
preprocessed text files
dictionary_path = 'D:/python
codes/Repositories/FInancialPhraseBank/all-data.csv' # path to
the training dictionary
model_path = 'D:/python
codes/Repositories/FInancialPhraseBank/best_model.pkl' # path to
the best model found using grid search

# Load the training dictionary
dictionary = pd.read_csv(dictionary_path, header=None,
names=['sentiment', 'text'], encoding='ISO-8859-1')

# Vectorize the training data
vectorizer = CountVectorizer(stop_words='english')
X = vectorizer.fit_transform(dictionary['text'])
y = dictionary['sentiment']

# Define the parameter grid for grid search
param_grid = {
    'n_estimators': [50, 100, 200],
    'max_depth': [10, 20, 30, None],
    'min_samples_split': [2, 5, 10],
    'min_samples_leaf': [1, 2, 4]
```

This code uses the **pandas** library to load the Financial PhraseBank training dictionary from the specified file path, and the **sklearn** library to vectorize the text data using the **CountVectorizer** class and train the Random Forest model using the **RandomForestClassifier** class. It then uses the **os** library to iterate through all the preprocessed text files in the specified folder and subfolders, reads each file using the **open** function, vectorizes the text data using the same **CountVectorizer** object used for training, and predicts the sentiment using the trained Random Forest model. The predicted sentiment is then printed to the console.

You can again fine-tune this mode using **GridSearchCV()** and apply the best model to our preprocessed text data using the code below:

```python
}

# Perform grid search to find the best hyperparameters
grid_search = GridSearchCV(RandomForestClassifier(), param_grid,
cv=5, n_jobs=-1)
grid_search.fit(X, y)

# Print the best hyperparameters and the corresponding mean
cross-validation score
print(f'Best parameters: {grid_search.best_params_}')
print(f'Best score: {grid_search.best_score_}')

# Save the best model found using grid search
joblib.dump(grid_search.best_estimator_, model_path)

# Load the best model
model = joblib.load(model_path)

# Predict sentiment for each preprocessed file in the folder and
subfolders
for root, dirs, files in os.walk(preprocessed_folder):
    for file in files:
        if file.endswith('.txt'):
            # Read the preprocessed file
            with open(os.path.join(root, file), 'r') as f:
                text = f.read()
            # Vectorize the preprocessed file
            X_test = vectorizer.transform([text])
            # Predict the sentiment using the trained model
            sentiment = model.predict(X_test)[0]
            print(f'{file}: {sentiment}')
```

```
import os
import pandas as pd
from sklearn.feature_extraction.text import TfidfVectorizer
from sklearn.ensemble import RandomForestClassifier
from sklearn.model_selection import train_test_split
from sklearn.decomposition import LatentDirichletAllocation

# Set the file paths
preprocessed_folder = 'D:/DATA/Preprocessed' # folder containing
preprocessed text files
dictionary_path = 'D:/python
codes/Repositories/FInancialPhraseBank/all-data.csv' # path to
the training dictionary

# Load the training dictionary
dictionary = pd.read_csv(dictionary_path, header=None,
names=['sentiment', 'text'], encoding='ISO-8859-1')

# Split the dictionary into training and testing sets
X_train, X_test, y_train, y_test =
train_test_split(dictionary['text'], dictionary['sentiment'],
test_size=0.2, random_state=42)

# Create a TfidfVectorizer object to convert text data to
numerical features
tfidf = TfidfVectorizer(stop_words='english')
X_train_tfidf = tfidf.fit_transform(X_train)

# Create a Latent Dirichlet Allocation (LDA) model to extract
topics from the text data
lda = LatentDirichletAllocation(n_components=10,
random_state=42)
lda.fit(X_train_tfidf)

# Save the LDA model
joblib.dump(lda, 'lda_model.pkl')

# Load the LDA model
lda = joblib.load('lda_model.pkl')

# Vectorize the test data and predict the topic probabilities
X_test_tfidf = tfidf.transform(X_test)
topic_probabilities = lda.transform(X_test_tfidf)

# Create a Random Forest model for sentiment analysis
```

In this code, we perform grid search to find the best hyperparameters for the Random Forest model using the **GridSearchCV** function, and save the best model found using **joblib.dump** to the specified path **model_path**. We then load the best model using **joblib.load** and use it to predict the sentiment of each preprocessed file in the specified folder and subfolders, printing the predicted sentiment to the console. You can modify this code to save the sentiments to csv file also.

Similarly, a topic modeling approach to random forest sentiment analysis would be:

```
rf = RandomForestClassifier()
rf.fit(X_train_tfidf, y_train)

# Predict the sentiment for each document and each topic
doc_sentiments = []
for i, prob in enumerate(topic_probabilities):
    topic_sentiments = []
    for j, topic_prob in enumerate(prob):
        if topic_prob >= 0.1: # consider only topics with probability >= 0.1
            # Extract the text corresponding to the topic from the test data
            topic_text = X_test[i][(lda.components_[j] * topic_prob).argsort()[::-1][:10]] # consider only the top 10 words in the topic
            # Predict the sentiment using the trained Random Forest model
            topic_sentiment = rf.predict(tfidf.transform([topic_text]))[0]
            topic_sentiments.append(topic_sentiment)
    # Aggregate the sentiments for the individual topics and calculate the overall sentiment for the document
    if len(topic_sentiments) > 0:
        doc_sentiment = sum([1 if sentiment == 'positive' else -1 for sentiment in topic_sentiments]) / len(topic_sentiments)
    else:
        doc_sentiment = 0 # no relevant topics found
    doc_sentiments.append(doc_sentiment)

# Calculate the aggregate sentiment for the entire document set
doc_sentiments = pd.Series(doc_sentiments)
doc_sentiments_agg = doc_sentiments.mean()
print(f'Aggregate sentiment for the document set: {doc_sentiments_agg:.2f}')
```

In this code, we first load the training dictionary and split it into training and testing sets. We create a **TfidfVectorizer** object to convert the text data to numerical features and fit it on the training data. We then create a **LatentDirichletAllocation** model to extract topics from the text data and fit it on the training.

- **Logistic Regression:**

Logistic regression is a linear classification algorithm that works by estimating the probability of each sentiment class given the words in the text. It uses a logistic function to map the output to a probability value between 0 and 1. Logistic regression is a simple and fast algorithm that can be easily trained on large datasets. Here's a code that creates the model, splits Financial Phrase-Bank dictionary randomly into training and text model and finetunes the model, and applies the best model to normalized text data:

```python
import os
import pandas as pd
from sklearn.feature_extraction.text import TfidfVectorizer
from sklearn.linear_model import LogisticRegression
from sklearn.model_selection import train_test_split

# Path to the preprocessed text files
folder_path = 'D:/DATA/Preprocessed'

# Load the Financial PhraseBank dataset
df = pd.read_csv('D:/python
codes/Repositories/FInancialPhraseBank/all-data.csv',
header=None, names=['sentiment', 'text'])

# Split the dataset into training and testing sets
X_train, X_test, y_train, y_test = train_test_split(df['text'],
df['sentiment'], test_size=0.2, random_state=42)

# Create a TfidfVectorizer object to convert text data to
numerical features
tfidf = TfidfVectorizer()

# Fit the vectorizer on the training data and transform both the
training and testing data
X_train_tfidf = tfidf.fit_transform(X_train)
X_test_tfidf = tfidf.transform(X_test)

# Create a Logistic Regression model and fit it on the training
data
lr = LogisticRegression()
lr.fit(X_train_tfidf, y_train)

# Evaluate the model on the testing data
accuracy = lr.score(X_test_tfidf, y_test)
print("Accuracy:", accuracy)

# Classify new text data using the trained model
for root, dirs, files in os.walk(folder_path):
    for file in files:
        file_path = os.path.join(root, file)
        with open(file_path, 'r') as f:
            text = f.read()
            text_tfidf = tfidf.transform([text])
sentiment = lr.predict(text_tfidf)
print(f"File: {file_path} - Sentiment: {sentiment}")
```

Here, we first load the Financial PhraseBank dataset and split it into training and testing sets. We then create a **TfidfVectorizer** object to convert the text data to numerical features and fit it on the training data. We create a **LogisticRegression** model and fit it on the training data. We evaluate the model on the testing data and print the accuracy. Finally, we classify the text files in the specified folder and subfolders using the trained model and print the predicted sentiment for each file.

These are just a few examples of the many machine learning algorithms that can be used for sentiment analysis. Each algorithm has its own strengths and weaknesses, and the choice of algorithm depends on the specific requirements and constraints of the project. It's important to evaluate the performance of different algorithms on a validation set before choosing the final algorithm for the project.

Sentiment analysis is an increasingly important tool in academic research, particularly in accounting and finance. With the explosion of digital data, sentiment analysis provides a powerful way to extract insights from text data and understand the sentiment of individuals or groups on various topics. In accounting and finance, sentiment analysis can be used to analyze the sentiment of financial news articles, earnings call transcripts, and social media discussions related to firms, industries, or financial markets. By identifying and categorizing the sentiment expressed in these texts, researchers can gain valuable insights into the impact of public sentiment on the financial markets and the behavior of investors.

Despite its potential, sentiment analysis is still underutilized in academic research, especially in accounting and finance. This may be due to the perception that sentiment analysis is a black box technique that lacks transparency and interpretability. However, recent advances in machine learning and deep learning algorithms have made sentiment analysis more accurate and transparent, and it is now possible to understand how the models make their predictions.

To fully realize the potential of sentiment analysis in academic research, it is important for researchers to become familiar with the various techniques and tools available and to carefully evaluate their performance and limitations. Researchers should also be aware of the ethical considerations involved in using sentiment analysis, such as the potential biases in the data or the potential impact on privacy.

Part V

Machine Learning and Predictive Analytics

18

Basic Regression

Regression analysis is a powerful statistical technique used to analyze the relationship between a dependent variable (also known as the response variable) and one or more independent variables (also known as predictor or explanatory variables). It helps to understand how changes in the independent variables affect the dependent variable and provides a mathematical model that can be used for prediction and forecasting purposes. The primary goal of regression analysis is to create an equation that best describes the relationship between the variables and minimizes the errors between the predicted and observed values.

Regression analysis serves as the main analytical tool for academic researchers in accounting and finance, playing an integral role in their research endeavors. It enables them to identify factors driving various financial metrics such as stock returns, earnings, and financial ratios. Moreover, it allows for the estimation of causal effects between variables, such as the impact of corporate governance structures on firm performance. Researchers also use regression analysis to assess the predictive power of financial models by comparing their forecasts with actual outcomes and to test hypotheses related to theoretical models in finance, such as the Capital Asset Pricing Model (CAPM) and the Fama–French Three-Factor Model. Additionally, it can be employed to evaluate the effectiveness of accounting and financial policies or regulatory changes. By leveraging regression analysis, academic researchers in accounting and finance can gain insights into the complex relationships among variables, make informed decisions, and develop evidence-based strategies that benefit various stakeholders, including investors, managers, and regulators.

Regressions can be categorized based on various factors such as the relationship between variables, the nature of the dependent variable, and the type of data being analyzed. Here are some common ways to categorize different regression techniques:

1. Based on the relationship between variables:

    a. Linear Regression: Assumes a linear relationship between the dependent variable and the independent variables. Examples include simple linear regression (one independent variable) and multiple linear regression (two or more independent variables).

    b. Non-linear Regression: Models a non-linear relationship between the dependent variable and the independent variables. Examples include polynomial regression, exponential regression, and logarithmic regression.

2. Based on the nature of the dependent variable:

    a. Continuous Dependent Variable: Assumes the dependent variable is continuous in nature. Examples include simple linear regression, multiple linear regression, and polynomial regression.

    b. Categorical or Binary Dependent Variable: Assumes the dependent variable is categorical or binary in nature. Examples include logistic regression, probit regression, and multinomial logistic regression.

    c. Count Dependent Variable: Assumes the dependent variable represents count data. Examples include Poisson regression and negative binomial regression.

3. Based on the type of data being analyzed:

    a. Cross-sectional Regression: Uses cross-sectional data, i.e., data collected at a single point in time. Examples include simple linear regression, multiple linear regression, and logistic regression.

    b. Time Series Regression: Uses time series data, i.e., data collected over multiple time periods. Examples include autoregressive models, distributed lag models, and vector autoregressive models.

    c. Panel Data Regression: Uses both cross-sectional and time series data. Examples include fixed effects models, random effects models, and dynamic panel models.

4. Based on other specialized criteria:

    a. Robust Regression: Techniques designed to handle outliers and influential observations, such as Huber regression and M-estimators.

b. Regularized Regression: Techniques that incorporate regularization to prevent overfitting and improve model performance, such as LASSO, Ridge, and Elastic Net regression.
c. Quantile Regression: Models the relationship between the dependent variable and the independent variables at different quantiles of the dependent variable's distribution.

These categorizations are not mutually exclusive, and many regression techniques can belong to multiple categories based on their characteristics and application. Researchers can choose the most appropriate regression method depending on the research question, data availability, and underlying assumptions.

A brief discussion of the regressions commonly used in academic accounting and finance research along with the Python implementation is provided below.

Linear Regressions

Linear regression is a fundamental method in statistical analysis, which is widely used to model the relationships between a dependent variable and one or more independent variables. However, linear regression has several assumptions and limitations that need to be considered before using this technique. The first assumption is linearity, which means that the relationship between the dependent and independent variables should be linear. The second assumption is independence, which assumes that the observations are independent of each other. The third assumption is homoscedasticity, which means that the variance of the error term is constant across all levels of the independent variable. The fourth assumption is normality, which assumes that the error term follows a normal distribution. These assumptions are important to ensure the accuracy and validity of the results obtained from simple linear regression. Failure to meet these assumptions can lead to inaccurate or biased results. Therefore, it is important to carefully consider these assumptions before using linear regression.

Simple Linear Regression

Simple linear regression is a method for modeling the relationship between a dependent variable (Y) and a single independent variable (X). The model takes the form of $Y = \beta_0 + \beta_1 X + \varepsilon$, where β_0 and β_1 are the regression coefficients, and ε is the error term.

For implementation of simple linear regression, we will use the 'statsmodels' library. A simple model that regresses price of BTC on ETH from D:/Data/my_data.csv is as follows:

```
import pandas as pd
import statsmodels.api as sm

data = pd.read_csv('D:/Data/my_data.csv')
X = data['ETH']
Y = data['BTC']

X = sm.add_constant(X)
model = sm.OLS(Y, X).fit()
print(model.summary())
```

This Python code uses the pandas library to read in a CSV file containing data on two variables, "ETH" and "BTC". It then uses the **statsmodels.api** library to perform a simple linear regression analysis on the data, where "ETH" is the independent variable and "BTC" is the dependent variable. The X variable is created by selecting the "ETH" column from the dataset and the Y variable is created by selecting the "BTC" column from the dataset. The X variable is then augmented with a constant term using the **add_constant** function from the **statsmodels.api** library. The OLS function from the **statsmodels.api** library is then used to fit the linear regression model with Y as the dependent variable and X as the independent variable, and the resulting model is stored in the "model" variable. Finally, the code prints out a summary of the regression analysis, which includes information about the model's coefficients, statistical significance, goodness-of-fit measures, and other important diagnostic statistics.

The result will look like:

```
                            OLS Regression Results
==============================================================================
Dep. Variable:                    BTC   R-squared:                       0.875
Model:                            OLS   Adj. R-squared:                  0.875
Method:                 Least Squares   F-statistic:                 2.215e+04
Date:                Wed, 03 May 2023   Prob (F-statistic):               0.00
Time:                        06:21:32   Log-Likelihood:                -31676.
No. Observations:                3171   AIC:                         6.336e+04
Df Residuals:                    3169   BIC:                         6.337e+04
Df Model:                           1
Covariance Type:            nonrobust
==============================================================================
                 coef    std err          t      P>|t|      [0.025      0.975]
------------------------------------------------------------------------------
const       2228.9228    104.490     21.332      0.000    2024.049    2433.797
ETH           14.9804      0.101    148.833      0.000      14.783      15.178
==============================================================================
Omnibus:                     1795.017   Durbin-Watson:                   0.011
Prob(Omnibus):                  0.000   Jarque-Bera (JB):            17946.581
Skew:                           2.535   Prob(JB):                         0.00
Kurtosis:                      13.495   Cond. No.                     1.16e+03
==============================================================================

Notes:
[1] Standard Errors assume that the covariance matrix of the
errors is correctly specified.
[2] The condition number is large, 1.16e+03. This might
indicate that there are
strong multicollinearity or other numerical problems.
```

The results also show the tests which normally are not available automatically from other statistical softwares while you carry out regression. Here, the Omnibus test and Jarque–Bera test indicate that the residuals are not normally distributed, and the Durbin-Watson test suggests that there may be some autocorrelation in the residuals. The condition number of 1160 suggests that there might be strong multicollinearity or other numerical problems. To interpret the results, we can also print the coefficients and intercept separately:

```
print('Intercept:', model.params[0])
print('Coefficients:', model.params[1])
```

This will print the parameters:

```
Intercept: 2228.9227940130877
Coefficients: 14.980444452692431
```

To see the scatterplot with regression line the following code will be useful:

```
import pandas as pd
import statsmodels.api as sm
import matplotlib.pyplot as plt
import seaborn as sns

data = pd.read_csv('D:/Data/my_data.csv')
X = data['ETH']
Y = data['BTC']

X = sm.add_constant(X)
model = sm.OLS(Y, X).fit()

# Scatterplot
plt.figure(figsize=(10, 6))
sns.scatterplot(x='ETH', y='BTC', data=data, label='Data
Points')

# Regression line
reg_line_x = X['ETH']
reg_line_y = model.params[0] + model.params[1] * reg_line_x
plt.plot(reg_line_x, reg_line_y, color='red',
label='Regression Line')

plt.xlabel('ETH')
plt.ylabel('BTC')
plt.legend()
plt.show()
```

This code imports the necessary libraries, loads the dataset, defines the dependent and independent variables, and fits an OLS regression model. It then creates a scatterplot using **seaborn.scatterplot** and adds the regression line to the plot using the model's parameters to create the plot. The output will look like:

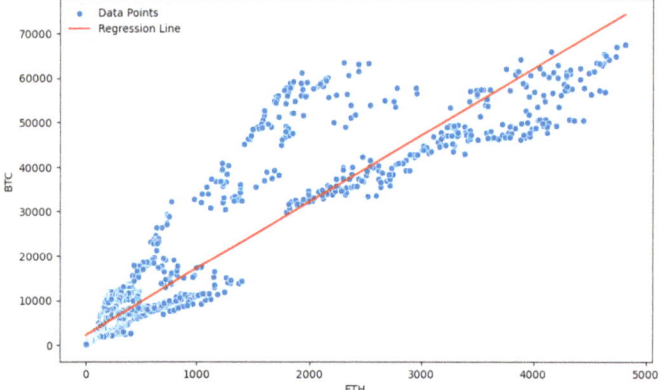

Multiple Regression

Multiple linear regression is a method for modeling the relationship between a dependent variable (Y) and a single independent variable (X). The model takes the form of $Y = \beta_0 + \Sigma \beta_n X_n + \varepsilon$, where β_0 and β_n are the regression coefficients, and ε is the error term.

For implementation of simple linear regression, we simply replace line 5 of the code to:

```
X = data[['ETH', 'BNB', 'ADA']]
```

The result will look like:

```
                            OLS Regression Results
==============================================================================
Dep. Variable:                    BTC   R-squared:                       0.879
Model:                            OLS   Adj. R-squared:                  0.878
Method:                 Least Squares   F-statistic:                     7635.
Date:                Wed, 03 May 2023   Prob (F-statistic):               0.00
Time:                        06:53:17   Log-Likelihood:                -31629.
No. Observations:                3171   AIC:                         6.327e+04
Df Residuals:                    3167   BIC:                         6.329e+04
Df Model:                           3
Covariance Type:            nonrobust
==============================================================================
                 coef    std err          t      P>|t|      [0.025      0.975]
------------------------------------------------------------------------------
const       2550.5781    108.286     23.554      0.000    2338.259    2762.897
ETH           10.8469      0.437     24.811      0.000       9.990      11.704
BNB           19.1904      2.506      7.659      0.000      14.278      24.103
ADA         2866.1216    587.033      4.882      0.000    1715.119    4017.125
==============================================================================
Omnibus:                     1718.829   Durbin-Watson:                   0.010
Prob(Omnibus):                  0.000   Jarque-Bera (JB):            14168.129
Skew:                           2.472   Prob(JB):                         0.00
Kurtosis:                      12.098   Cond. No.                     6.66e+03
==============================================================================

Notes:
[1] Standard Errors assume that the covariance matrix of the
errors is correctly specified.
[2] The condition number is large, 6.66e+03. This might
indicate that there are
strong multicollinearity or other numerical problems.
```

Alternately we can use statsmodels library also. An implementation using statmodels library is as follows:

```
import pandas as pd
import statsmodels.api as sm
from sklearn.linear_model import LinearRegression

data = pd.read_csv('D:/Data/my_data.csv')
X = data[['ETH', 'BNB', 'ADA']]
Y = data['BTC']

model = LinearRegression().fit(X, Y)
print('Intercept:', model.intercept_)
print('Coefficients:', model.coef_)
```

You can export the results in a publication quality table in a word file with the code below:

```
import pandas as pd
import statsmodels.api as sm
import os
from docx import Document
from docx.shared import Inches

# Read the data
data = pd.read_csv('D:/Data/my_data.csv')
X = data[['ETH', 'BNB', 'ADA']]
Y = data['BTC']

# Fit the OLS model
X = sm.add_constant(X)
model = sm.OLS(Y, X).fit()

# Create a new Word document
doc = Document()
```

```python
# Add a title to the document
doc.add_heading('Regression Model Summary', level=1)

# Create a table to hold the summary results
table = doc.add_table(rows=5, cols=2)
table.style = 'Table Grid'

# Add the regression model results to the table
table.cell(0, 0).text = 'Adjusted R-squared:'
table.cell(0, 1).text = str(round(model.rsquared_adj, 4))
table.cell(1, 0).text = 'F-statistic:'
table.cell(1, 1).text = str(round(model.fvalue, 4))
table.cell(2, 0).text = 'p-value (F-statistic):'
table.cell(2, 1).text = str(round(model.f_pvalue, 4))
table.cell(3, 0).text = 'Number of observations:'
table.cell(3, 1).text = str(model.nobs)
table.cell(4, 0).text = 'Degrees of freedom:'
table.cell(4, 1).text = str(model.df_resid)

# Add a table to hold the regression coefficients and
statistics
table = doc.add_table(rows=len(X.columns)+1, cols=7)
table.style = 'Table Grid'

# Add the header row to the table
hdr_cells = table.rows[0].cells
hdr_cells[0].text = 'Variable'
hdr_cells[1].text = 'Coefficient'
hdr_cells[2].text = 'Std. Error'
hdr_cells[3].text = 't-statistic'
hdr_cells[4].text = 'p-value'
hdr_cells[5].text = 'Lower 95%'
hdr_cells[6].text = 'Upper 95%'


# Add the regression model results to the table
conf_int_df = pd.DataFrame(model.conf_int(alpha=0.05).T,
columns=X.columns)
for i, var in enumerate(X.columns):
    row_cells = table.add_row().cells
    row_cells[0].text = var
    row_cells[1].text = str(round(model.params[i], 4))
    row_cells[2].text = str(round(model.bse[i], 4))
    row_cells[3].text = str(round(model.tvalues[i], 4))
    row_cells[4].text = str(round(model.pvalues[i], 4))
    row_cells[5].text = str(round(conf_int_df[var][0], 4))
    row_cells[6].text = str(round(conf_int_df[var][1], 4))

# Save the document in the same folder as the data file
file_dir =
os.path.dirname(os.path.abspath('D:/Data/my_data.csv'))
doc.save(file_dir + '/regression_results.docx')
```

In multiple regression, we often have a large number of predictors, and some of them may be highly correlated or irrelevant for predicting the outcome variable. This can lead to overfitting, which means the model fits the training data too well, but performs poorly on new data. Regularization methods can help address this issue by shrinking the coefficients of the predictors and reducing their variance.

- LASSO (Least Absolute Shrinkage and Selection Operator): LASSO is a regularization technique that adds a penalty term to the least squares loss function. The penalty term is the absolute value of the sum of the regression coefficients. This penalty tends to force some of the coefficients to zero, effectively performing variable selection and producing a sparse model. Implementation of LASSO regression with alpha = 0.1 will be:

```
lasso_model = Lasso(alpha=0.1).fit(X, Y)
print('LASSO Coefficients:', lasso_model.coef_)
```

- Ridge Regression: Ridge regression is a regularization technique that adds a penalty term to the least squares loss function. The penalty term is the sum of the squared regression coefficients, multiplied by a tuning parameter lambda. This penalty tends to shrink the coefficients toward zero, but does not force them to zero. Ridge regression can be useful when there are many predictors with small to moderate effects. Implementation of Ridge regression with alpha = 0.1 will be:

```
ridge_model = Ridge(alpha=0.1).fit(X, Y)
print('Ridge Coefficients:', ridge_model.coef_)
```

- Elastic Net: Elastic net is a combination of LASSO and Ridge regression. It adds a penalty term that is a mixture of the absolute and squared values of the regression coefficients. The mixing parameter alpha determines the balance between the LASSO and Ridge penalties. Elastic net can be useful when there are many correlated predictors. Implementation of Elastic Net regression with alpha = 0.1 and l1_ratio = 0.5 will be:

```
elastic_net_model = ElasticNet(alpha=0.1, l1_ratio=0.5).fit(X, Y)
print('Elastic Net Coefficients:', elastic_net_model.coef_)
```

To visualize the regression in a 3-dimensional picture, the following code would be useful:

```
import pandas as pd
import numpy as np
from sklearn.linear_model import LinearRegression
from mpl_toolkits.mplot3d import Axes3D
import matplotlib.pyplot as plt

data = pd.read_csv('D:/Data/my_data.csv')
X = data[['ETH', 'BNB', 'ADA']]
Y = data['BTC']
model = LinearRegression().fit(X, Y)
print('Intercept:', model.intercept_)
print('Coefficients:', model.coef_)

# 3D scatterplot
fig = plt.figure(figsize=(10, 8))
ax = fig.add_subplot(111, projection='3d')

ax.scatter(data['ETH'], data['BNB'], data['ADA'], c='blue',
marker='o', alpha=1)

# Regression plane
xx, yy = np.meshgrid(data['ETH'].unique(),
data['BNB'].unique())
zz = model.intercept_ + model.coef_[0] * xx + model.coef_[1] *
yy

ax.plot_surface(xx, yy, zz, color='red', alpha=0.5)

ax.set_xlabel('ETH')
ax.set_ylabel('BNB')
ax.set_zlabel('ADA')
plt.show()
```

This code a linear regression model using **LinearRegression** from the **sklearn.linear_model** library creates a 3D scatterplot using the **scatter** method of an **Axes3D** object and adds the regression plane to the plot using the model's intercept and coefficients. Note that this plot includes only ETH and BNB on the *x* and *y*-axes, with ADA as the dependent variable. The output will look like:

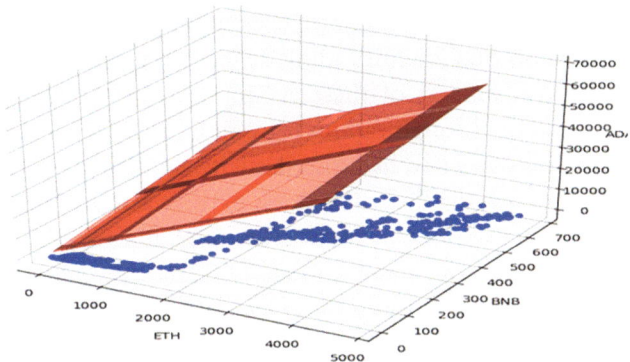

For such plots a 3-dimensional plot that can be manipulated to view from various angles would be more useful, which can be obtained using the following code:

```
import pandas as pd
import numpy as np
from sklearn.linear_model import LinearRegression
import plotly.express as px
import plotly.graph_objects as go

data = pd.read_csv('D:/Data/my_data.csv')
X = data[['ETH', 'BNB', 'ADA']]
Y = data['BTC']

model = LinearRegression().fit(X, Y)
print('Intercept:', model.intercept_)
print('Coefficients:', model.coef_)

# 3D scatterplot
fig = px.scatter_3d(data, x='ETH', y='BNB', z='ADA')

# Regression plane
xx, yy = np.meshgrid(data['ETH'].unique(),
data['BNB'].unique())
zz = model.intercept_ + model.coef_[0] * xx + model.coef_[1] * yy

fig.add_trace(go.Surface(x=xx, y=yy, z=zz,
colorscale='Inferno', opacity=0.5))

fig.update_layout(scene=dict(xaxis_title='ETH',
yaxis_title='BNB', zaxis_title='ADA'))
fig.show()
```

Regression Diagnostics

Linear regression relies on several key assumptions that must be true for the results to be meaningful and accurate. If any of these assumptions are incorrect, the results of the regression may be misleading or incorrect. Therefore, it's important to check the truthfulness of these assumptions before using linear regression. This can be done through various regression diagnostic techniques, such as linearity, multicollinearity, heteroscedasticity, autocorrelation, and normality of residuals. By verifying the truthfulness of these assumptions, we can ensure that the results of the linear regression are valid and reliable.

Before diving into the diagnostics, let's load the data D:/Data/my_data.csv and perform a multiple linear regression using the given data.

```python
import pandas as pd
import statsmodels.api as sm

# Load the data
data = pd.read_csv("D:/Data/my_data.csv")

# Define the dependent variable
y = data["BTC"]
# Define the independent variables
X = data[["ETH", "BNB", "SOL", "ADA"]]
X = sm.add_constant(X)

# Fit the model
model = sm.OLS(y, X).fit()
```

Linearity Check

To test the linearity assumption, we can visually inspect the relationship between the dependent variable and each independent variable using scatter plots.

```python
import matplotlib.pyplot as plt

for column in X.columns[1:]:
    plt.scatter(X[column], y)
    plt.xlabel(column)
    plt.ylabel("BTC")
    plt.show()
```

The output will show the scatter plots of all the X variables and Y variables separately. If the relationships appear to be non-linear, you can consider transforming the independent variables using functions such as logarithmic, square root, or inverse transformations.

Multicollinearity

Multicollinearity occurs when two or more independent variables are highly correlated. It can lead to unstable parameter estimates and reduce the interpretability of the regression coefficients. We can use the Variance Inflation Factor (VIF) to detect multicollinearity.

```python
vif = pd.DataFrame()
vif["Features"] = X.columns[1:]
vif["VIF"] = [variance_inflation_factor(X.values[:, 1:], i)
              for i in range(X.shape[1] - 1)]

print(vif)
```

The output will look like:

```
   Features        VIF
0       ETH  23.538970
1       BNB  14.001344
2       SOL   4.176452
3       ADA  12.946235
```

A VIF value greater than 10 indicates high multicollinearity, as is seen here. If multicollinearity is present, you can consider dropping one of the correlated variables or use dimensionality reduction techniques such as Principal Component Analysis (PCA).

Heteroscedasticity

Heteroscedasticity refers to the unequal variance of the residuals across the range of the predicted values. It can lead to inefficient parameter estimates. To test for heteroscedasticity, we can use the Breusch-Pagan test from the **statsmodels** library.

```
from statsmodels.stats.diagnostic import het_breuschpagan

bp_test = het_breuschpagan(model.resid, X)
print("Breusch-Pagan test p-value:", bp_test[1])
```

The output will show Breusch–Pagan test p-value. A p-value less than 0.05 indicates the presence of heteroscedasticity. If heteroscedasticity is present, you can consider using robust standard errors or weighted least squares regression.

Autocorrelation

Autocorrelation occurs when the residuals are correlated across time. It can lead to biased parameter estimates. To test for autocorrelation, we can use the Durbin–Watson test from the **statsmodels** library.

```
from statsmodels.stats.stattools import durbin_watson

dw = durbin_watson(model.resid)
print("Durbin-Watson statistic:", dw)
```

The output will show Durbin-Watson statistic. A Durbin-Watson statistic close to 2 indicates no autocorrelation, while values significantly below 2 suggest positive autocorrelation. If autocorrelation is present, you can consider using time series regression techniques such as autoregressive integrated moving average (ARIMA) or vector autoregression (VAR) models, which are specifically designed to handle time series data.

Normality of Residuals

The normality assumption states that the residuals should follow a normal distribution. To test this assumption, we can use the Shapiro–Wilk test from the **scipy.stats** library and visually inspect the residuals using a Q–Q plot.

```
from scipy.stats import shapiro
import statsmodels.graphics.gofplots as gofplots

# Perform the Shapiro-Wilk test
shapiro_test = shapiro(model.resid)
print("Shapiro-Wilk test p-value:", shapiro_test[1])

# Create a Q-Q plot
gofplots.qqplot(model.resid)
plt.show()
```

A p-value less than 0.05 from the Shapiro–Wilk test indicates non-normal residuals. If the residuals are not normally distributed, you can consider transforming the dependent variable or using non-linear regression techniques such as generalized linear models (GLMs) or machine learning algorithms.

Regression diagnostics are crucial for validating the assumptions of the regression model and ensuring its accuracy. By carefully examining linearity, multicollinearity, heteroscedasticity, autocorrelation, and the normality of residuals, you can improve the quality of your regression analysis and enhance the reliability of your research findings.

Linear regression is a powerful and widely used technique in accounting and finance research. This chapter has provided a comprehensive overview of both simple and multiple linear regression and its implementation using Python. Regression diagnostics were also discussed to ensure the validity of the models and conclusions. The use of Python for implementing linear regression and advanced regression techniques provides researchers with a flexible and accessible toolkit for their research endeavors. The upcoming chapters will delve into more advanced regression techniques accompanied by Python implementations, allowing researchers to confidently apply them to their research projects.

19

Logistic Regression

Logistic regression is a statistical method used for analyzing the relationship between one or more predictor variables and a binary outcome or dependent variable. It is particularly useful for researchers in the fields of accounting, finance, and other business-related disciplines, as it can help predict the probability of events or outcomes, such as financial distress, fraud, or bankruptcy, based on relevant predictor variables. The key advantage of logistic regression is its ability to handle categorical response variables, making it suitable for analyzing binary classification problems.

The logistic regression equation is derived from the logistic function, also known as the sigmoid function. The logistic function takes the form: $P(Y = 1) = 1/(1 + e^{(-z)})$, where $P(Y = 1)$ represents the probability of the outcome variable Y taking the value 1 (the event of interest), and z is the linear combination of predictor variables $X_1, X_2, \ldots, X_n$, and their corresponding coefficients $\beta_1, \beta_2, \ldots, \beta_n$, along with an intercept term β_0. The linear combination of predictor variables can be represented as: $z = \beta_0 + \beta_1 X_1 + \beta_2 X_2 + \cdots + \beta_n X_n$. Substituting this back into the logistic function, we get the logistic regression equation: $P(Y = 1) = 1/(1 + e^{-(\beta_0 + \beta_1 X_1 + \beta_2 X_2 + \cdots + \beta_n X_n)})$.

Logistic regression is based on the concept of odds, which is the ratio of the probability of the event of interest ($Y = 1$) to the probability of its complement ($Y = 0$). The odds are defined as: $\text{Odds}(Y = 1) = P(Y = 1)/P(Y = 0)$. Using the logistic regression equation, we can relate the odds to the predictor variables and their coefficients: $\text{Odds}(Y = 1) = e^{(\beta_0 + \beta_1 X_1 + \beta_2 X_2 + \cdots + \beta_n X_n)}$. Taking the natural logarithm of both sides, we obtain the log-odds (also known as logit) representation:

$Logit(Y = 1) = ln(Odds(Y = 1)) = \beta_0 + \beta_1 X_1 + \beta_2 X_2 + \cdots + \beta_n X_n.$
The logit is a linear function of the predictor variables, and the coefficients $\beta_1, \beta_2, \ldots, \beta_n$ represent the change in the log-odds for a one-unit increase in the corresponding predictor variable, while holding all other variables constant. This makes the interpretation of the coefficients more straightforward in terms of the odds ratios.

Implementing Logistic Regression in Python

An implementation of logistic regression using **statsmodels** library and binomial data in D:/Data/my_data1.csv is as follows:

```
import pandas as pd
import statsmodels.api as sm

# Load the dataset
data = pd.read_csv('D:/Data/my_data1.csv')

# Define the dependent variable (Y) and independent variables
(X)
Y = data['BTC']
X = data[['ETH', 'BNB', 'SOL', 'ADA']]

# Add a constant to the independent variables
X = sm.add_constant(X)
# Fit the logistic regression model
logit_model = sm.Logit(Y, X)
result = logit_model.fit()
# Print the results summary
print(result.summary())
```

This code snippet imports the necessary libraries and loads the **my_data1.csv** dataset into a DataFrame. It then defines the dependent variable (Y) as the 'BTC' column and the independent variables (X) as the 'ETH', 'BNB', 'SOL', and 'ADA' columns. After adding a constant term to the independent variables, the code fits a logistic regression model using the **Logit** function from the **statsmodels** library and prints a summary of the results. The output will look like:

```
Optimization terminated successfully.
         Current function value: 0.692409
         Iterations 3
                        Logit Regression Results
==============================================================================
Dep. Variable:                    BTC   No. Observations:                 3171
Model:                          Logit   Df Residuals:                     3166
Method:                           MLE   Df Model:                            4
Date:                Wed, 03 May 2023   Pseudo R-squ.:                0.001059
Time:                        07:58:32   Log-Likelihood:                -2195.6
converged:                       True   LL-Null:                       -2198.0
Covariance Type:            nonrobust   LLR p-value:                    0.3248
==============================================================================
                 coef    std err          z      P>|z|      [0.025      0.975]
------------------------------------------------------------------------------
const         -0.1085      0.079     -1.372      0.170      -0.263       0.046
ETH            0.1249      0.071      1.755      0.079      -0.015       0.264
BNB           -0.0343      0.071     -0.483      0.629      -0.174       0.105
SOL            0.0662      0.071      0.931      0.352      -0.073       0.206
ADA            0.0481      0.071      0.676      0.499      -0.091       0.188
==============================================================================
```

The logistic regression results show that the model has a relatively low pseudo R-squared value of 0.001059, indicating a weak explanatory power. Among the independent variables, only the coefficient of ETH is significant at the 10% level (p-value $= 0.079$), suggesting a positive association between ETH and the log-odds of BTC being equal to 1. The coefficients of BNB, SOL, and ADA are not statistically significant (p-values > 0.1), indicating that there is no strong evidence to support a relationship between these variables and the log-odds of BTC being equal to 1 in this model.

To save the output of the logistic regression model in a publication quality table in Microsoft Word format, you can use the **stargazer** package from the **stargazer** library. First, you'll need to install the **stargazer** library by running **!pip install stargazer**. Then, you can modify your code as follows:

```python
import pandas as pd
import statsmodels.api as sm
from stargazer.stargazer import Stargazer

# Load the dataset
data = pd.read_csv('D:/Data/my_data1.csv')

# Define the dependent variable (Y) and independent variables
(X)
Y = data['BTC']
X = data[['ETH', 'BNB', 'SOL', 'ADA']]

# Add a constant to the independent variables
X = sm.add_constant(X)

# Fit the logistic regression model
logit_model = sm.Logit(Y, X)
result = logit_model.fit()

# Save the result summary to a Word document
stargazer = Stargazer([result])
stargazer.title("Logistic Regression Results")
stargazer.custom_columns(['Model'], [1])
stargazer.render_latex()

# Save the latex output to a .tex file in the D:/Data folder
with open("D:/Data/output_table.tex", "w") as f:
    f.write(stargazer.render_latex())

# Convert the .tex file to a Word document in the D:/Data folder
!pandoc D:/Data/output_table.tex -o D:/Data/output_table.docx
```

This code will generate a .tex file with the LaTeX code for the table and then convert it to a Word document (.docx) using **pandoc**. If you don't have **pandoc** installed, you can download and install it from the official Pandoc website. Sometime after installing you may have to add pandoc as system PATH variable also. The output table will be saved as **output_table.docx** in the same folder as your script. You can use the code similar to the one we used for linear regression also, but pandoc gives the tables in the format typically used in accounting and finance literature.

Evaluating the performance of a logistic regression model is crucial to understanding its predictive power and accuracy. Several metrics can be used to assess the model, including the confusion matrix, ROC curve, AUC, precision, recall, and F1-score.

Confusion Matrix

A confusion matrix is a table that represents the number of true positive (TP), true negative (TN), false positive (FP), and false negative (FN) predictions made by a binary classifier. It helps to visualize the performance of the model in terms of its correct and incorrect predictions. A confusion matrix for this regression can be calculated as:

```
from sklearn.metrics import confusion_matrix
from sklearn.model_selection import train_test_split
from sklearn.linear_model import LogisticRegression

# Split the data into train and test sets
X_train, X_test, Y_train, Y_test = train_test_split(X, Y,
test_size=0.3, random_state=42)

# Fit the logistic regression model using sklearn
lr_model = LogisticRegression()
lr_model.fit(X_train, Y_train)

# Make predictions on the test set
Y_pred = lr_model.predict(X_test)

# Calculate the confusion matrix
cm = confusion_matrix(Y_test, Y_pred)
print("Confusion Matrix:\n", cm)
```

The output will look like:

```
Confusion Matrix:
 [[251 241]
  [213 247]]
```

The confusion matrix shows that out of 952 instances, the model correctly predicted 498 instances (251 true positives and 247 true negatives), while it incorrectly predicted 454 instances (241 false positives and 213 false negatives). The model's accuracy can be calculated as $(251 + 247)/952 = 0.525$, while the misclassification rate can be calculated as $(241 + 213)/952 = 0.475$.

ROC Curve and AUC

The Receiver Operating Characteristic (ROC) curve is a graphical representation of the model's sensitivity (true positive rate) against its false positive rate (1—specificity) at various threshold levels. The area under the ROC curve (AUC) is a single value that summarizes the overall performance of the model. A higher AUC value indicates better classification performance.

```python
import pandas as pd
import statsmodels.api as sm
import matplotlib.pyplot as plt
from sklearn.metrics import roc_curve, roc_auc_score

# Load the dataset
data = pd.read_csv('D:/Data/my_data1.csv')

# Define the dependent variable (Y) and independent variables (X)
Y = data['BTC']
X = data[['ETH', 'BNB', 'SOL', 'ADA']]

# Add a constant to the independent variables
X = sm.add_constant(X)

# Fit the logistic regression model
logit_model = sm.Logit(Y, X)
result = logit_model.fit()

# Get predicted probabilities for the test set
Y_prob = result.predict(X)

# Calculate the false positive rate, true positive rate, and thresholds
fpr, tpr, thresholds = roc_curve(Y, Y_prob)

# Calculate the AUC
auc_score = roc_auc_score(Y, Y_prob)

print("AUC Score:", auc_score)

# Plot the ROC curve
plt.plot(fpr, tpr, color='darkorange', label='ROC curve (area = %0.2f)' % auc_score)
plt.plot([0, 1], [0, 1], color='navy', linestyle='--')
plt.xlabel('False Positive Rate')
plt.ylabel('True Positive Rate')
plt.title('Receiver operating characteristic')
plt.legend(loc="lower right")
plt.show()
```

The given code performs logistic regression on a dataset using **statsmodels** package in Python. It first loads the dataset and separates the dependent and independent variables. It then adds a constant to the independent variables and fits the logistic regression model using **logit()** function. It predicts the target variable using **predict()** function and calculates the false positive rate, true positive rate, and thresholds for the ROC curve using **roc_curve()** function. Finally, it calculates the AUC using **roc_auc_score()** function and plots the ROC curve using the matplotlib library. The output will look like:

```
Optimization terminated successfully.
         Current function value: 0.692409
         Iterations 3
AUC Score: 0.522284677717709
```

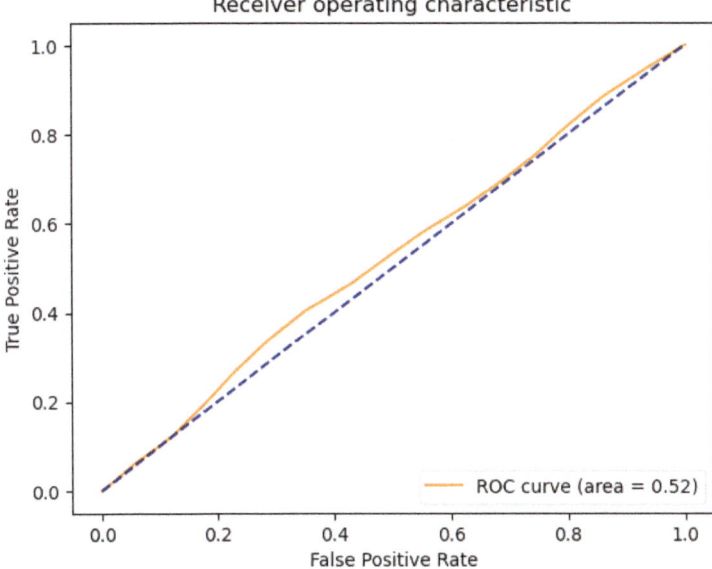

The output indicates that the logistic regression model was optimized successfully with a log-likelihood value of 0.692409. The algorithm required only 3 iterations to converge. The AUC score of 0.522284677717709 indicates that the model's discriminatory power is only slightly better than random guessing, as a score of 0.5 corresponds to a random classifier. In general, an AUC score of 0.6–0.7 is considered poor, 0.7–0.8 is considered fair, 0.8–0.9 is considered good, and 0.9–1.0 is considered excellent. Therefore, the AUC score of 0.522284677717709 suggests that the model's discriminatory power is poor, and the model may not be suitable for practical applications where accurate classification is important.

Precision, Recall, and F1-Score

Precision is the proportion of true positive predictions among all positive predictions made by the model. Recall (sensitivity or true positive rate) is the proportion of true positive predictions among all actual positive instances. F1-score is the harmonic mean of precision and recall and is used to measure the trade-off between precision and recall. An implementation in this case would be:

```python
import pandas as pd
import statsmodels.api as sm
from sklearn.metrics import classification_report

# Load the dataset
data = pd.read_csv('D:/Data/my_data1.csv')

# Define the dependent variable (Y) and independent variables
(X)
Y = data['BTC']
X = data[['ETH', 'BNB', 'SOL', 'ADA']]

# Add a constant to the independent variables
X = sm.add_constant(X)

# Fit the logistic regression model
logit_model = sm.Logit(Y, X)
result = logit_model.fit()

# Get predicted classes for the test set
Y_pred = result.predict(X)
Y_pred = (Y_pred > 0.5).astype(int)

# Print classification report
print(classification_report(Y, Y_pred))
```

Here, we use the **classification_report()** function from **sklearn.metrics** to calculate the Precision, Recall, and F1-score. We first fit the logistic regression model on the training data and predict the target variable using the **predict()** function. We then convert the predicted probabilities to predicted classes by setting a threshold of 0.5 and finally calculate the Precision, Recall, and F1-score using **classification_report()** function and print the classification report. The output will look like:

```
Optimization terminated successfully.
         Current function value: 0.692409
         Iterations 3
              precision    recall  f1-score   support

           0       0.52      0.51      0.51      1590
           1       0.51      0.53      0.52      1581

    accuracy                           0.52      3171
   macro avg       0.52      0.52      0.52      3171
weighted avg       0.52      0.52      0.52      3171
```

The first three lines indicate that the optimization of the logistic regression model was successful, and the algorithm required only three iterations to converge.

The classification report shows the Precision, Recall, and F1-score for both the positive (class 1) and negative (class 0) classes. The Precision of class 0 is 0.52, meaning that 52% of the predicted positive values are true positives, and the Precision of class 1 is 0.51, meaning that 51% of the predicted negative values are true negatives. The Recall of class 0 is 0.51, meaning that 51% of the actual positive values are correctly predicted, and the Recall of class 1 is 0.53, meaning that 53% of the actual negative values are correctly predicted. The F1-score of class 0 is 0.51, which is the harmonic mean of Precision and Recall, and the F1-score of class 1 is 0.52.

The accuracy of the model is 0.52, which means that it correctly predicted 52% of the test set. The macro average of Precision, Recall, and F1-score is also 0.52, which is the average of the scores for both the classes. The weighted average of Precision, Recall, and F1-score is also 0.52, which is the average of the scores weighted by the number of samples in each class.

In summary, the classification report suggests that the logistic regression model has poor performance in distinguishing between the two classes, with an accuracy of 0.52 and F1-scores of 0.51 and 0.52 for the two classes.

By leveraging Python's extensive libraries and tools, academic researchers can efficiently implement and analyze logistic regression models, gaining a deeper understanding of the relationships between predictor variables and binary response variables. The integration of Python in this field not only streamlines the research process but also enhances the accuracy and interpretability of the results, ultimately leading to more robust and impactful findings in accounting and finance research.

20

Probit and Logit Regression

The probit model and logit model are both types of generalized linear models (GLMs) used to analyze the relationship between a binary dependent variable and one or more independent variables. While they are similar in many aspects, they differ in the link function they use to model the relationship between the dependent and independent variables.

Probit Regression

Probit regression is a statistical technique used to analyze the relationship between a binary dependent variable and one or more independent variables, that may or may not be binary. It is commonly employed in accounting and finance research to study events with binary outcomes, such as bankruptcy or fraud, which can be represented as either a 0 or 1. Probit regression is a useful alternative to logistic regression when the underlying assumption of the logistic model may not hold.

The probit regression equation is based on the cumulative distribution function (CDF) of the standard normal distribution, also known as the probit function:

$$\Phi(z) = P(Z \leq z).$$

Here, Z is a standard normal random variable, and z is a specific value from the standard normal distribution. The function $\Phi(z)$ gives the probability that Z is less than or equal to z.

In probit regression, we model the probability of the binary outcome variable Y as a function of independent variables X:

$$P(Y = 1|X) = \Phi(X\beta).$$

Here, X is the matrix of independent variables, β is a vector of coefficients to be estimated, and $\Phi(X\beta)$ is the standard normal CDF.

The probit regression model is based on the latent variable approach. We assume that there is an unobservable continuous variable Y^* that is determined by a linear combination of the independent variables X and an error term ε:

$$Y* = X\beta + \varepsilon.$$

The error term ε follows a standard normal distribution with a mean of 0 and a variance of 1. The binary outcome variable Y is defined based on Y^*:

$$Y = 1 \text{ if } Y* > 0 \quad Y = 0 \text{ if } Y* \leq 0.$$

The probit model estimates the probability of the binary outcome variable Y being equal to 1, given the values of the independent variables X. This probability is given by the CDF of the standard normal distribution evaluated at $X\beta$:

$$P(Y = 1|X) = \Phi(X\beta).$$

The coefficients in a probit regression model represent the change in the z-score for a one-unit increase in the corresponding independent variable, holding all other variables constant. A positive coefficient indicates that an increase in the independent variable is associated with an increase in the probability of the binary outcome variable being equal to 1, while a negative coefficient suggests a decrease in this probability.

It is important to note that the coefficients in a probit regression model are not directly interpretable as changes in probabilities, unlike in a linear probability model. To interpret the effects of the independent variables on the probability of the binary outcome variable, one can calculate marginal effects or use other techniques such as average partial effects.

Probit model is based on the cumulative distribution function (CDF) of the standard normal distribution. This can make it easier to interpret the coefficients, as they can be directly related to the standard normal distribution. The normal distribution is also a well-known and widely used statistical distribution, which means that the underlying assumptions and properties of the model are more familiar to many researchers. However, a disadvantage of the probit model is that the coefficients are not as easily interpretable as odds ratios, which are a natural output of the logit model. Odds ratios can provide a more intuitive understanding of the relationship between the independent variables and the probability of the event occurring.

An implementation of probit on a data my_data3.csv in D:/Data folder can be done using the code below. Here, BTC is the dependent variable and independent variables are ETH, BNB, SOL, and ADA. All independent variables are continuous, except SOL, which is binary. BTC is also binary.

```
import pandas as pd
import statsmodels.api as sm
import statsmodels.formula.api as smf

# Read the data
data = pd.read_csv("D:/Data/my_data3.csv")

# Define the dependent variable and independent variables
dependent_variable = "BTC"
independent_variables = ["ETH", "BNB", "SOL", "ADA"]

# Add a constant to the independent variables
data["const"] = 1

# Define the probit regression model
formula = f"{dependent_variable} ~ {' + '.join(independent_variables)}"
probit_model = smf.probit(formula, data=data)

# Fit the probit regression model
probit_results = probit_model.fit()

# Print the regression results
print(probit_results.summary())

# Calculate marginal effects
marginal_effects = probit_results.get_margeff(at="mean", method="dydx")
print(marginal_effects.summary())
```

This code uses the **statsmodels** library. The code reads the data from the CSV file, adds a constant to the independent variables, defines the probit regression model using the specified dependent variable and independent variables, fits the model, and prints the results. The output will look like:

```
Optimization terminated successfully.
         Current function value: 0.689319
         Iterations 4
                          Probit Regression Results
==============================================================================
Dep. Variable:                    BTC   No. Observations:                 1554
Model:                         Probit   Df Residuals:                     1549
Method:                           MLE   Df Model:                            4
Date:                Sun, 07 May 2023   Pseudo R-squ.:                0.002171
Time:                        07:29:46   Log-Likelihood:                -1071.2
converged:                       True   LL-Null:                       -1073.5
Covariance Type:            nonrobust   LLR p-value:                    0.3238
==============================================================================
                 coef    std err          z      P>|z|      [0.025      0.975]
------------------------------------------------------------------------------
Intercept      0.0506      0.048      1.065      0.287      -0.043       0.144
ETH            0.0001      0.000      0.884      0.377      -0.000       0.000
BNB           -0.0010      0.001     -1.559      0.119      -0.002       0.000
SOL            0.1234      0.082      1.502      0.133      -0.038       0.284
ADA           -0.0161      0.144     -0.112      0.911      -0.298       0.266
==============================================================================
         Probit Marginal Effects
=====================================
Dep. Variable:                    BTC
Method:                          dydx
At:                              mean
==============================================================================
                dy/dx    std err          z      P>|z|      [0.025      0.975]
------------------------------------------------------------------------------
ETH          4.141e-05   4.69e-05      0.884      0.377   -5.04e-05       0.000
BNB            -0.0004      0.000     -1.559      0.119      -0.001       0.000
SOL             0.0491      0.033      1.502      0.133      -0.015       0.113
ADA            -0.0064      0.057     -0.112      0.911      -0.118       0.106
==============================================================================
```

The results can be exported to a word file in a table using the following code:

```python
import os
from docx import Document
from docx.shared import Pt

# Create a new Word document
doc = Document()

# Add a title to the document
doc.add_heading("Probit Regression Results", level=1)

# Convert the regression results summary to a string
summary_str = probit_results.summary().as_text()

# Create a new paragraph style with a monospace font
style = doc.styles.add_style("MonospaceStyle",
doc.styles["Normal"].type)
style.font.name = "Courier New"
style.font.size = Pt(10)

# Add the regression results summary to the document using the
new style
paragraph = doc.add_paragraph()
paragraph.style = style
paragraph.add_run(summary_str)

# Save the Word document in the same folder as the data file
output_filename =
os.path.join(os.path.dirname("D:/Data/my_data3.csv"),
"probit_results.docx")
doc.save(output_filename)
```

A publication quality table can be formed using the following code:

```python
import os
import pandas as pd
from docx import Document
from docx.shared import Pt

# Read the data
data = pd.read_csv("D:/Data/my_data3.csv")

# Define the dependent variable and independent variables
dependent_variable = "BTC"
independent_variables = ["ETH", "BNB", "SOL", "ADA"]

# Add a constant to the independent variables
data["const"] = 1

# Define the probit regression model
formula = f"{dependent_variable} ~ {' + '.join(independent_variables)}"
probit_model = smf.probit(formula, data=data)

# Fit the probit regression model
probit_results = probit_model.fit()

# Create a new Word document
doc = Document()

# Add a title to the document
doc.add_heading("Probit Regression Results", level=1)

# Create a new table with the regression results
num_rows = len(independent_variables) + 4
num_columns = 4
table = doc.add_table(rows=num_rows, cols=num_columns)

# Add the headers
header_cells = table.rows[0].cells
header_cells[0].text = "Variable"
header_cells[1].text = "Coefficient"
header_cells[2].text = "t-stat"
header_cells[3].text = "p-value"

# Add the dependent variable row
dep_var_cells = table.rows[1].cells
dep_var_cells[0].text = "Dependent Variable"
dep_var_cells[1].text = dependent_variable
dep_var_cells[2].text = ""
dep_var_cells[3].text = ""
```

```
# Add the Pseudo R2 row
r2_cells = table.rows[2].cells
r2_cells[0].text = "Pseudo R2"
r2_cells[1].text = f"{probit_results.prsquared:.4f}"
r2_cells[2].text = ""
r2_cells[3].text = ""

# Add the coefficient rows
for i, variable in enumerate(["Intercept"] +
independent_variables):
    var_cells = table.rows[i+3].cells
    var_cells[0].text = variable
    var_cells[1].text = f"{probit_results.params[variable]:.4f}"
    var_cells[2].text =
f"{probit_results.tvalues[variable]:.4f}"
    var_cells[3].text =
f"{probit_results.pvalues[variable]:.4f}"

# Apply formatting to the table
for row in table.rows:
    for cell in row.cells:
        paragraphs = cell.paragraphs
        for paragraph in paragraphs:
            paragraph_format = paragraph.paragraph_format
            paragraph_format.space_before = Pt(0)
            paragraph_format.space_after = Pt(0)

# Save the Word document in the same folder as the data file
output_filename =
os.path.join(os.path.dirname("D:/Data/my_data3.csv"),
"probit_results.docx")
doc.save(output_filename)

print("Regression results saved successfully.")
```

To assess the quality of a probit regression model, following criteria can be considered in addition to pseudo R-squared, coefficients and p-values, confidence intervals, and residual analysis:

Akaike Information Criterion (AIC) and Bayesian Information Criterion (BIC)

These are information criteria that help to compare different models. Lower values of AIC and BIC indicate a better model. These criteria can be especially useful when comparing probit models with different sets of independent variables, and it is often useful to compare the AIC and BIC of different models to determine which model has the best fit to the data. AIC and BIC can be calculated during the regression adding the following code after fitting the model:

```
# Print the AIC and BIC
print(f"\nAIC: {probit_results.aic:.4f}")
print(f"BIC: {probit_results.bic:.4f}")
```

Model Diagnostics

Model diagnostics help in identifying potential issues in your probit regression model. Here are some diagnostics that can be used:

- Multicollinearity: Multicollinearity occurs when two or more independent variables are highly correlated, making it difficult to determine the individual effect of each variable. You can use the Variance Inflation Factor (VIF) to detect multicollinearity. A VIF greater than 10 may indicate a high degree of multicollinearity, and you may need to address it by removing or transforming variables.

```
from statsmodels.stats.outliers_influence import
variance_inflation_factor

# Calculate VIF
X = data[independent_variables]
vif = pd.DataFrame()
vif["VIF Factor"] = [variance_inflation_factor(X.values, i) for
i in range(X.shape[1])]
vif["feature"] = X.columns
print(vif)
```

- Link test: The link test is used to check for model misspecification, i.e., whether important variables or higher-order terms have been omitted from the model. The test creates two additional variables: the predicted values of the dependent variable ($\widehat{Y}$) and the squared predicted values ($\widehat{Y}^2$). Then, a new probit regression model is fitted using $\widehat{Y}$ and $\widehat{Y}^2$ as independent variables. If the coefficient for $\widehat{Y}^2$ is significant, it suggests that the original model might be misspecified. Breusch-Pagan test is most commonly used link testy, which is implemented as below:

```
import statsmodels.api as sm
from statsmodels.stats.diagnostic import het_breuschpagan

# Define the dependent variable and independent variables
dependent_variable = "BTC"
independent_variables = ["ETH", "BNB", "SOL", "ADA"]

# Read the data
data = pd.read_csv("D:/Data/my_data3.csv")

# Add a constant to the independent variables
data["const"] = 1

# Fit the OLS regression model
ols_model = sm.OLS(data[dependent_variable],
data[independent_variables + ["const"]])
ols_results = ols_model.fit()

# Perform the Breusch-Pagan test
link_test = het_breuschpagan(ols_results.resid,
ols_results.model.exog)
print(link_test)
```

The **het_breuschpagan** function returns four statistics. The output will look like:

(4.609828115548001,   0.3297225985432579,   1.1521668138469945, 0.3303018227688186)

The first statistic (4.61 in this case) is the LM test statistic, and the second statistic (0.33 here) is the *p*-value for the test. The null hypothesis of the Breusch–Pagan test is that there is no heteroskedasticity in the residuals, and the alternative hypothesis is that there is heteroskedasticity. Since the *p*-value of the test (0.33) is greater than the typical significance level of 0.05, we cannot reject the null hypothesis, and we conclude that there is no significant evidence of heteroskedasticity in the residuals. The third statistic (1.15 in this case) is the degrees of freedom used in the LM test, and the fourth statistic (0.33 here) is the *p*-value for the *F*-test of the null hypothesis that all of the coefficients on the squared terms are zero. Since this test is not significant (*p*-value > 0.05), we conclude that the squared terms do not contribute significantly to the model.

- Hosmer–Lemeshow test: This test is used to assess the goodness-of-fit of the model by dividing the data into groups based on the predicted probabilities and comparing the observed and expected frequencies of events in each group. A non-significant p-value indicates that the model fits the data well.

```
from sklearn.metrics import log_loss
from sklearn.calibration import calibration_curve

# Create a logistic regression model
logit_model = sm.Logit(data[dependent_variable],
data[independent_variables])

# Fit the model
logit_results = logit_model.fit()

# Compute the predicted probabilities
y_pred_prob = logit_results.predict(data[independent_variables])

# Compute the log loss
log_loss_val = log_loss(data[dependent_variable], y_pred_prob)
print("Log Loss: {:.4f}".format(log_loss_val))

# Compute the calibration curve
true_prob, pred_prob =
calibration_curve(data[dependent_variable], y_pred_prob,
n_bins=10)

# Plot the calibration curve
plt.plot([0, 1], [0, 1], linestyle='--')
plt.plot(pred_prob, true_prob, marker='.')
plt.title('Calibration Curve')
plt.xlabel('Predicted Probability')
plt.ylabel('True Probability')
plt.show()
```

The code will give the following outputs:

```
Optimization terminated successfully.
         Current function value: 0.689682
         Iterations 4
Log Loss: 0.6897
```

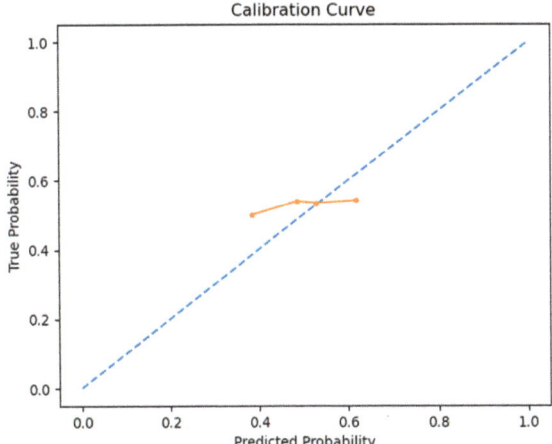

Calibration Curve

A good calibration curve should show the predicted probabilities closely matching the actual outcomes. Ideally, the calibration curve should be a 45-degree line, where the predicted probabilities are equal to the actual probabilities. However, in practice, a small amount of deviation from the 45-degree line is acceptable. If the curve deviates substantially from the 45-degree line, it suggests that the model is poorly calibrated and the predicted probabilities may not be reliable.

Logit Model

The logit model, also known as logistic regression, uses the logit link function. The logit link function is the natural logarithm of the odds, which is the ratio of the probability of the event (success) to the probability of the non-event (failure). In mathematical terms, the logit model can be expressed as:

$$\log(p/(1-p)) = \beta_0 + \beta_1 X_1 + \beta_2 X_2 + \cdots + \beta_n X_n$$

where p is the probability of success (the dependent variable taking the value 1), $X_1, X_2, \ldots, X_n$ are the independent variables, and $\beta_0, \beta_1, \ldots, \beta_n$ are the coefficients to be estimated.

While both probit and logit models are used to estimate the probability of an event occurring, they differ in the way they model the relationship between the dependent variable and independent variables. One of the primary differences between the two models is the link function used to transform the linear predictor into a probability. The logit model uses the logit link function, while the probit model uses the probit link function. The functional forms of the two models differ because the logistic distribution used in the logit model has heavier tails than the standard normal distribution used in the probit model. Moreover, the interpretation of coefficients differs between the two models. In the logit model, the coefficients represent the change in the log-odds of success for a one-unit increase in the independent variable, while in the probit model, the coefficients represent the change in the z-score for a one-unit increase in the independent variable.

Despite these differences, the choice between the two models often comes down to personal preference or the specific research question being addressed. Both models can produce similar results, especially when the relationship between the dependent and independent variables is not extreme. Ultimately, the choice of which model to use will depend on the researcher's familiarity with each method and which model is best suited for the research question at hand.

Python implementation of logit (logistic regression) is quite similar to the probit model. You can use the **statsmodels** library for both models, and the process of fitting the model and interpreting the results is almost identical. The only difference is that you use **smf.logit()** instead of **smf.probit()** to define the logit model. Here's an example of how to implement a logit model in Python using the same dataset from the previous example:

```python
import pandas as pd
import statsmodels.formula.api as smf

# Read the data
data = pd.read_csv("D:/Data/my_data3.csv")

# Define the dependent variable and independent variables
dependent_variable = "BTC"
independent_variables = ["ETH", "BNB", "SOL", "ADA"]

# Define the logit regression model
formula = f"{dependent_variable} ~ {' + '.join(independent_variables)}"
logit_model = smf.logit(formula, data=data)

# Fit the logit regression model
logit_results = logit_model.fit()

# Print the regression results
print(logit_results.summary())
```

The methods to export the results and to assess the quality of regression also remain the same.

The logit model has the advantage of producing coefficients that can be easily transformed into odds ratios, which can be more intuitive for researchers to understand and interpret. The logit model is based on the logistic function, which has an S-shaped curve and is symmetrical around the inflection point. This property can be useful when modeling probabilities, as it ensures that the predicted probabilities are always bounded between 0 and 1. However, logit model assumes that the relationship between the independent variables and the log-odds of the event occurring is linear. This assumption may not always hold in practice, and it can be more difficult to assess the goodness-of-fit and model diagnostics for the logit model compared to the probit model.

Probit and logit regression models are indispensable tools for academic accounting and finance research. As binary choice models, they provide a robust and flexible framework for analyzing relationships between binary dependent variables and multiple independent variables. Both models have been widely employed in various areas of accounting and finance research, including fraud detection, bankruptcy prediction, corporate governance, earnings management, and financial market participation, among others. These models enable researchers to delve deeper into the complex relationships that exist within financial data, providing valuable insights that can be used to inform decision-making, policy design, and theoretical development in accounting and finance. Both models offer unique advantages and disadvantages, but ultimately, the choice between probit and logit models is often determined by the research question, the nature of the data, and the preferences of the researcher.

21

Polynomial Regression

Polynomial regression is a type of regression analysis in which the relationship between the independent variable (X) and the dependent variable (Y) is modeled as an nth-degree polynomial. In accounting and finance research, polynomial regression can be used to capture more complex relationships between variables, which may not be adequately represented by a linear model. For instance, the relationship between stock returns and market conditions could be better modeled using a polynomial function to capture the various fluctuations and changes in trends.

The equation for a polynomial regression of degree n is given by:

$$Y = B_0 + B_1 X + B_2 X^2 + \ldots + B_n X^n + \varepsilon,$$

where

- Y is the dependent variable;
- X is the independent variable;
- $\beta_0, \beta_1, \ldots, \beta_n$ are the regression coefficients;
- n is the degree of the polynomial;
- ε is the error term.

The objective of polynomial regression is to estimate the values of the regression coefficients (β_0, β_1, ..., β_n) that minimize the sum of squared errors (SSE) between the observed and predicted values of the dependent variable.

Polynomial regression extends the simple linear regression model by allowing for higher-degree polynomial terms in the equation. The linear regression model is limited in its ability to capture complex relationships between variables because it assumes a straight-line relationship between them. Polynomial regression, on the other hand, allows for curved relationships, which can better model real-world phenomena.

The decision to use polynomial regression over linear regression depends on the nature of the data and the underlying relationship between the independent and dependent variables. If a scatter plot of the data points suggests a non-linear relationship, then polynomial regression may be more appropriate. However, it is important to note that using a high-degree polynomial can lead to overfitting the model. Overfitting occurs when a model becomes too complex and captures not only the underlying relationship but also the noise in the data. This can lead to poor predictions when the model is applied to new data. To avoid overfitting, researchers should use cross-validation techniques or regularization methods, such as Lasso or Ridge regression.

Here is an example of a third degree polynomial regression on a data in D:/Data/my_data.csv. Dependent variable is BTC and independent variable is ETH.

```
import pandas as pd
import numpy as np
import statsmodels.api as sm
from sklearn.model_selection import train_test_split
from sklearn.preprocessing import PolynomialFeatures

# Load the data
data = pd.read_csv("D:/Data/my_data.csv")

# Assign the independent variable (X) and dependent variable (Y)
X = data['ETH'].values.reshape(-1, 1)
Y = data['BTC'].values.reshape(-1, 1)

# Split the data into training and testing sets
X_train, X_test, Y_train, Y_test = train_test_split(X, Y,
test_size=0.2, random_state=0)

# Create and fit the polynomial features
poly_features = PolynomialFeatures(degree=3)
X_train_poly = poly_features.fit_transform(X_train)

# Add a constant term to the independent variables matrix for
the intercept
X_train_poly = sm.add_constant(X_train_poly)

# Create and fit the OLS model using statsmodels
model = sm.OLS(Y_train, X_train_poly).fit()

# Print the summary of the regression results
print(model.summary())

# Extract the coefficients, t-statistics, and number of
observations
coefficients = model.params
t_stats = model.tvalues
n_obs = model.nobs
```

This code uses the **statsmodels** library to create and fit an ordinary least squares (OLS) regression model with the third degree polynomial features. It then prints a summary of the regression results, including the coefficients, t-statistics, number of observations, and other important statistics such as R-squared and adjusted R-squared.

The output will look like:

```
                            OLS Regression Results
==============================================================================
Dep. Variable:                      y   R-squared:                       0.915
Model:                            OLS   Adj. R-squared:                  0.915
Method:                 Least Squares   F-statistic:                     9112.
Date:                Sat, 06 May 2023   Prob (F-statistic):               0.00
Time:                        09:37:54   Log-Likelihood:                -24852.
No. Observations:                2536   AIC:                         4.971e+04
Df Residuals:                    2532   BIC:                         4.974e+04
Df Model:                           3
Covariance Type:            nonrobust
==============================================================================
                 coef    std err          t      P>|t|      [0.025      0.975]
------------------------------------------------------------------------------
const        467.9759    116.990      4.000      0.000     238.569     697.383
x1            27.0842      0.584     46.395      0.000      25.940      28.229
x2            -0.0045      0.000    -11.503      0.000      -0.005      -0.004
x3          2.965e-07   6.62e-08      4.480      0.000    1.67e-07    4.26e-07
==============================================================================
Omnibus:                      674.324   Durbin-Watson:                   1.961
Prob(Omnibus):                  0.000   Jarque-Bera (JB):             4866.455
Skew:                           1.062   Prob(JB):                         0.00
Kurtosis:                       9.446   Cond. No.                     1.91e+10
==============================================================================
```

Notes:
[1] Standard Errors assume that the covariance matrix of the errors is correctly specified.
[2] The condition number is large, 1.91e+10. This might indicate that there are strong multicollinearity or other numerical problems.

To export the output as a picture on Word document as a publication quality table the following code will be useful:

```
import pandas as pd
import numpy as np
import statsmodels.api as sm
from sklearn.model_selection import train_test_split
from sklearn.preprocessing import PolynomialFeatures
import matplotlib.pyplot as plt
from PIL import Image
import os

# Load the data
data = pd.read_csv("D:/Data/my_data.csv")

# Assign the independent variable (X) and dependent variable (Y)
X = data['ETH'].values.reshape(-1, 1)
Y = data['BTC'].values.reshape(-1, 1)

# Split the data into training and testing sets
X_train, X_test, Y_train, Y_test = train_test_split(X, Y, test_size=0.2, random_state=0)
```

```
# Create and fit the polynomial features
poly_features = PolynomialFeatures(degree=3)
X_train_poly = poly_features.fit_transform(X_train)

# Add a constant term to the independent variables matrix for
the intercept
X_train_poly = sm.add_constant(X_train_poly)

# Create and fit the OLS model using statsmodels
model = sm.OLS(Y_train, X_train_poly).fit()

# Save the summary output as an image
fig = plt.figure(figsize=(12, 8))
txt = str(model.summary())
plt.figtext(0.1, 0.05, txt, fontsize=12, family='monospace')
plt.axis('off')
fig.savefig('D:/Data/regression_summary.png', dpi=300,
bbox_inches='tight')

# Open the Word document and insert the image
doc = Document()
doc.add_heading('Regression Summary:', level=1)
doc.add_picture('D:/Data/regression_summary.png',
width=Inches(6))
doc.save('D:/Data/regression_summary.docx')

# Display the image in a window (optional)
img = Image.open('D:/Data/regression_summary.png')
img.show()
```

This code creates a Word document with the title "Regression Summary:" and inserts the image using the **docx** module. Note that you may need to adjust the file path for saving the image and the document, based on your system configuration.

If you need to export a publication quality table that can directly be used in a paper, the following code will be useful:

```
import pandas as pd
import numpy as np
import statsmodels.api as sm
from sklearn.model_selection import train_test_split
from sklearn.preprocessing import PolynomialFeatures
from docx import Document
from docx.shared import Inches
```

```python
# Load the data
data = pd.read_csv("D:/Data/my_data.csv")

# Assign the independent variable (X) and dependent variable (Y)
X = data['ETH'].values.reshape(-1, 1)
Y = data['BTC'].values.reshape(-1, 1)

# Split the data into training and testing sets
X_train, X_test, Y_train, Y_test = train_test_split(X, Y,
test_size=0.2, random_state=0)

# Create and fit the polynomial features
poly_features = PolynomialFeatures(degree=3)
X_train_poly = poly_features.fit_transform(X_train)

# Add a constant term to the independent variables matrix for
the intercept
X_train_poly = sm.add_constant(X_train_poly)

# Create and fit the OLS model using statsmodels
model = sm.OLS(Y_train, X_train_poly).fit()

# Extract the coefficients, t-statistics, adjusted R-squared,
and number of observations
coefficients = model.params[1:]
t_stats = model.tvalues[1:]
n_obs = model.nobs
r2 = model.rsquared_adj

# Create a list of lists containing the regression results
results = []
for i in range(len(coefficients)):
    results.append([i+1, coefficients[i], t_stats[i]])

# Create a table and add the header row
doc = Document()
doc.add_heading('Polynomial Regression Summary', level=1)
doc.add_paragraph('Dependent Variable: BTC\nIndependent
Variable: ETH')
table = doc.add_table(rows=1, cols=3)
hdr_cells = table.rows[0].cells
hdr_cells[0].text = 'Degree'
hdr_cells[1].text = 'Coefficient'
hdr_cells[2].text = 't-statistic'

# Add the regression results to the table
for result in results:
```

```
    row_cells = table.add_row().cells
    row_cells[0].text = str(result[0])
    row_cells[1].text = str(round(result[1], 4))
    row_cells[2].text = str(round(result[2], 4))

# Add the number of observations and adjusted R-squared to the
table
table.add_row()
obs_row = table.add_row().cells
obs_row[0].text = 'Observations'
obs_row[1].text = str(n_obs)
obs_row[2].text = ''
r2_row = table.add_row().cells
r2_row[0].text = 'Adjusted R-squared'
r2_row[1].text = str(round(r2, 4))
r2_row[2].text = ''

# Save the Word document
doc.save('D:/Data/regression_summary.docx')
```

This code creates a table in the Word document with three columns: variable, coefficient, and t-statistic. The code extracts these values from the regression model and adds them to the table, along with the number of observations and the adjusted R-squared. Note that you may need to adjust the file path for saving the document, based on your system configuration.

Choosing the appropriate degree for the polynomial regression model is critical for accurate predictions and avoiding overfitting. Several model selection and validation techniques can help researchers determine the optimal degree of the polynomial.

1. **Visual Inspection of Scatter Plots**

Before applying polynomial regression, researchers can create scatter plots of the independent and dependent variables. If the plot suggests a non-linear relationship, polynomial regression may be more appropriate than linear regression. However, the scatter plot may not provide a clear indication of the optimal degree for the polynomial.

2. **Cross-Validation**

Cross-validation is a popular technique for estimating the model's performance on new, unseen data. In k-fold cross-validation, the data is divided into k subsets, and the model is trained on $k-1$ subsets and tested on the remaining subset. This process is repeated k times, each time using a different subset for testing. The average error across all k iterations is used to evaluate the model's performance. By comparing the cross-validation errors for different polynomial degrees, researchers can choose the degree that yields the lowest error. Cross-validation scores can be calculated by inserting the following code in main code:

```
# Perform k-fold cross-validation
scores = cross_val_score(lin_reg, X_poly, Y, cv=5)
```

The whole code for separately calculating the cross-validation score is as follows:

```
import pandas as pd
import numpy as np
import statsmodels.api as sm
from sklearn.model_selection import train_test_split,
cross_val_score
from sklearn.preprocessing import PolynomialFeatures
from sklearn.linear_model import LinearRegression

# Load the data
data = pd.read_csv("D:/Data/my_data.csv")

# Assign the independent variable (X) and dependent variable (Y)
X = data['ETH'].values.reshape(-1, 1)
Y = data['BTC'].values.reshape(-1, 1)

# Create and fit the polynomial features
poly_features = PolynomialFeatures(degree=3)
X_poly = poly_features.fit_transform(X)

# Split the data into training and testing sets
X_train, X_test, Y_train, Y_test = train_test_split(X_poly, Y,
test_size=0.2, random_state=0)

# Create and fit the linear regression model using scikit-learn
lin_reg = LinearRegression()
lin_reg.fit(X_train, Y_train)

# Perform k-fold cross-validation
scores = cross_val_score(lin_reg, X_poly, Y, cv=5)

# Print the cross-validation scores and their mean
print('Cross-validation scores:', scores)
print('Mean cross-validation score:', np.mean(scores))
```

This code uses **LinearRegression** from scikit-learn to create a linear regression model that implements the **fit** method. The model is created and fitted using
    lin_reg.fit(X_train, Y_train).

To perform cross-validation, the code calls **cross_val_score** with the **lin_ reg** estimator. The output will look like:

```
Cross-validation scores: [-5.64837924e+02 -6.23929554e-01 -
3.45478086e+00 -4.39766254e+00
 -1.66073712e-01]
Mean cross-validation score: -114.69607409916057
```

The **cross_val_score** function returns an array of scores, one for each fold, which represents the negative MSE for that fold. A negative MSE indicates that the model performed worse than the average of the target values, which means that the model did not fit the data well for that fold. A lower negative MSE indicates a better fit.

In the output you provided, the cross-validation scores are mostly negative, with one exception. The mean cross-validation score is also negative, which indicates that the model did not fit the data well overall.

It's worth noting that the interpretation of the MSE depends on the scale of the target variable, and it may not be immediately clear how well the model is performing based on the scores alone. It's often useful to compare the cross-validation scores to the mean and standard deviation of the target variable to get a sense of how well the model is doing relative to the variability in the data.

3. Regularization Techniques

Regularization techniques, such as Lasso or Ridge regression, can be applied to polynomial regression models to prevent overfitting. These techniques add a penalty term to the objective function, which reduces the magnitude of the regression coefficients. This helps prevent the model from becoming overly complex and fitting the noise in the data. By incorporating regularization, researchers can select the optimal degree of the polynomial while avoiding overfitting.

Lasso can be implemented on a polynomial regression as follows:

```
import pandas as pd
import numpy as np
from sklearn.model_selection import train_test_split
from sklearn.preprocessing import PolynomialFeatures
from sklearn.linear_model import Lasso
from sklearn.metrics import mean_squared_error

# Load the data
data = pd.read_csv("D:/Data/my_data.csv")

# Assign the independent variable (X) and dependent variable (Y)
```

In this code, we first load and split the data into training and testing sets. We then create polynomial features using **PolynomialFeatures**, and fit a Lasso regression model using **Lasso**. The **alpha** parameter controls the regularization strength, and the **max_iter** parameter controls the maximum number of iterations. We then use the trained model to predict the test set results and calculate the mean squared error. Note that Lasso regression can be sensitive to the choice of **alpha**. You may need to experiment with different values of **alpha** to find the best one for your data. The output will look like:

```
Mean squared error: 18797519.140195906
Coefficients:
Degree 1: 0.0
Degree 2: 27.0842
Degree 3: -0.0045
Degree 4: 0.0
```

Mean squared error is a measure of the difference between the actual values and the predicted values of the test set. In this case, the mean squared error is quite high at 18,797,519.14. This indicates that the Lasso model is not a very good fit for the data. The coefficients indicate the relative importance of each feature in predicting the dependent variable. In this case, the Lasso model has assigned zero weight to the first and fourth degree polynomial features, indicating that they are not contributing to the model. The second degree polynomial feature (i.e., **ETH^2**) has a positive weight of 27.0842, indicating that it has a positive relationship with the dependent variable. The third degree polynomial feature (i.e., **ETH^3**) has a negative weight of -0.0045, indicating that it has a negative relationship with the dependent variable. It's worth noting that the Lasso model is a type of regularized regression that can help prevent overfitting, by penalizing the coefficients of features that are not important. However, in this case, the mean squared error is quite high, indicating that the Lasso model is not a very good fit for the data. You may need to try other types of regression or adjust the hyperparameters of the Lasso model to improve its performance.

On the same data, ridge regression would be implemented as follows:

```python
import pandas as pd
import numpy as np
import statsmodels.api as sm
from sklearn.model_selection import train_test_split
from sklearn.preprocessing import PolynomialFeatures, StandardScaler
from sklearn.linear_model import Ridge
from docx import Document
from docx.shared import Inches

# Load the data
data = pd.read_csv("D:/Data/my_data.csv")

# Assign the independent variable (X) and dependent variable (Y)
X = data['ETH'].values.reshape(-1, 1)
Y = data['BTC'].values.reshape(-1, 1)

# Split the data into training and testing sets
X_train, X_test, Y_train, Y_test = train_test_split(X, Y, test_size=0.2, random_state=0)

# Create and fit the polynomial features
poly_features = PolynomialFeatures(degree=3)
X_train_poly = poly_features.fit_transform(X_train)

# Normalize the input features
scaler = StandardScaler()
```

```python
X_train_poly_scaled = scaler.fit_transform(X_train_poly)

# Fit the Ridge regression model
ridge_model = Ridge(alpha=1.0)
ridge_model.fit(X_train_poly_scaled, Y_train)

# Extract the coefficients, number of observations, and adjusted
R-squared
coefficients = ridge_model.coef_[0][1:]
n_obs = X_train.shape[0]
r2 = ridge_model.score(X_train_poly_scaled, Y_train)

# Create a list of lists containing the regression results
results = []
for i in range(len(coefficients)):
    results.append([i+1, coefficients[i]])

# Create a table and add the header row
doc = Document()
doc.add_heading('Polynomial Ridge Regression Summary', level=1)
doc.add_paragraph('Dependent Variable: BTC\nIndependent
Variable: ETH')
table = doc.add_table(rows=1, cols=2)
hdr_cells = table.rows[0].cells
hdr_cells[0].text = 'Degree'
hdr_cells[1].text = 'Coefficient'

# Add the regression results to the table
for result in results:
    row_cells = table.add_row().cells
    row_cells[0].text = str(result[0])
    row_cells[1].text = str(round(result[1], 4))

# Add the number of observations and adjusted R-squared to the
table
table.add_row()
obs_row = table.add_row().cells
obs_row[0].text = 'Observations'
obs_row[1].text = str(n_obs)
r2_row = table.add_row().cells
r2_row[0].text = 'Adjusted R-squared'
r2_row[1].text = str(round(r2, 4))

# Save the Word document
doc.save('D:/Data/ridge_regression_summary.docx')
```

The output will be saved in a publication quality table. This code creates a table in the Word document with two columns: degree and coefficient. The code fits a Ridge regression model with an alpha value of 1.0, and extracts the coefficients, number of observations, and adjusted R-squared.

The main advantage of polynomial regression is that it can capture non-linear relationships between the independent and dependent variables, which cannot be modeled using linear regression. This makes polynomial regression a useful tool in fields such as physics, engineering, and finance, where non-linear relationships between variables are common. However, one major disadvantage of polynomial regression is that it can easily overfit the data, especially when the degree of the polynomial is high. Overfitting occurs when the model fits the noise in the data, rather than the underlying relationship between the variables. This can lead to poor performance when the model is applied to new, unseen data. Moreover, interpreting the coefficients of a polynomial regression model can be challenging, especially when the degree of the polynomial is high. This can make it difficult to understand the effect of the independent variables on the dependent variable.

Polynomial regression is a powerful tool for academic researchers in accounting and finance. It allows for more flexible modeling of non-linear relationships between variables, which can be especially important when studying complex financial phenomena. By incorporating polynomial regression into their research methodologies, scholars can better understand the underlying dynamics of financial markets and improve the accuracy of their predictions. Ultimately, this can lead to more informed decision-making and more effective policy recommendations in the accounting and finance sectors.

22

Quantile Regression

Quantile regression is an extension of linear regression that allows researchers to estimate the conditional quantiles of a response variable given a set of predictor variables. Unlike linear regression, which focuses on estimating the conditional mean of the response variable, quantile regression provides a more comprehensive understanding of the relationship between the response and predictor variables across different quantiles. It enables researchers to analyze the impact of predictor variables on various parts of the response variable's distribution. This is especially important when the relationship between the predictor and response variables is heterogeneous or when the distribution of the response variable is asymmetric or skewed.

The quantile regression equation is similar to that of linear regression, but with a crucial difference. In linear regression, we estimate the conditional mean of the response variable (Y) given the predictor variables (X). In quantile regression, we estimate the conditional quantile (τ) of the response variable, which ranges between 0 and 1.

The quantile regression equation can be written as:

$$Q\_Y(\tau|X) = X\beta(\tau),$$

where

- $Q\_Y(\tau|X)$ represents the τ-th quantile of the response variable Y conditional on the predictor variables X;
- $\beta(\tau)$ is a vector of coefficients to be estimated, which varies across different quantiles τ;

- X is the matrix of predictor variables;

Quantile regression is based on the concept of minimizing the sum of asymmetrically weighted absolute residuals. The asymmetric weights are determined by the choice of quantile τ. The objective function for quantile regression can be expressed as:

$$\min\_\beta(\tau) \sum \left(\rho\_\tau (y\_i - x\_i' \beta(\tau)) \right),$$

where

- $\rho\_\tau(u) = u(\tau - I(u < 0))$ is the check function;
- y_i is the response variable for the i-th observation;
- x_i' is the transposed vector of predictor variables for the i-th observation;
- $I(u < 0)$ is an indicator function that equals 1 if $u < 0$ and 0 otherwise.

The minimization of this objective function results in the estimation of the $\beta(\tau)$ coefficients. This process is performed for each desired quantile τ.

An implementation of quantile regression using Python using the data in D:/Data/my_data2.csv is as follows:

```python
import pandas as pd
import numpy as np
import statsmodels.api as sm
import statsmodels.formula.api as smf
data = pd.read_csv('D:/Data/my_data2.csv')
X = data[['sex', 'dex', 'lex', 'kwit', 'job_tenure', 'censored']]
X = sm.add_constant(X)
y = data['y']
quantiles = [0.25, 0.5, 0.75]
for quantile in quantiles:
    model = sm.QuantReg(y, X).fit(q=quantile)
    print(f"Results for {quantile} Quantile Regression:\n")
    print(model.summary())
    print("\n\n")
```

This code snippet demonstrates how to perform quantile regression using the **statsmodels** library in Python. It first imports the necessary libraries, including pandas, numpy, and statsmodels. The data is then read from a CSV file into a pandas DataFrame. The predictor variables 'sex', 'dex', 'lex', 'kwit', 'job_tenure', and 'censored' are selected, and a constant term is added to the matrix of predictor variables. The dependent variable 'y' is also selected from the data. The code defines a list of quantiles (0.25, 0.5, and 0.75) for which the quantile regression models will be estimated. A for loop is used to iterate through each quantile, fit the quantile regression model using the **QuantReg** class from the **statsmodels** library, and print the results, including the estimated coefficients for each quantile. You can modify the **quantiles** list to include any desired quantiles between 0 and 1. The code will give an output for all the quantiles. In this case output for 0.25 quantile is as follows:

```
Results for 0.25 Quantile Regression:
                     QuantReg Regression Results
==============================================================================
Dep. Variable:                      y   Pseudo R-squared:               0.2050
Model:                       QuantReg   Bandwidth:                      0.5631
Method:                 Least Squares   Sparsity:                        3.404
Date:                Sat, 06 May 2023   No. Observations:                  683
Time:                        15:45:49   Df Residuals:                      676
                                        Df Model:                            6
==============================================================================
                 coef    std err          t      P>|t|      [0.025      0.975]
------------------------------------------------------------------------------
const         10.3025      0.819     12.573      0.000       8.694      11.912
sex           -0.8047      0.118     -6.833      0.000      -1.036      -0.573
dex            0.1098      0.009     11.895      0.000       0.092       0.128
lex           -0.0693      0.058     -1.184      0.237      -0.184       0.046
kwit          -0.0166      0.145     -0.114      0.909      -0.302       0.269
job_tenure    -0.0003      0.000     -1.782      0.075      -0.001    3.19e-05
censored       0.1063      0.154      0.689      0.491      -0.197       0.410
==============================================================================

The condition number is large, 7.31e+03. This might indicate that
there are strong multicollinearity or other numerical problems.
```

The condition number is a measure of the sensitivity of a function's output to its input, which is often used to diagnose multicollinearity and numerical instability in regression models. A large condition number indicates that small changes in the input (i.e., predictor variables) could lead to significant changes in the output (i.e., coefficients), suggesting that the model might be unstable or have multicollinearity issues. Generally, condition numbers above 30 are considered large, and the condition number in this case is 7.31e + 03, which is quite high.

Determining the quality of a quantile regression model involves assessing its goodness-of-fit, stability, and validity of the underlying assumptions. Here are some ways to evaluate the quality of a quantile regression model:

1. **Pseudo R-Squared**

Pseudo R-squared is a measure of goodness-of-fit for quantile regression models, analogous to R-squared in linear regression. It helps assess the proportion of variation in the dependent variable that the model explains. Higher values of pseudo R-squared indicate a better fit. However, keep in mind that pseudo R-squared values are not directly comparable to R-squared values from linear regression, and a high pseudo R-squared is not always indicative of a good model.

2. Residual Analysis

Analyze the residuals (differences between the observed values and the predicted values) to assess the model's performance. Ideally, the residuals should be randomly distributed and not show any patterns, trends, or heteroscedasticity. You can create residual plots, such as scatter plots of residuals versus predicted values or histograms of residuals, to visually assess the distribution and behavior of the residuals. A residual plot for this 0.25 quantile regression can be created as follows:

```python
import matplotlib.pyplot as plt
# Fit the 0.25 quantile regression model
quantile_025 = 0.25
model_025 = sm.QuantReg(y, X).fit(q=quantile_025)
# Calculate predicted values
y_pred_025 = model_025.predict(X)

# Calculate residuals
residuals_025 = y - y_pred_025

# Create teh residual plot
plt.scatter(y_pred_025, residuals_025, alpha=0.5)
plt.axhline(y=0, color='r', linestyle='--')
plt.xlabel('Predicted Values')
plt.ylabel('Residuals')
plt.title('Residual Plot for 0.25 Quantile Regression')
plt.show()
```

The output will show the scatterplot of residuals for different predicted values of Y:

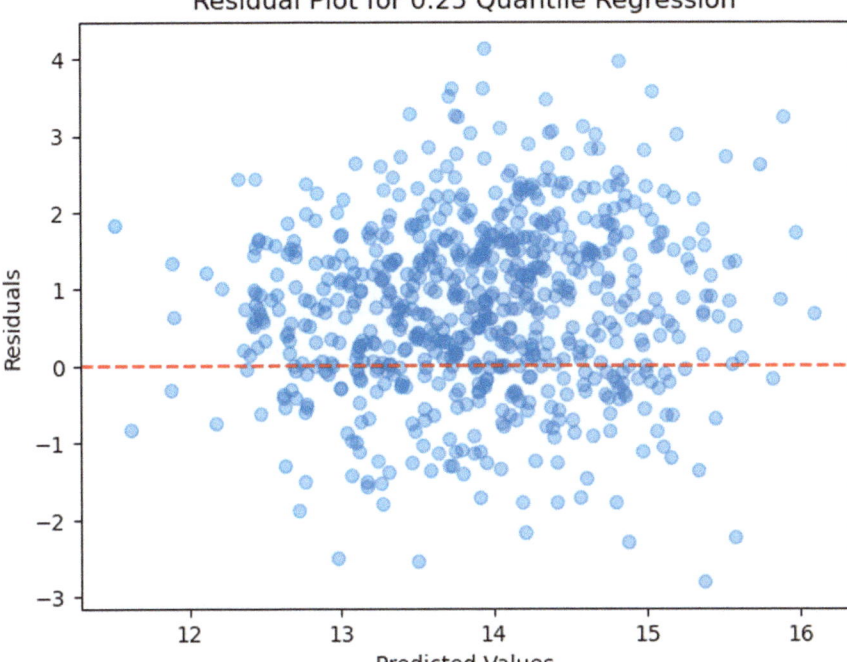

A good residual plot for a regression model, whether it's linear or quantile regression, should display the following characteristics:

- Randomly Distributed Points: The residuals should be scattered randomly around the horizontal axis ($y = 0$) without any clear patterns or trends. This indicates that the model has captured the underlying relationship between the independent and dependent variables and that the error term is random.
- No Heteroscedasticity: The spread of the residuals should be approximately constant across the range of predicted values. If the spread increases or decreases systematically with the predicted values, it suggests that the model suffers from heteroscedasticity, which may lead to unreliable coefficient estimates and incorrect inferences.
- No Autocorrelation: In time series data, the residuals should not exhibit any autocorrelation. The presence of autocorrelation indicates that the model has not captured the full structure of the time series data, and the coefficient estimates may be biased.
- Normally Distributed Residuals: For linear regression models, the residuals should be approximately normally distributed. This can be assessed visually

using a histogram or a Q-Q plot. For quantile regression, this assumption is not required, as quantile regression is robust to non-normal error distributions.

3. Coefficient Significance

Examine the p-values and confidence intervals of the estimated coefficients to assess their statistical significance. Smaller *p*-values (typically below 0.05) and confidence intervals not containing zero indicate that the corresponding predictor variables have a statistically significant effect on the dependent variable at the specified quantile.

4. Stability and Multicollinearity

Check the stability of the model by examining the condition number (as already discussed above), as high condition numbers may indicate multicollinearity or numerical instability. You can also assess multicollinearity by analyzing the correlation matrix or the variance inflation factor (VIF) of the predictor variables. Addressing multicollinearity issues may improve the quality of the model. VIF can be calculated as:

```
# Import necessary libraries
import pandas as pd
from statsmodels.stats.outliers_influence import variance_inflation_factor

#Load Data
data = pd.read_csv('D:/Data/my_data2.csv')

#Prepare data for VIF Calculation
X = data[['sex', 'dex', 'lex', 'kwit', 'job_tenure', 'censored']]
X = X.dropna()   # Drop rows with missing values (optional, if needed)

#Calculate VIF for each predictor variable
vif_data = pd.DataFrame()
vif_data["feature"] = X.columns
vif_data["VIF"] = [variance_inflation_factor(X.values, i) for i in range(X.shape[1])]

#Print VIF Values
print(vif_data)
```

This will print the VIF for all the independent variables, which in this would be:

```
     feature        VIF
0        sex   2.429564
1        dex  38.462426
2        lex  42.388046
3       kwit   1.689730
4  job_tenure   2.957368
5   censored   5.402124
```

Generally, a VIF value greater than 10 indicates potential multicollinearity issues, while some researchers suggest a stricter threshold of 5.

5. Out-of-Sample Predictive Performance

Assess the model's performance on new, unseen data by using techniques like cross-validation or a hold-out validation set. Good out-of-sample performance indicates that the model generalizes well to new data, which is important in practical applications.

6. Comparison with Other Models

Compare the quantile regression model with other models, such as linear regression or alternative quantile regression specifications, to see if it provides a better fit or more meaningful insights into the data. This can help you choose the most appropriate model for your research question.

Quantile regression offers several advantages over traditional linear regression, making it particularly well-suited for accounting and finance research. One of the key benefits is its robustness to outliers. Unlike linear regression, which minimizes squared residuals, quantile regression minimizes absolute residuals. This makes it less sensitive to outliers and more robust when dealing with skewed or heavy-tailed distributions, which are common in financial data. Another advantage of quantile regression is its ability to estimate heterogeneous relationships between predictor and response variables across different quantiles. This is especially useful in cases where the impact of predictor variables varies across different levels of the response variable. By considering the relationships at various quantiles, researchers can gain a more detailed understanding of the underlying processes and interactions, which can lead to more accurate and insightful models. Quantile regression enables researchers to perform a more comprehensive analysis of the distribution of the response variable by estimating multiple quantiles. This provides important insights into the behavior of financial variables, such as tail risk and extreme events. By examining the entire distribution, rather than just the mean, researchers can develop a deeper understanding of the data and better identify patterns and trends that may not be apparent with traditional linear regression methods.

Despite its advantages, quantile regression also has some limitations and drawbacks. One of the main disadvantages is its computational complexity. Quantile regression often requires more computational resources and time to estimate the model parameters compared to linear regression, particularly when dealing with large datasets or multiple quantiles. This can be a limiting factor for researchers working with extensive data or constrained computational resources. Another limitation of quantile regression is the difficulty in interpreting the results, especially for non-experts or those unfamiliar with the technique. Unlike linear regression, which provides a single coefficient for each predictor variable that can be easily interpreted, quantile regression produces different coefficients for each quantile. This can make the interpretation of the results more complex and may require additional effort to convey the findings to a broader audience. Quantile regression does not provide a direct measure of goodness-of-fit like the R-squared value in linear regression. Instead, it relies on pseudo R-squared, which can be harder to interpret and compare with the goodness-of-fit values from other models. This makes model selection and evaluation more challenging for researchers using quantile regression. The presence of multicollinearity can still be an issue in quantile regression, similar to linear regression. While quantile regression is robust to outliers and can handle heteroskedasticity, it does not inherently address the problems caused by multicollinearity. Researchers should still be cautious when dealing with highly correlated predictor variables and consider appropriate techniques to mitigate the potential issues.

Quantile regression has emerged as an essential tool for academic accounting and finance research due to its ability to model complex relationships, robustness to outliers, and comprehensive distributional analysis capabilities. While it has some limitations, such as computational complexity and challenges in interpretation, the advantages it offers make it a valuable alternative to traditional linear regression methods. By employing quantile regression, researchers can gain deeper insights into the underlying processes, better understand the behavior of financial variables across different quantiles, and ultimately contribute to the development of more accurate and informative models. Embracing quantile regression in academic accounting and finance research can help researchers uncover important findings that may have been overlooked by conventional methods, thereby advancing the understanding of financial markets and their dynamics.

23

Advanced Regressions

In the field of accounting and finance research, the complex and diverse nature of data often requires the utilization of a diverse range of analytical tools. Standard linear regression models, such as ordinary least squares regression and logistic regression, may not always be suitable due to their assumptions not aligning with the characteristics of the data or research question at hand. This necessitates researchers to explore alternative statistical methods that respect the specific properties of the data and cater to the unique needs of their research. In this chapter, we will discuss some specific regression technique that are suitable for some specific situations.

Tobit Regression

Tobit regression is a statistical method specifically designed to estimate relationships between variables when there is either left- or right-censoring in the dependent variable. It is a type of generalized linear model that combines elements of both linear regression and probit models, making it particularly well-suited to deal with scenarios in which the range of outcomes is mechanically constrained. In accounting and finance research, Tobit regression is particularly beneficial in situations where the dependent variable is censored. Censoring occurs when the value of the dependent variable is only partially known; for instance, when the value of an observation falls below or above a certain threshold, it is not exactly observable. A classic example in finance is the situation where a firm's liabilities exceed its assets, but the exact amount of negative equity is not observed or reported, making it a case of left-censoring at zero.

The Tobit model can be formally written as follows:

$$y_i* = \beta' x_i + \varepsilon_i,$$

$$y_i = \max(0, y_i*),$$

where y_i* is the latent (or unobservable) variable, x_i is a vector of independent variables, β is a vector of parameters to be estimated, and ε_i is an error term. y_i is the observed variable. When the latent variable y_i* is less than zero, we observe y_i as zero (in the case of left-censoring).

Tobit regression is particularly valuable in accounting and finance research because of the prevalence of censored and truncated data. For example, the analysis of corporate financial distress (where negative equity is censored) and the evaluation of investment decisions (where certain investment returns are either not realized or not observed) can be suitably handled by Tobit regression models. The Tobit model's power lies in its ability to simultaneously model the probability of observing a non-censored outcome and the expectation of the outcome, given it is not censored.

The key advantage of Tobit regression lies in its ability to provide unbiased and consistent estimators when dealing with censored or limited dependent variables. Unlike ordinary least squares (OLS) regression, which can yield biased and inefficient estimates under censoring, Tobit models can account for the specific distributional features of the dependent variable, leading to more accurate inferences. On the other hand, Tobit regression models require a strong assumption that the censoring mechanism and the process determining the actual values of the dependent variable (when it is not censored) are driven by the same process. If this assumption is violated, Tobit models can lead to biased and inconsistent parameter estimates. Moreover, Tobit models are more complex and computationally intensive than standard linear models, which could be a limitation in some settings.

An example of Tobit model implementation on D:/Data/my_data4.csv where dependent variable is BTC_returns and independent variables are ETH _returns, BNB _returns, and ADA _returns is provided below. The model censors negative observations for dependent variable. This is just an example, and the Tobit model may actually not be suitable in this case.

```python
import pandas as pd
import numpy as np
from scipy import stats, linalg
from scipy.optimize import minimize
from scipy.stats import norm

class TobitModel:
    def __init__(self, y, X, sigma=1, left=0):
        self.y = y
        self.X = X
        self.n, self.k = X.shape
        self.sigma = sigma
        self.left = left

    def nll(self, params):
        X, y, left = self.X, self.y, self.left
        beta = params[:-1]
        sigma = params[-1]
        mu = np.matmul(X, beta)
        # likelihood for uncensored observations
        nll_obs = stats.norm.logpdf(y, loc=mu, scale=sigma)
        # likelihood for censored observations
        nll_cens = stats.norm.logcdf(y, loc=mu, scale=sigma)
        return -np.sum(np.where(y > left, nll_obs, nll_cens))

    def fit(self):
        x0 = np.concatenate([np.ones(self.k), [self.sigma]])  # Use a different initial guess
        res = minimize(self.nll, x0, method='BFGS')  # Use BFGS optimizer which also computes the Hessian
```

```
        self.params_ = res.x
        self.sigma_ = res.x[-1]
        self.beta_ = res.x[:-1]

        # Compute standard errors using the inverse of the Hessian
        self.cov_ = res.hess_inv
        self.stderr_ = np.sqrt(np.diag(self.cov_))
        self.tvalues_ = self.params_ / self.stderr_
        self.pvalues_ = 2 * (1 - norm.cdf(np.abs(self.tvalues_)))   # two-sided pvalue = Prob(abs(t)>tt)
        return self

    def summary(self):
        results = pd.DataFrame({
            'Variable': ['intercept', 'ETH_returns', 'BNB_returns', 'ADA_returns', 'sigma'],
            'Estimate': self.params_,
            'Standard Error': self.stderr_,
            't-statistic': self.tvalues_,
            'p-value': self.pvalues_
        })
        return results


# Load the dataset
df = pd.read_csv('D:/Data/my_data4.csv')

# Make sure the Date column is in datetime format
df['Date'] = pd.to_datetime(df['Date'])

# Sort values by Date
df = df.sort_values(by='Date')

# Define the dependent variable (y) and the independent variables (X)
X = df[['ETH_returns', 'BNB_returns', 'ADA_returns']].values
y = df['BTC_returns'].values

# Add constant to the independent variables matrix
X = np.concatenate([np.ones((X.shape[0], 1)), X], axis=1)

# Fit the Tobit model
model = TobitModel(y, X).fit()

# Print the summary
print(model.summary())
```

The output will look like:

```
    Variable      Estimate  Standard Error    t-statistic  p-value
0  intercept  -2177.241218        0.785767   -2770.849870      0.0
1  ETH_returns    34.552806        1.000004      34.552665      0.0
2  BNB_returns    32.702362        1.000004      32.702217      0.0
3  ADA_returns    33.906818        1.000002      33.906751      0.0
4        sigma -1897.621215        0.685535   -2768.087278      0.0
```

As the Tobit model is not directly supported by any major Python statistical package (though some custom solutions by user community do exist), a custom solution needs to be created to compute additional statistics in the code above. The following code will directly export the regression table in a format typically used in accounting and finance papers to a word file.

```python
from docx import Document
from docx.shared import Inches

class TobitModel:
    def __init__(self, Y, X, sigma=1, left=0):
        self.Y = Y
        self.X = X
        self.n, self.k = X.shape
        self.sigma = sigma
        self.left = left

    def nll(self, params):
        X, Y, left = self.X, self.Y, self.left
        beta = params[:-1]
        sigma = params[-1]
        mu = np.matmul(X, beta)
        # likelihood for uncensored observations
        nll_obs = stats.norm.logpdf(Y, loc=mu, scale=sigma)
        # likelihood for censored observations
        nll_cens = stats.norm.logcdf(Y, loc=mu, scale=sigma)
        return -np.sum(np.where(Y > left, nll_obs, nll_cens))

    def fit(self):
        x0 = np.concatenate([np.ones(self.k), [self.sigma]])  # Use a different initial guess
        res = minimize(self.nll, x0, method='BFGS')  # Use BFGS optimizer which also computes the Hessian
        self.params_ = res.x
        self.sigma_ = res.x[-1]
        self.beta_ = res.x[:-1]
```

```python
        # Compute standard errors using the inverse of the Hessian
        self.cov_ = res.hess_inv
        self.stderr_ = np.sqrt(np.diag(self.cov_))
        self.tvalues_ = self.params_ / self.stderr_
        self.pvalues_ = 2 * (1 - norm.cdf(np.abs(self.tvalues_)))  # two-sided pvalue = Prob(abs(t)>tt)
        return self

    def summary(self):
        results = pd.DataFrame({
            'Variable': ['intercept', 'ETH_returns', 'BNB_returns', 'ADA_returns', 'sigma'],
            'Estimate': self.params_,
            'Standard Error': self.stderr_,
            't-statistic': self.tvalues_,
            'p-value': self.pvalues_
        })
        return results

    def export_to_word(self, doc_path):
        # Create a new Document
        doc = Document()

        # Create a table in the document
        table = doc.add_table(rows=1, cols=5)

        # Add the header row
        for i, column_name in enumerate(['Variable', 'Estimate', 'Standard Error', 't-statistic', 'p-value']):
            table.cell(0, i).text = column_name

        # Add the data rows
        for i in range(self.k + 1):
            cells = table.add_row().cells
            cells[0].text = ['intercept', 'ETH_returns', 'BNB_returns', 'ADA_returns', 'sigma'][i]
            cells[1].text = str(self.params_[i])
            cells[2].text = str(self.stderr_[i])
            cells[3].text = str(self.tvalues_[i])
            cells[4].text = str(self.pvalues_[i])

        # Save the document
        doc.save(doc_path)
```

```python
# Load the dataset
df = pd.read_csv('D:/Data/my_data4.csv')

# Make sure the Date column is in datetime format
df['Date'] = pd.to_datetime(df['Date'])

# Sort values by Date
df = df.sort_values(by='Date')

# Define the dependent variable (Y) and the independent
variables (X)
X = df[['ETH_returns', 'BNB_returns', 'ADA_returns']].values
Y = df['BTC_returns'].values

# Add constant to the independent variables matrix
X = np.concatenate([np.ones((X.shape[0], 1)), X], axis=1)

# Fit the Tobit model
model = TobitModel(Y, X).fit()

# Export the summary to a Word document
model.export_to_word('D:/Data/summary.docx')
```

The script begins by importing necessary modules and defining the Tobit model class. The class constructor initializes the dependent variable (Y), the independent variables (X), and some optional parameters, including the standard deviation of the error term (sigma) and a left-censoring limit. The class includes methods for fitting the model and summarizing or exporting the results. The method **nll** calculates the negative log-likelihood of the model given a set of parameters, differentiating between censored and uncensored observations. The **fit** method uses the Broyden-Fletcher-Goldfarb-Shanno (BFGS) algorithm to find the set of parameters that minimizes the negative log-likelihood, computes standard errors using the inverse of the Hessian matrix (an approximation of the second derivative of the likelihood function), and calculates the t-statistics and two-sided p-values for the parameters. The **summary** method returns a DataFrame containing the estimated parameters, their standard errors, *t*-statistics, and *p*-values. The **export_to_word** method generates a Word document with a table that includes the same information as the summary.

The script then loads a dataset, converts the Date column to datetime format, and sorts the observations by date. It defines the dependent variable as 'BTC_returns' and the independent variables as 'ETH_returns', 'BNB_returns', and 'ADA_returns'. It adds a constant to the matrix of independent variables to account for the intercept term in the regression model. After preparing the data, the script fits the Tobit model and exports the summary of the results to a Word document. This implementation provides a comprehensive analysis of the relationship between Bitcoin returns and the returns of three other cryptocurrencies under the assumption that Bitcoin returns are left-censored at zero.

This was a censoring model. If we need to implement a trimming model, then **TobitModel** class in the above code needs to be modified as:

```python
from scipy.stats import norm
import numpy as np
import pandas as pd
from scipy.optimize import minimize

class TobitModel:
    def __init__(self, y, X, trim_percent=0.1):
        self.y = y
        self.X = X
        self.n, self.k = X.shape
        self.trim_percent = trim_percent

    def nll(self, params):
        X, y, trim_percent = self.X, self.y, self.trim_percent
        beta = params[:-1]
        sigma = params[-1]
        mu = np.matmul(X, beta)
        resid = y - mu

        # Compute the trimmed indices
        trim_count = int(trim_percent * self.n)
        sorted_resid = np.sort(resid)
        trim_indices = np.concatenate([np.arange(trim_count),
np.arange(-trim_count, 0)])
        trimmed_resid = sorted_resid[trim_indices]

        # Compute the negative log-likelihood using the trimmed residuals
        nll_obs = norm.logpdf(trimmed_resid, scale=sigma)
        return -np.sum(nll_obs)

    def fit(self):
        x0 = np.concatenate([np.ones(self.k), [1]])  # Use a different initial guess
        res = minimize(self.nll, x0, method='BFGS')
        self.params_ = res.x
        self.sigma_ = res.x[-1]
        self.beta_ = res.x[:-1]
        return self

    def summary(self):


        results = pd.DataFrame({
            'Variable': ['intercept'] + ['X{}'.format(i) for i in range(1, self.k + 1)],
            'Estimate': self.params_[:-1],
            'Standard Error': np.nan,
            't-statistic': np.nan,
            'p-value': np.nan
        })
        return results
```

Poisson Regression

Poisson regression is a generalized linear model form of regression analysis used to model count data and contingency tables. The Poisson regression model allows us to examine the relationship between a set of predictor variables and a count-dependent variable. In a Poisson regression model, the response variable Y is assumed to follow a Poisson distribution, and the logarithm of its expected value can be modeled by a linear combination of unknown parameters. The canonical form of the Poisson regression model is expressed as:

$$\log(E[Y|X]) = \beta_0 + \beta_1 X_1 + \beta_2 X_2 + \ldots + \beta_p * X_p,$$

where

- $E[Y|X]$ denotes the expected (mean) response of Y given the predictor variables X;
- $\beta_0, \beta_1, \ldots, \beta_p$ are the parameters of the model;
- $X_1, X_2, \ldots, X_p$ are the predictor variables.

Poisson regression is a versatile tool in accounting and finance academic research, specifically when dealing with count-based data. The world of finance and accounting is replete with situations where the data points are discrete counts: the number of times a firm was audited, the number of bankruptcies in a certain industry over a period, the count of financial transactions in a day, etc. For instance, Poisson regression could be used to understand the drivers of corporate bankruptcies, by treating the number of bankruptcies as the dependent variable and potential predictors as firm size, leverage, profitability, and industry competition. Furthermore, Poisson regression models have been used to study the number of times a company releases financial reports, the number of defaults on a loan, or the count of insurance claims, among other applications.

Poisson regression is a valuable tool for analyzing count-based data due to its applicability, flexibility, and interpretability. It is specifically designed for datasets where the dependent variable represents counts or rates, making it well-suited for such scenarios. The model allows for the inclusion of numerous predictors, accommodating non-linear relationships and enabling a comprehensive understanding of the factors influencing the count-based outcome. Coefficients in Poisson regression are interpretable, providing insights into the relationship between predictor variables and the response. However, Poisson regression has certain limitations to consider. Violation of the assumption of equal mean and variance can lead to overdispersion, producing biased estimates. Moreover, zero-inflation, common in accounting and finance applications, can pose challenges for the model. Additionally, the assumption of independence among observed counts must be met for accurate inferences.

An implementation of Poisson regression on D:/Data/my_data5.csv where dependent variable is BTC_Poisson and independent variables are ETH_returns, BNB_returns, and ADA_returns is as follows:

```
# Import the necessary libraries
import pandas as pd
import statsmodels.api as sm

# Load the data
data = pd.read_csv('D:/Data/my_data5.csv')

# Define the dependent variable
y = data['BTC_Poisson']

# Define the independent variables
X = data[['ETH_returns', 'BNB_returns', 'ADA_returns']]

# Add a constant to the independent variables matrix
X = sm.add_constant(X)

# Fit the Poisson regression model
poisson_model = sm.GLM(y, X, family=sm.families.Poisson()).fit()

# Print the model summary
print(poisson_model.summary())
```

Negative Binomial Regression

Negative Binomial Regression, an integral tool within the statistical toolbox of the accounting and finance researcher, is a type of generalized linear model (GLM) that is particularly useful when dealing with count data. This specialized regression technique is invoked when the variance of the count data significantly exceeds its mean, a condition known as overdispersion, which is not uncommon in the realm of financial and accounting data. The basic model of Negative Binomial Regression is expressed as follows:

$$y = \log(\mu) = \beta_0 + \beta_1 * x_1 + \beta_2 * x_2 + \ldots + \beta_n * x_n.$$

Here, 'y' is the log of the expected count 'μ', and 'x1, x2, ..., x_n' are the predictor variables. '$\beta_0, \beta_1, \ldots, \beta_n$' are the parameters to be estimated, reflecting the impact of respective predictors on the expected log count.

One of the reasons Negative Binomial Regression holds importance in academic accounting and finance is its flexibility in modeling count data that exhibits overdispersion. Compared to the Poisson regression, which assumes equality of the mean and variance, Negative Binomial Regression incorporates an additional parameter to model the overdispersion, thus providing more accurate and reliable estimates. For instance, in accounting research, if one aims to study the number of accounting restatements by firms, which can be conceptualized as count data, the Negative Binomial Regression can be an appropriate choice. Similarly, in finance, it may be employed to model the number of times a company issues bonds within a certain period.

The primary advantages of the Negative Binomial Regression include its ability to handle overdispersed count data and to estimate the extra-Poisson variation. By modeling this overdispersion, researchers can produce more efficient and unbiased estimates than would be possible with Poisson regression. In addition, Negative Binomial Regression does not require the dependent variable to be normally distributed, which is often a stringent assumption in many types of research scenarios in accounting and finance. It also accommodates offset terms, allowing for exposure or at-risk periods, which can be highly pertinent in financial risk assessment studies. Nonetheless, Negative Binomial Regression requires the dependent variable to be a non-negative integer, which can limit its applicability. It also assumes that observations are independent, an assumption that might not always hold in time series or panel data scenarios common in finance and accounting research. Furthermore, the interpretation of coefficients in Negative Binomial Regression, being on a logarithmic scale, can be somewhat less intuitive, often requiring transformation into incidence rate ratios for clearer interpretability.

In Python, the **statsmodels** library is used for implementing Negative Binomial Regression. Here is a code that implements Negative Binomial Regression on D:/Data/my_data5.csv where dependent variable is BTC_Neg_Bin and independent variables are ETH_returns, BNB_returns, ADA_returns.

```
import pandas as pd
import statsmodels.api as sm

# Load the data
data = pd.read_csv('D:/Data/my_data5.csv')

# Define dependent variable
Y = data['BTC_Neg_Bin']

# Define independent variables
X = data[['ETH_returns', 'BNB_returns', 'ADA_returns']]

# Add a constant to the independent variables matrix
X = sm.add_constant(X)

# Fit Negative Binomial Regression Model
nb_model = sm.GLM(Y, X,
family=sm.families.NegativeBinomial()).fit()

# Print model summary
print(nb_model.summary())
```

Instrumental Variables (IV) Regression

Instrumental variable (IV) regression is a statistical technique employed to address endogeneity issues in econometric models. Endogeneity occurs when explanatory variables in a regression model correlate with the error term, resulting in biased and inconsistent parameter estimates. IV regression resolves this problem by introducing instrumental variables, which are exogenous variables correlated with the endogenous variables but not directly associated with the error term.

The IV regression model can be represented as follows:

$$Y = \beta_0 + \beta_1 X + \varepsilon.$$

Here, Y represents the dependent variable, X denotes the endogenous variable, β_0 and β_1 are the parameters of interest, and ε represents the error term. However, in IV regression, an instrumental variable Z is introduced, which correlates with X:

$$X = \gamma_0 + \gamma_1 Z + u.$$

The parameters γ_0 and γ_1 reflect the relationship between Z and X, while u denotes the error term. By substituting the original equation's X with its instrumental variable representation, the IV regression equation is obtained:

$$Y = \beta_0 + \beta_1(\gamma_0 + \gamma_1 Z + u) + \varepsilon.$$

IV regression finds widespread usage in accounting and finance research, particularly for addressing endogeneity concerns. It enables researchers to establish causal relationships between variables by mitigating biases arising from endogenous regressors. This technique is particularly valuable when direct manipulation of the endogenous variables is infeasible or ethically challenging. IV regression provides a framework for identifying causal effects by leveraging the correlation between instrumental variables and endogenous variables.

The advantages of IV regression encompass its capacity to provide consistent estimates of causal effects in the presence of endogeneity. By employing instrumental variables, researchers can effectively address issues such as omitted variable bias and measurement error, thereby yielding more reliable and interpretable results. IV regression empowers researchers to make causal claims regarding variable relationships, thereby enhancing the rigor of empirical studies. However, IV regression also exhibits certain limitations. Identifying valid instrumental variables satisfying the assumptions of relevance and exogeneity can be challenging. Researchers must exercise careful selection and justification of their chosen instruments to ensure the validity of IV estimates. Also, compared to standard regression models, IV regression requires a larger sample size due to the estimation of additional parameters. Moreover, if the instruments are weak or there are numerous weak instruments, the IV estimates may lack precision, resulting in large standard errors and reduced statistical power.

```
import pandas as pd
import statsmodels.api as sm

# Load the data
data = pd.read_csv("D:/Data/my_dataX.csv")

# Assign variables
endogenous_variable = data[''Dep_Var'']
exogenous_variables = data[['Indep_Var1', ' Indep_Var2', '
Indep_Var3', 'Indep_Var4', 'Indep_Var5']]
instrumental_variable = data['Instrumental_variable']

# Add a constant term to the exogenous variables
exogenous_variables = sm.add_constant(exogenous_variables)

# Fit the IV regression model
model = sm.OLS(endogenous_variable, exogenous_variables)
iv_model = model.fit(cov_type='robust',
iv_instrument=instrumental_variable)

# Print the summary of the IV regression results
print(iv_model.summary())
```

Two-Stage Least Squares (2SLS) Regression

The 2SLS regression model is designed to tackle endogeneity, which arises when explanatory variables are correlated with the error term. This correlation undermines the assumption of exogeneity, leading to biased parameter estimates and invalid inference. By utilizing instrumental variables, the 2SLS technique allows researchers to overcome endogeneity concerns.

The basic idea behind 2SLS regression is to estimate the relationship between an endogenous explanatory variable and the dependent variable by replacing it with a proxy variable. This proxy variable is constructed by using instrumental variables (IVs) that are assumed to be correlated with the endogenous variable but not directly correlated with the error term. By using these IVs, the endogeneity problem can be mitigated, allowing for consistent estimation of the parameters of interest.

The 2SLS regression model can be expressed as follows:
Stage 1: $Y = X\Pi + Z\Gamma + UX = W\Omega + V$.

Stage 2: $Y = X\beta + \varepsilon$.

In Stage 1, the endogenous variable X is regressed on a set of instrumental variables Z. The fitted values from this stage, denoted as $\bar{Y}$, are obtained. In Stage 2, the outcome variable Y is regressed on $\bar{Y}$ and other exogenous variables X, yielding unbiased and consistent estimates of the coefficients β.

The application of 2SLS regression in academic research within the accounting and finance domain is wide-ranging. It is commonly employed in studies that investigate the impact of certain accounting practices, financial policies, or corporate governance mechanisms on firm performance, investment decisions, or market outcomes. Moreover, 2SLS regression is frequently utilized in examining the effects of financial regulations, mergers and acquisitions, or capital structure choices on firm valuation and financial performance.

The importance of 2SLS regression lies in its ability to provide researchers with a rigorous method to address endogeneity problems that may arise in empirical studies. By incorporating instrumental variables and employing a two-stage estimation procedure, 2SLS allows researchers to obtain consistent and unbiased estimates of causal relationships, thereby enhancing the validity of their findings. This is particularly crucial in empirical research where establishing causal relationships is of paramount importance. However, 2SLS relies on the availability of valid instrumental variables. Finding suitable instruments that meet the necessary criteria, such as relevance and exogeneity, can be challenging and may introduce additional sources of bias if improperly chosen. Moreover, 2SLS regression assumes that there are no measurement errors in the instrumental variables, which may not always hold in practice.

A python implementation of 2SLS would be as follows:

```
import pandas as pd
import statsmodels.api as sm

# Load the data
data = pd.read_csv("D:/Data/my_dataX.csv")

# Assign variables
endogenous_variable = data['Dep_Var']
exogenous_variables = data[['Indep_Var1', 'Indep_Var2',
'Indep_Var3', 'Indep_Var4', 'Indep_Var5']]
instrumental_variable = data['Instrumental_variable']

# Stage 1: Estimate the predicted values of the endogenous
variable
exogenous_variables_stage1 =
sm.add_constant(exogenous_variables)  # Add a constant term
stage1_model = sm.OLS(endogenous_variable,
exogenous_variables_stage1)
stage1_results = stage1_model.fit()

# Obtain the predicted values of the endogenous variable
predicted_endogenous = stage1_results.predict()

# Stage 2: Estimate the 2SLS model
exogenous_variables_stage2 =
sm.add_constant(predicted_endogenous)  # Use predicted
endogenous as an additional exogenous variable
stage2_model = sm.OLS(endogenous_variable,
exogenous_variables_stage2)
tsls_results = stage2_model.fit(cov_type='robust')

# Print the summary of the 2SLS regression results
print(tsls_results.summary())
```

Advanced regression techniques, such as Tobit, Poisson, Negative Binomial regressions, Instrumental Variable (IV) regression, and Two-Stage Least Squares (2SLS) regression, hold immense importance in academic research within accounting and finance. These techniques offer robust tools to address specific challenges encountered in analyzing financial and accounting datasets, including modeling non-negative numerical data, handling censoring, analyzing count data, and addressing endogeneity issues. Their judicious use, underpinned by a sound understanding of their assumptions and applicability, enables scholars to unearth valuable insights, strengthening the empirical robustness of academic discourse. Therefore, these methods are not just statistical techniques; they are instrumental in the academic pursuit of knowledge, contributing to the elucidation of intricate financial phenomena and the development of accounting theory.

24

Time Series Analysis

Time series data is a type of data that is collected over time at regular intervals. It is a sequence of data points that are ordered by time, where each data point represents a measurement or observation made at a specific point in time. Time series data can be collected from a wide range of sources, including sensors, financial markets, social media, weather stations, and more. Time series data is unique in that it contains both temporal and structural dependencies, meaning that each observation is dependent on previous observations and that the data points have a specific order. This makes time series data useful for analyzing trends, patterns, and behaviors over time. Time series analysis techniques can be used to identify trends and seasonality, make predictions and forecasts, and detect anomalies or outliers. Examples of time series data include annual financial statements, daily stock prices, hourly weather data, monthly sales figures, and annual GDP growth rates. Time series data is commonly used in fields such as accounting, finance, economics, engineering, and environmental science, among others, where understanding patterns and trends over time is crucial for making informed decisions.

Time series analysis is a set of statistical techniques used to analyze and draw insights from data points collected over time. In accounting and finance research, time series analysis plays a crucial role in understanding the dynamics of financial markets, firm performance, and economic indicators. By analyzing time series data, researchers can identify patterns, detect trends, and make forecasts to support informed decision-making.

To effectively analyze time series data, researchers should understand the following basic concepts:

1. **Components of a time series**: Time series data can often be decomposed into three main components:
   - **Trend**: The long-term movement of a time series, which represents the underlying growth or decline in the data.
   - **Seasonality**: Regular and predictable fluctuations that recur over a specific period, such as daily, weekly, or yearly patterns.
   - **Noise**: The irregular or random component of a time series, which includes any unpredictable fluctuations that are not explained by the trend or seasonality.
2. **Stationarity**: A time series is stationary if its statistical properties, such as mean and variance, remain constant over time. Stationarity is important for time series analysis because many forecasting models assume that the data is stationary. If a time series is non-stationary, researchers may need to apply transformations, such as differencing or log transformation, to achieve stationarity before applying forecasting techniques.
3. **Autocorrelation and Partial Autocorrelation**: Autocorrelation is the correlation between a time series and a lagged version of itself, measuring the dependence between observations over time. Partial autocorrelation measures the correlation between a time series and a lagged version of itself after accounting for the effects of any shorter lags. These concepts are important for identifying the appropriate order of autoregressive (AR) and moving average (MA) components in time series forecasting models.

Before analyzing time series data, it's essential to preprocess the data to ensure it's in the correct format and has the desired properties.

1. **Importing and Handling Time Series Data with Pandas**

First, let's import the necessary libraries and load the data from the specified file path:

```
import pandas as pd

# Load the data
file_path = "D:/Data/my_data4.csv"
data = pd.read_csv(file_path)

# Set 'Date' as the index and convert it to datetime format
data['Date'] = pd.to_datetime(data['Date'])
data.set_index('Date', inplace=True)

print(data.head())
```

2. Resampling and Interpolation Techniques

Resampling is useful for converting the data's frequency to a different time scale, such as daily, weekly, or monthly. Interpolation is used to fill missing values in the data. Here's an example of how to resample and interpolate the data:

```
# Resample the data to a weekly frequency
weekly_data = data.resample('W').mean()

# Interpolate missing values using linear interpolation
interpolated_data = weekly_data.interpolate(method='linear')

print(interpolated_data.head())
```

3. Checking for Stationarity and Performing Transformations

To check for stationarity, you can use the Augmented Dickey-Fuller (ADF) test provided by the statsmodels library. If the data is non-stationary, you can apply transformations like differencing or log transformation to achieve stationarity.

```
import numpy as np
from statsmodels.tsa.stattools import adfuller

# Define a function to perform the ADF test
def adf_test(timeseries):
    result = adfuller(timeseries)
    print('ADF Statistic: {}'.format(result[0]))
    print('p-value: {}'.format(result[1]))
    print('Critical Values:')
    for key, value in result[4].items():
        print('\t{}: {}'.format(key, value))

# Check stationarity for the dependent variable (BTC)
print("Results of ADF test for BTC:")
adf_test(interpolated_data['BTC'])

# Apply differencing to achieve stationarity (if needed)
interpolated_data['BTC_diff'] =
interpolated_data['BTC'].diff().dropna()
print("Results of ADF test for differenced BTC:")
adf_test(interpolated_data['BTC_diff'].dropna())

# Apply log transformation to achieve stationarity (if needed)
interpolated_data['BTC_log'] = np.log(interpolated_data['BTC'])
print("Results of ADF test for log-transformed BTC:")
adf_test(interpolated_data['BTC_log'].dropna())
```

This code defines a function called **adf_test** to perform the Augmented Dickey-Fuller (ADF) test. The code then applies the ADF test to the original

BTC data, the differenced BTC data (created by taking the first difference), and the log-transformed BTC data to determine if any of these transformations result in a stationary time series. The results will look like:

```
Results of ADF test for BTC:
ADF Statistic: -0.4283402641607983
p-value: 0.9052583521073132
Critical Values:
    1%: -3.4621857592784546
    5%: -2.875537986778846
    10%: -2.574231080806213
Results of ADF test for differenced BTC:
ADF Statistic: -4.669125404631494
p-value: 9.613914488683148e-05
Critical Values:
    1%: -3.4623415245233145
    5%: -2.875606128263243
    10%: -2.574267439846904
Results of ADF test for log-transformed BTC:
ADF Statistic: -0.9652852765849609
p-value: 0.7657142553972955
Critical Values:
    1%: -3.4602906385073884
    5%: -2.874708679520702
    10%: -2.573788599127782
```

In this example, for the original BTC time series, the ADF statistic is -0.428, and the p-value is 0.905, which suggests that the data is non-stationary and that there is a high probability that it has a unit root (i.e., a trend). This is supported by the critical values, which are all greater than the ADF statistic. For the differenced BTC time series, which is the original series after differencing (i.e., taking the first difference), the ADF statistic is -4.669, and the p-value is 9.61e−05, which suggests that the data is now stationary and that there is a low probability that it has a unit root. This is supported by the critical values, which are all less than the ADF statistic. For the log-transformed BTC time series, which is the original series after taking the natural logarithm, the ADF statistic is -0.965, and the p-value is 0.766, which suggests that the data is still non-stationary and that there is a high probability that it has a unit root. This is supported by the critical values, which are all greater than the ADF statistic. ADF test results suggest that the original BTC time series data is non-stationary and has a unit root, which means that it has a trend that needs to be removed.

For our research question, if the data calls for a difference transformation, it can be achieved as follows:

```
# Assuming 'interpolated_data' is the DataFrame containing your
time series data
differenced_data = interpolated_data.diff().dropna()

print(differenced_data.head())
```

This code will apply the first-order difference transformation to all columns in the **interpolated_data** DataFrame and store the result in a new DataFrame called **differenced_data**. The **dropna()** method is used to remove any rows with missing values that result from the differencing process.

4. **Autocorrelation and Partial Autocorrelation**

Calculating autocorrelation and partial autocorrelation functions (ACF and PACF) is an important step in analyzing a time series data and building a time series regression model. These functions help identify the appropriate lag order for the model and provide information on the strength and nature of the relationships between the variables. The ACF shows the correlation between a time series and its own lagged values. It helps determine the lag order for the model by identifying the number of lags at which the autocorrelation coefficients become close to zero. The PACF, on the other hand, shows the correlation between a time series and its own lagged values, but only after the effects of the intermediate lags have been removed. It helps identify the direct relationship between a variable and its lagged values. By analyzing the ACF and PACF, one can determine the appropriate lag order for the time series regression model. ACF and PACF plots can also provide information on whether the time series is stationary, has seasonality, or has other features that may need to be accounted for in the regression model. The following code gives the ACF and PACF plots:

```
import matplotlib.pyplot as plt
from statsmodels.graphics.tsaplots import plot_acf, plot_pacf

# Assuming 'differenced_data' is the DataFrame containing your
differenced time series data
btc_diff = differenced_data['BTC']

# Plot ACF
plt.figure(figsize=(12, 6))
plot_acf(btc_diff, lags=20, title='Autocorrelation Function
(ACF) for BTC')
plt.show()

# Plot PACF
plt.figure(figsize=(12, 6))
plot_pacf(btc_diff, lags=20, title='Partial Autocorrelation
Function (PACF) for BTC')
plt.show()
```

The output will be two plots:

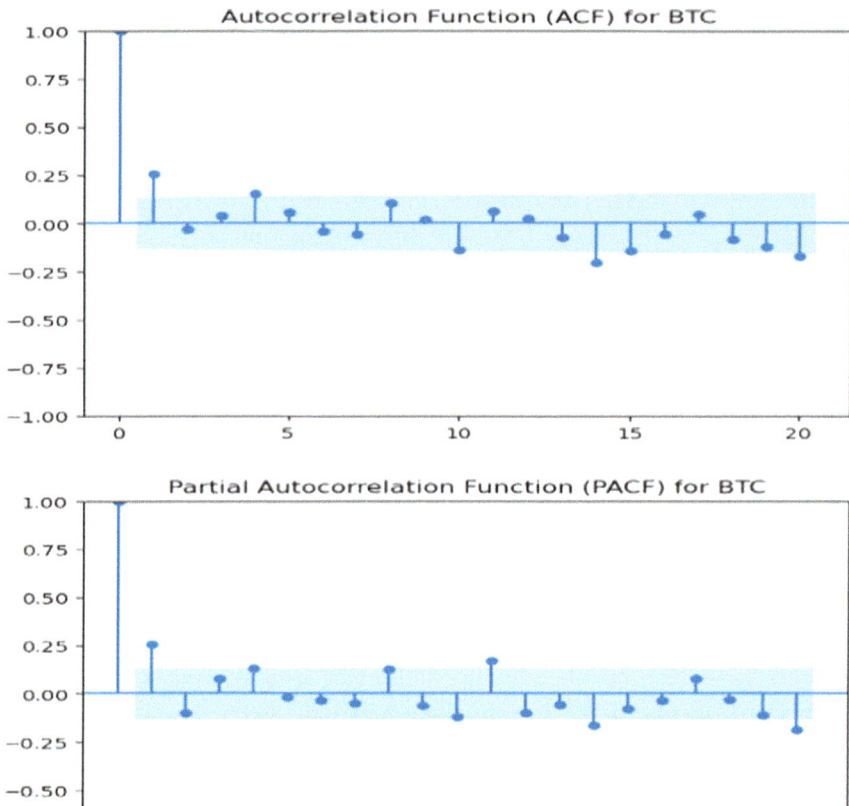

When interpreting Autocorrelation Function (ACF) and Partial Autocorrelation Function (PACF) plots, several aspects must be considered. The ACF plot displays the autocorrelation of the time series at different lags, measuring the linear dependence between the time series and its lagged version. If the ACF plot shows a gradual decrease or exponential decay and cuts off after a certain lag, it suggests an Autoregressive (AR) process of that order. Conversely, if the ACF plot displays a sudden cut-off after a certain lag, it indicates a Moving Average (MA) process of that order. If the ACF plot exhibits a mixture of gradual decrease and sudden cut-offs, it signifies a mix of AR and MA components.

The PACF plot showcases the partial autocorrelation of the time series at different lags. The partial autocorrelation gauges the linear dependence between the time series and its lagged version, accounting for the effects of any shorter lags. If the PACF plot demonstrates a gradual decrease or exponential decay and cuts off after a specific lag, it suggests a Moving Average (MA) process of that order. If the PACF plot displays a sudden cut-off after a certain lag, it indicates an Autoregressive (AR) process of that order. If the PACF plot exhibits a mixture of gradual decrease and sudden cut-offs, it signifies a mix of AR and MA components.

It is also essential to consider the confidence intervals. The confidence intervals are shown as horizontal lines or shaded areas around zero. If the autocorrelation or partial autocorrelation is within the confidence interval, it is not statistically significant, and you can consider it as zero. If the autocorrelation or partial autocorrelation is outside the confidence interval, it is statistically significant.

There are several types of time series models, each with its own assumptions and applications. Some of the most commonly used time series models include:

- Autoregressive (AR) model.
- Moving Average (MA) model.
- Autoregressive Moving Average (ARMA) model.
- Autoregressive Integrated Moving Average (ARIMA) model.
- Seasonal Autoregressive Integrated Moving Average (SARIMA) model.
- Seasonal Autoregressive Integrated Moving Average with Exogenous Regressors (SARIMAX) model.
- Vector Autoregression (VAR) model.
- Vector Error Correction Model (VECM).
- Generalized Autoregressive Conditional Heteroskedasticity (GARCH) model.

Each of these models has different assumptions and is suited for different types of time series data. AR, MA, and ARMA models are useful for analyzing stationary time series, while ARIMA, SARIMA, and SARIMAX models can handle non-stationary and seasonal data. VAR and VECM models are useful for analyzing the relationships between multiple time series variables, while GARCH models are used to model volatility in financial time series data. We'll discuss the most important time series models in greater detail.

Autoregressive (AR) Model

An autoregressive (AR) model is a widely used time series forecasting technique that leverages the linear relationship between a variable's past values and its future values. In an AR model, the value of a variable at time t is predicted as a linear combination of its previous values at time $t-1$, $t-2$, ..., $t-p$, where p denotes the order of the autoregressive model (i.e., the number of lagged terms included in the model). The general form of an AR(p) model can be expressed as:

$$X_t = c + \alpha_1 * X_{(t-1)} + \alpha_2 X_{(t-2)} + \ldots + \alpha_p X_{(t-p)} + \varepsilon_t.$$

In this equation:

- X_t represents the value of the variable at time t.
- c is a constant term.
- $\alpha_1, \alpha_2, \ldots, \alpha_p$ are the autoregressive coefficients.
- ε_t is the error term, which is assumed to be white noise.

To fit an AR model to your data in Python, you can use the **statsmodels.tsa.ar_model.AutoReg** class from the **statsmodels** library. Here's the complete code to import the data, preprocess it, check for stationarity, and fit a third order AR model to the differenced BTC data:

24 Time Series Analysis

```python
import pandas as pd
import numpy as np
import matplotlib.pyplot as plt
from statsmodels.tsa.stattools import adfuller
from statsmodels.tsa.ar_model import AutoReg

# Load the data
file_path = "D:/Data/my_data4.csv"
data = pd.read_csv(file_path)

# Set 'Date' as the index and convert it to datetime format
data['Date'] = pd.to_datetime(data['Date'])
data.set_index('Date', inplace=True)

# Resample the data to a weekly frequency
weekly_data = data.resample('W').mean()

# Interpolate missing values using linear interpolation
interpolated_data = weekly_data.interpolate(method='linear')

# Apply differencing transformation to the data
differenced_data = interpolated_data.diff().dropna()

# Define a function to perform the ADF test
def adf_test(timeseries):
    result = adfuller(timeseries)
    print('ADF Statistic: {}'.format(result[0]))
    print('p-value: {}'.format(result[1]))
    print('Critical Values:')
    for key, value in result[4].items():
        print('\t{}: {}'.format(key, value))

# Check stationarity for the differenced BTC data
print("Results of ADF test for differenced BTC.")
adf_test(differenced_data['BTC'])
# Fit an AR model to the differenced BTC data
btc_diff = differenced_data['BTC']
model = AutoReg(btc_diff, lags=3)
results = model.fit()

# Print the model summary
print(results.summary())
```

This code will import the data from the specified file path, preprocess it by setting the index to the 'Date' column and resampling to weekly frequency, interpolate missing values, apply the difference transformation, check stationarity using the ADF test, and fit an AR(1) model to the differenced BTC data. If the AR model is to be employed on the main dataset i.e., my_data4.csv, then the code has to be modified accordingly. Keep in mind that the choice of the AR model's order should be informed by the ACF and PACF plots, as well as model selection criteria such as Akaike Information Criterion (AIC) and Bayesian Information Criterion (BIC). You can export the results in a publication quality table in word file as follows:

```python
import pandas as pd
import numpy as np
import matplotlib.pyplot as plt
from statsmodels.tsa.stattools import adfuller
from statsmodels.tsa.ar_model import AutoReg
from docx import Document
from docx.enum.text import WD_ALIGN_PARAGRAPH
from docx.shared import Pt

# Load the data
file_path = "D:/Data/my_data4.csv"
data = pd.read_csv(file_path)

# Set 'Date' as the index and convert it to datetime format
data['Date'] = pd.to_datetime(data['Date'])
data.set_index('Date', inplace=True)

# Resample the data to a weekly frequency
weekly_data = data.resample('W').mean()

# Interpolate missing values using linear interpolation
interpolated_data = weekly_data.interpolate(method='linear')

# Apply differencing transformation to the data
differenced_data = interpolated_data.diff().dropna()

# Define a function to perform the ADF test
def adf_test(timeseries):
    result = adfuller(timeseries)
```

```python
        print('ADF Statistic: {}'.format(result[0]))
        print('p-value: {}'.format(result[1]))
        print('Critical Values:')
        for key, value in result[4].items():
            print('\t{}: {}'.format(key, value))

# Check stationarity for the differenced BTC data
print("Results of ADF test for differenced BTC:")
adf_test(differenced_data['BTC'])

# Fit an AR model to the differenced BTC data
btc_diff = differenced_data['BTC']
model = AutoReg(btc_diff, lags=3)
results = model.fit()

# Create a Word document
doc = Document()

# Add a title to the document
title = doc.add_heading("AR Model Results", level=1)
title.alignment = WD_ALIGN_PARAGRAPH.CENTER

# Add the model summary as a table to the document
table = doc.add_table(rows=len(results.params) + 2, cols=4)
table.style = "TableGrid"
# Set the header row
header_row = table.rows[0].cells
header_row[0].text = "Dependent Variable"
header_row[1].text = "BTC"
header_row[2].text = ""
header_row[3].text = ""

# Set the independent variable row
row_cells = table.rows[1].cells
row_cells[0].text = "Independent Variable"
row_cells[1].text = "Coefficient"
row_cells[2].text = "t-stat"
row_cells[3].text = "p-value"

# Fill in the data rows
for i in range(len(results.params)):
    row_cells = table.rows[i + 2].cells
    row_cells[0].text = f"Lag {i}" if i > 0 else "Intercept"
    row_cells[1].text = str(round(results.params[i], 4))
    row_cells[2].text = str(round(results.tvalues[i], 4))
    row_cells[3].text = str(round(results.pvalues[i], 4))


# Change the font size of the table
for row in table.rows:
    for cell in row.cells:
        for paragraph in cell.paragraphs:
            for run in paragraph.runs:
                font = run.font
                font.size = Pt(12)

# Save the Word document
doc.save("D:/Data/ar_model_results.docx")
```

Moving Average (MA) Model

A Moving Average (MA) model is a time series forecasting technique that uses the weighted average of past values to make predictions. The model assumes that the current observation is a linear combination of previous errors or random shocks. The order of an MA model, denoted as q, represents the number of lagged error terms included in the model. An MA(q) model can be written as:

$$Y_t = \mu + \varepsilon_t + \theta_1 \varepsilon_{(t-1)} + \theta_2 \varepsilon_{(t-2)} + \ldots + \theta_q \varepsilon_{(t-q)},$$

where Y_t is the observed value at time t, μ is the mean of the time series, ε_t is the error term at time t, and θ_i is the parameter associated with the error term at $t-i$.

The MA model is useful for modeling time series with short-term dependencies or when the underlying process generating the data is best described by moving averages of past errors. A second-order MA model can be implemented on D:/Data/my_data4.csv as follows:

```python
import pandas as pd
import matplotlib.pyplot as plt
from statsmodels.graphics.tsaplots import plot_acf
from statsmodels.tsa.arima.model import ARIMA

# Step 1: Load the data
file_path = "D:/Data/my_data4.csv"
data = pd.read_csv(file_path)

# Step 2: Preprocess the data
data['Date'] = pd.to_datetime(data['Date'])
data.set_index('Date', inplace=True)
data = data.resample('D').mean().interpolate()

# Step 3: Determine the order of the MA model (q) using the ACF plot
column_name = 'BTC'  # Replace with the desired column name
plot_acf(data[column_name], lags=20)
plt.show()
# Based on the ACF plot, choose an appropriate value for q
q = 2  # Replace with the appropriate value based on the ACF plot

# Step 4: Fit the MA model
ma_model = ARIMA(data[column_name], order=(0, 0, q))
ma_result = ma_model.fit()

print(ma_result.summary())

# Based on the ACF plot, choose an appropriate value for q
q = 2  # Replace with the appropriate value based on the ACF plot

# Step 4: Fit the MA model
ma_model = ARIMA(data[column_name], order=(0, 0, q))
ma_result = ma_model.fit()

print(ma_result.summary())
```

The result will look like:

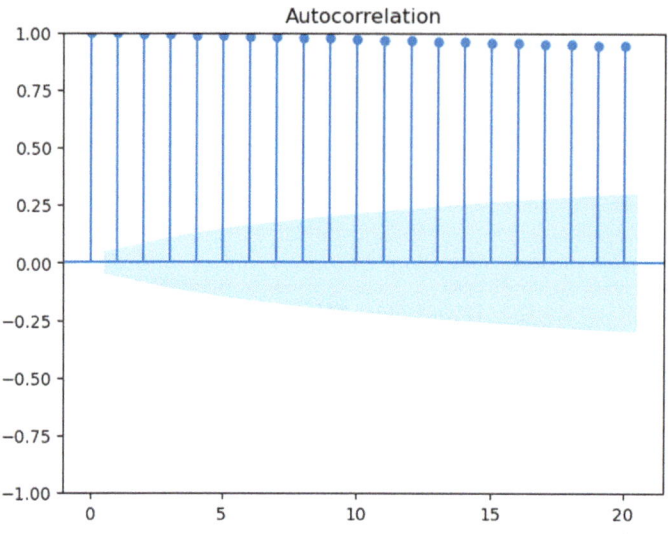

```
                               SARIMAX Results
==============================================================================
Dep. Variable:                    BTC   No. Observations:                 1554
Model:                 ARIMA(0, 0, 2)   Log Likelihood              -15744.331
Date:                Sun, 07 May 2023   AIC                          31496.662
Time:                        10:59:56   BIC                          31518.056
Sample:                    10-01-2017   HQIC                         31504.618
                         - 01-01-2022
Covariance Type:                  opg
==============================================================================
                 coef    std err          z      P>|z|      [0.025      0.975]
------------------------------------------------------------------------------
const       1.787e+04    561.039     31.851      0.000    1.68e+04    1.9e+04
ma.L1          1.4355      0.010    142.757      0.000       1.416       1.455
ma.L2          0.9999      0.014     71.448      0.000       0.972       1.027
sigma2      3.663e+07      0.320   1.14e+08      0.000    3.66e+07    3.66e+07
===================================================================================
Ljung-Box (L1) (Q):                 370.45   Jarque-Bera (JB):              2025.11
Prob(Q):                              0.00   Prob(JB):                         0.00
Heteroskedasticity (H):               6.10   Skew:                             1.99
Prob(H) (two-sided):                  0.00   Kurtosis:                         6.94
===================================================================================

Warnings:
[1] Covariance matrix calculated using the outer product of
gradients (complex-step).
[2] Covariance matrix is singular or near-singular, with
condition number 3.25e+22. Standard errors may be unstable.
```

Ljung-Box (L1) (Q) and Prob(Q): The Ljung-Box test is a statistical test used to check for autocorrelation in the residuals of the model. A high test statistic value (Q) and a low *p*-value (Prob(Q)) indicate that there is significant autocorrelation in the residuals, suggesting that the model may not have captured all the underlying patterns in the data. In this case, the test statistic is 370.45, and the *p*-value is 0.00, indicating evidence of significant autocorrelation in the residuals. Jarque–Bera (JB) and Prob(JB): The Jarque–Bera test is a statistical test used to determine whether the residuals are normally distributed. A high test statistic value (JB) and a low p-value (Prob(JB)) indicate that the residuals are not normally distributed. In this case, the test statistic is 2025.11, and the p-value is 0.00, suggesting that the residuals do not follow a normal distribution.

ARMA, ARIMA, SARIMA, and SARIMAX

ARMA (Autoregressive Moving Average) is a linear model that combines the autoregressive (AR) and moving average (MA) components. It assumes that the current value of the time series depends on its previous values (AR component) and past errors or random shocks (MA component). ARMA models are suitable for stationary time series without trends or seasonality.

In an ARMA(p, q) model, p represents the order of the autoregressive (AR) component, and q represents the order of the moving average (MA) component. The ARMA model combines both the AR and MA components. The general equation for an ARMA(p, q) model can be expressed as:

$$Y_t = c + \Sigma(\phi_i * Y_{t-i}) + \varepsilon_t + \Sigma(\theta_j * \varepsilon_{\{t-j\}}),$$

where

- Y_t is the value of the time series at time t;
- c is a constant term;
- ϕ_i is the autoregressive coefficient for lag i ($i = 1, 2, ..., p$);
- $Y_{\{t-i\}}$ is the value of the time series at time $t-i$ (i.e., i lags before the current observation);
- E_t is the error term (residual or random shock) at time t;
- θ_j is the moving average coefficient for lag j ($j = 1, 2, ..., q$);
- $\varepsilon_{\{t-j\}}$ is the error term at time $t-j$ (i.e., j lags before the current observation).

In simpler terms, the ARMA model represents the current value of the time series as a linear combination of its past values (autoregressive component), past error terms (moving average component), and a constant term. The model aims to capture both the dependence on previous values and the effect of random shocks in the data.

The ARIMA model is a generalization of the Autoregressive (AR) and MA models, combining their strengths to better capture the dynamics of time series data. ARIMA models can also incorporate differencing to handle non-stationary data. An ARIMA model is defined by three parameters: p, d, and q. The p parameter refers to the order of the AR component, the d parameter represents the degree of differencing, and the q parameter indicates the order of the MA component. An ARIMA(p, d, q) model can be written as:

$$\Phi(B)(1-B)^d Y_t = \mu + \Theta(B)\varepsilon_t,$$

where $\Phi(B)$ and $\Theta(B)$ are the lag polynomials for the AR and MA components, respectively, B is the backshift operator, and $(1-B)^d$ represents the differencing operator applied d times.

ARIMA (Autoregressive Integrated Moving Average) models are particularly useful for forecasting time series data with complex patterns, such as trends, seasonality, and a combination of short- and long-term dependencies. By choosing the appropriate p, d, and q parameters, researchers can capture a wide range of time series behaviors using ARIMA models.

SARIMA (Seasonal Autoregressive Integrated Moving Average) is an extension of the ARIMA model that specifically addresses seasonality. In addition to the AR, differencing, and MA components, SARIMA models include seasonal components for each of these features. This allows the model to capture complex seasonal patterns in the time series data.

SARIMAX (Seasonal Autoregressive Integrated Moving Average with Exogenous Regressors) is an extension of the ARMA model that incorporates both seasonality and exogenous variables, making it more versatile and capable of capturing complex patterns in time series data, especially when there are seasonal effects or additional external factors influencing the observations. This enhanced modeling capability allows SARIMAX to provide more accurate forecasts and insights compared to the simpler ARMA model.

There is no universally "best" model among ARMA, ARIMA, SARIMA, and SARIMAX, as the choice of the most suitable model depends on the specific characteristics of the time series data being analyzed. The optimal model for a given dataset may vary based on factors such as trend, seasonality, and the presence of exogenous variables.

To select the best model for your data, consider the following:

- If your time series is stationary and has no trend or seasonality, an ARMA model might be sufficient.
- If your time series has a trend but no seasonality, consider using an ARIMA model.

- If your time series exhibits both trend and seasonality, a SARIMA model may be the most appropriate choice.
- If your time series has trend, seasonality, and is affected by external factors or exogenous variables, a SARIMAX model is likely the best option.

It is generally recommended to test multiple models and compare their performance using metrics such as the Akaike Information Criterion (AIC), Bayesian Information Criterion (BIC), or Mean Absolute Error (MAE) to determine the model that best captures the underlying patterns in your data.

To implement an SARIMAX model on your data in the file **D:/Data/my_data4.csv**, you can use the **ARIMA** class from the **statsmodels.tsa.arima.model** module.

```python
import pandas as pd
from statsmodels.tsa.arima.model import ARIMA

# Load the data
file_path = "D:/Data/my_data4.csv"
data = pd.read_csv(file_path)

# Set 'Date' as the index and convert it to datetime format
data['Date'] = pd.to_datetime(data['Date'])
data.set_index('Date', inplace=True)

# Assuming you have chosen the AR, differencing, and MA orders
# as p, d, and q
p = 2
d = 1
q = 3

# Choose the dependent variable, e.g., 'BTC'
dependent_variable = 'BTC'

# Fit the ARIMA model
arima_model = ARIMA(data[dependent_variable], order=(p, d, q))
arima_fit = arima_model.fit()

# Print model summary
print(arima_fit.summary())
```

The output will look like:

```
                               SARIMAX Results
==============================================================================
Dep. Variable:                    BTC   No. Observations:                 1554
Model:                 ARIMA(2, 1, 3)   Log Likelihood              -12914.550
Date:                Sun, 07 May 2023   AIC                          25841.100
Time:                        11:27:37   BIC                          25873.187
Sample:                             0   HQIC                         25853.033
                               - 1554
Covariance Type:                  opg
==============================================================================
                 coef    std err          z      P>|z|      [0.025      0.975]
------------------------------------------------------------------------------
ar.L1          0.7486      0.033     22.721      0.000       0.684       0.813
ar.L2         -0.9021      0.034    -26.280      0.000      -0.969      -0.835
ma.L1         -0.7920      0.036    -22.196      0.000      -0.862      -0.722
ma.L2          0.9124      0.040     22.678      0.000       0.834       0.991
ma.L3         -0.0092      0.017     -0.533      0.594      -0.043       0.025
sigma2      9.793e+05   1.45e+04     67.450      0.000    9.51e+05    1.01e+06
===================================================================================
Ljung-Box (L1) (Q):                   0.03   Jarque-Bera (JB):              7231.20
Prob(Q):                              0.86   Prob(JB):                         0.00
Heteroskedasticity (H):               0.08   Skew:                             0.16
Prob(H) (two-sided):                  0.00   Kurtosis:                        13.57
===================================================================================

Warnings:
[1] Covariance matrix calculated using the outer product of
    gradients (complex-step).
```

The model can be employed for forecasting as follows:

```
# Forecast the next 10 time steps
forecast_steps = 10
forecast = arima_fit.forecast(steps=forecast_steps)
print(forecast)
```

Vector Autoregression (VAR) Model

Vector Autoregression (VAR) is a multivariate time series model that captures the linear interdependencies among multiple time series. It is an extension of the univariate autoregressive (AR) model to a system of multiple variables. VAR models are particularly useful in studying the dynamic relationships among a set of variables and in predicting the behavior of multiple time series simultaneously. The VAR model represents each variable in the system as a linear function of its own lagged values and the lagged values of all other variables in the system. For example, consider a system with two time series, $Y1$ and $Y2$, and a VAR model with one lag:

$$Y1(t) = c1 + a11 * Y1(t-1) + a12 * Y2(t-1) + e1(t),$$

$$Y2(t) = c2 + a21 * Y1(t-1) + a22 * Y2(t-1) + e2(t).$$

Here, $Y1(t)$ and $Y2(t)$ represent the values of the two time series at time t, $c1$, and $c2$ are constants, $a11$, $a12$, $a21$, and $a22$ are the coefficients to be estimated, and $e1(t)$ and $e2(t)$ are the error terms.

VAR models can be estimated using ordinary least squares (OLS) on a per-equation basis. To determine the optimal number of lags to include in the model, researchers can use information criteria, such as the Akaike Information Criterion (AIC) or the Bayesian Information Criterion (BIC).

Some key features of VAR models include:

- **Impulse Response Functions (IRFs)**: IRFs show the response of each variable to a one-time shock in another variable while holding all other shocks constant. They help researchers analyze the dynamic effects of shocks to the system and understand the transmission of shocks across variables.
- **Variance Decomposition**: Variance decomposition measures the proportion of the forecast error variance of each variable that is attributable to shocks in other variables in the system. It provides insights into the relative importance of each variable in explaining the forecast error variance of the other variables.
- **Granger Causality Tests**: Granger causality tests are used to determine whether one variable can predict another variable better than using the past values of the dependent variable alone. Granger causality does not imply true causality but can provide evidence of a predictive relationship between variables.

To implement a VAR model in Python, researchers can use the **VAR** class from the **statsmodels.tsa.vector_ar.var_model** module, which provides tools for estimating, analyzing, and forecasting VAR models. An implementation of VAR model on D:/Data/my_data4.csv with BTC as $Y1$ and ETH as $Y2$, assuming that the series are stationary is as follows:

```python
import pandas as pd
from statsmodels.tsa.vector_ar.var_model import VAR
from statsmodels.tsa.stattools import adfuller

# Load the data
file_path = "D:/Data/my_data4.csv"
data = pd.read_csv(file_path)

# Set 'Date' as the index and convert it to datetime format
data['Date'] = pd.to_datetime(data['Date'])
data.set_index('Date', inplace=True)

# Select the columns of interest (BTC and ETH)
selected_data = data[['BTC', 'ETH']]

print("Results of ADF test for BTC:")
adf_test(selected_data['BTC'])
print("\nResults of ADF test for ETH:")
adf_test(selected_data['ETH'])

# Fit the VAR model
model = VAR(selected_data)

# Determine the optimal lag order using AIC
lags = model.select_order(maxlags=10)
optimal_lag_order = lags.aic
print(f"Optimal lag order (AIC): {optimal_lag_order}")

# Fit the VAR model with the optimal lag order
results = model.fit(optimal_lag_order)

# Print the VAR model summary
print(results.summary())
```

This code loads time series data of Bitcoin and Ethereum prices from a CSV file, sets the date column as the index, selects the columns of interest, and performs an Augmented Dickey-Fuller (ADF) test for each of the selected columns to check for stationarity. A vector autoregression (VAR) model is then fitted to the selected data, and the optimal lag order is determined using the Akaike information criterion (AIC). Finally, the VAR model is fit with the optimal lag order, and a summary of the model results is printed. This code provides a basic example of how to perform time series analysis using the VAR model in Python.

If the series are non-stationary, we need to include the differencing also. The code in this case would be:

```
lags = model.select_order(maxlags=10)
optimal_lag_order = lags.aic
print(f"Optimal lag order (AIC): {optimal_lag_order}")

# Fit the VAR model with the optimal lag order
results = model.fit(optimal_lag_order)

# Print the VAR model summary
print(results.summary())
```

One of the main advantages of the VAR model is its ability to capture the linear interdependencies among multiple time series, which makes it particularly useful for analyzing the behavior of interconnected economic and financial variables. The model is easy to estimate using ordinary least squares (OLS) and provides a straightforward way to examine the dynamic effects of shocks to the system through impulse response functions (IRFs). Additionally, VAR models can be used for forecasting multiple time series simultaneously, making them suitable for a variety of applications.

However, VAR model requires a large number of parameters to be estimated, which can lead to overfitting, especially in cases where the number of time series and lags is high. This problem can be mitigated by using model selection criteria, such as the Akaike Information Criterion (AIC) or the Bayesian Information Criterion (BIC), to determine the optimal number of lags. Another disadvantage of VAR models is that they are purely linear, which means they may not accurately capture complex, non-linear relationships between time series. This limitation can be addressed by employing non-linear models, such as the Nonlinear Vector Autoregression (NVAR) model or the Vector Autoregression Moving Average (VARMA) model. VAR models are not suitable for analyzing time series with structural breaks or regime shifts, as the model assumes constant coefficients throughout the entire sample period. In such cases, more advanced techniques, such as the Markov-switching VAR model or the time-varying parameter VAR model, may be more appropriate. VAR model is essentially a reduced-form model, which means it does not provide a structural interpretation of the relationships among the variables. Researchers who are interested in identifying causal relationships or understanding the underlying economic mechanisms may need to employ structural models, such as structural VAR models, which impose identifying restrictions based on economic theory.

Vector Error Correction Model (VECM)

Vector Error Correction Model (VECM) is a multivariate extension of the Error Correction Model (ECM) used for modeling and analyzing the long-run and short-run dynamics of multiple cointegrated time series. VECM is particularly useful when examining the relationships between multiple non-stationary variables that share a long-run equilibrium relationship.

The VECM framework can be expressed as follows:

$$\Delta X_t = \Pi X_{(t-1)} + \Gamma_1 \Delta X_{(t-1)} + \Gamma_2 \Delta X_{(t-2)} + \ldots + \Gamma_{(p-1)} \Delta X_{(t-p+1)} + \varepsilon_t,$$

where

- ΔX_t represents the first difference of the vector of time series variables at time t;
- Π is the matrix of long-run relationships, also known as the cointegration matrix;
- $X_{(t-1)}$ is the lagged level of the vector of time series variables;
- Γ_i represents the matrices of short-run dynamics (i=1, 2, ..., p−1);
- p is the number of lags;
- ε_t is the vector of error terms.

The cointegration matrix (Π) captures the long-run equilibrium relationships between the time series variables, while the short-run dynamics are represented by the Γ_i matrices. If there are cointegrating relationships among the variables, the rank of the Π matrix will be greater than zero and less than the number of variables in the system.

Advantages of VECM

- **Long-run and Short-run Dynamics:** VECM allows for the simultaneous analysis of long-run equilibrium relationships and short-run adjustments among multiple time series variables, providing a comprehensive understanding of their interdependencies.
- **Cointegration:** By considering the cointegration relationships between non-stationary variables, VECM avoids the problem of spurious regression, which can lead to misleading results when analyzing non-stationary data.

- **Impulse Response Analysis**: VECM can be used to perform impulse response analysis, which helps researchers understand the effects of shocks to one variable on other variables in the system.
- **Forecasting**: VECM can be used for forecasting multiple time series, taking into account both the long-run equilibrium relationships and short-run dynamics among the variables.

To implement VECM in Python, you can use the **VECM** class from the **statsmodels.tsa.vector_ar.vecm** module. Before applying VECM, you'll need to determine the appropriate lag order (p) and the rank of the cointegration matrix (Π) using model selection criteria or tests for cointegration, such as the Johansen test, as follows:

```python
import pandas as pd
from statsmodels.tsa.vector_ar.vecm import VECM, select_order, coint_johansen

# Load the data
file_path = "D:/Data/my_data4.csv"
data = pd.read_csv(file_path)

# Set 'Date' as the index and convert it to datetime format
data['Date'] = pd.to_datetime(data['Date'])
data.set_index('Date', inplace=True)

# Remove any missing values (if necessary)
data = data.dropna()

# Determine the optimal lag order using model selection criteria
model_selection = select_order(data, maxlags=10,
deterministic='ci', seasons=0)

# Find the optimal lag order based on the Akaike Information
Criterion (AIC)
lag_order = model_selection.aic
print("Optimal lag order (p):", lag_order)

# Perform the Johansen test to determine the cointegration rank
johansen_test = coint_johansen(data, det_order=0,
k_ar_diff=lag_order)

# Check the trace statistic against the critical values
rank = 0
for trace_stat, crit_value in zip(johansen_test.lr1,
johansen_test.cvt[:, 1]):
    if trace_stat > crit_value:
        rank += 1
    else:
        break

print("Cointegration rank:", rank)
```

The output will be:

```
Optimal lag order (p):10
Cointegration rank:2
```

Now we can implement the VECM as follows:

```
from statsmodels.tsa.vector_ar.vecm import VECM
# Fit the VECM model
vecm_model = VECM(data, k_ar_diff=lag_order, coint_rank=rank,
deterministic='ci')
vecm_fit = vecm_model.fit()
# Print the summary of the VECM model
print(vecm_fit.summary())
```

The VECM output provides with the loading coefficients (alpha) for each variable in the system, as well as the cointegration relations (beta) for each cointegrating vector.

1. Loading Coefficients (alpha):

These coefficients represent the speed of adjustment of each variable in response to deviations from the long-run equilibrium. The lower the value, the slower the variable adjusts back to equilibrium. You can find the alpha coefficients for each variable (Year, BTC, ETH, BNB, ADA) in the output. The significance of each alpha coefficient can be determined by looking at its p-value ($P > |z|$). A p-value lower than 0.05 indicates that the alpha coefficient is statistically significant.

2. Cointegration Relations (beta):

These coefficients represent the long-run relationships between the variables in the system. The beta coefficients show how each variable in the system contributes to the cointegrating relationship. The constant term in the cointegrating equation represents the equilibrium value around which the relationship holds.

Vector Error Correction Model (VECM) is a widely used econometric technique for modeling the long-run dynamics of time series data. One of the key advantages of VECM is its ability to capture both the short-term and long-term relationships between multiple non-stationary time series. Unlike traditional regression analysis, VECM accounts for the possibility of cointegration, which implies that the variables share a common stochastic trend. This makes it suitable for analyzing economic data that may have complex interrelationships.

However, one of the disadvantages of VECM is its sensitivity to the choice of lag length and the number of cointegrating relationships. Incorrect specification of these parameters can lead to biased parameter estimates and inefficient inference. Furthermore, VECM assumes that the errors are normally distributed and homoscedastic, which may not be the case in practice. Finally, interpretation of VECM results can be challenging, as the coefficients represent the dynamic adjustment process between the variables, rather than direct causal relationships.

GARCH Model

The Generalized Autoregressive Conditional Heteroskedasticity (GARCH) model is a time series model commonly used to model volatility clustering, where large price changes in a time series tend to be followed by more large price changes, and small price changes tend to be followed by more small price changes. This phenomenon is often observed in financial data, where sudden market shocks can lead to prolonged periods of high volatility.

The GARCH model is based on the idea that the variance of a time series is a function of its own past values and the past values of its own squared residuals, which are the deviations of the actual values from the predicted values. The model assumes that the squared residuals follow an Autoregressive (AR) process and a Moving Average (MA) process, with the variance of the time series being a function of these past squared residuals.

The GARCH(p, q) model can be represented by the following equation:

$$\sigma_t^2 = \omega + \sum \left(\alpha_i \varepsilon_{\{t-i\}}^2 \right) + \sum \left(\beta_j \sigma_{\{t-j\}}^2 \right),$$

where σ^2_t represents the conditional variance at time t, ω is a constant term, α_i and β_j are the GARCH parameters to be estimated, $\varepsilon^2_{\{t-i\}}$ represents the squared residual from the mean equation at time $t-i$, and $\sigma^2_{\{t-j\}}$ is the lagged conditional variance at time $t-j$. The model order is determined by p and q, which are the number of ARCH and GARCH lags, respectively.

GARCH models are commonly used in finance to forecast the volatility of financial time series data, such as stock returns, exchange rates, and commodity prices. These models have applications in various research areas, including accounting, finance, and other business disciplines. For example, GARCH models can be used to estimate the risk and manage portfolios, price options, calculate Value-at-Risk (VaR), and analyze market efficiency.

One of the key advantages of GARCH models is their flexibility in capturing a wide range of volatility patterns in financial time series data. GARCH models are also known for their superior forecasting performance, especially in the presence of time-varying volatility. Additionally, GARCH models provide consistent and unbiased estimates of volatility, even when the underlying data-generating process is not known. However, GARCH models are complex and can be computationally intensive, especially when estimating higher-order models or dealing with large datasets. GARCH models may also be prone to overfitting due to the large number of parameters, potentially leading to poor out-of-sample forecasting performance. Another limitation is that GARCH models may not accurately capture sudden jumps or structural breaks in volatility.

Most common GARCH model is (1,1), which can be implemented as follows:

```python
import pandas as pd
import numpy as np
from arch import arch_model
# Load the dataset
data = pd.read_csv("D:/Data/my_data4.csv")

# Convert the 'Date' column to datetime format
data["Date"] = pd.to_datetime(data["Date"])

# Set the 'Date' column as the index
data.set_index("Date", inplace=True)

# Calculate the returns for each cryptocurrency
returns = np.log(data / data.shift(1)).dropna()

# Extract BTC returns
btc_returns = returns["BTC"]

# Specify and fit the GARCH(1,1) model
garch_model = arch_model(btc_returns, vol="Garch", p=1, q=1)
garch_fit = garch_model.fit()

# Print the model summary
print(garch_fit.summary())
```

The output will look like:

```
Iteration:      1,     Func. Count:      6,    Neg. LLF: 7.092418139243091e+16
Iteration:      2,     Func. Count:     17,    Neg. LLF: 350402.7834263507
Iteration:      3,     Func. Count:     28,    Neg. LLF: 996426.506882519
Iteration:      4,     Func. Count:     39,    Neg. LLF: 1.2735325559267968e+16
Iteration:      5,     Func. Count:     50,    Neg. LLF: 423572095.13441086
Iteration:      6,     Func. Count:     62,    Neg. LLF: -2863.189282785077
Optimization terminated successfully    (Exit mode 0)
            Current function value: -2863.1892755314398
            Iterations: 10
            Function evaluations: 62
            Gradient evaluations: 6
                    Constant Mean - GARCH Model Results
==============================================================================
Dep. Variable:                    BTC   R-squared:                       0.000
Mean Model:             Constant Mean   Adj. R-squared:                  0.000
Vol Model:                      GARCH   Log-Likelihood:                 2863.19
Distribution:                  Normal   AIC:                           -5718.38
Method:            Maximum Likelihood   BIC:                           -5696.99
                                        No. Observations:                 1553
Date:                Sun, May 07 2023   Df Residuals:                     1552
Time:                        13:58:43   Df Model:                            1
                                  Mean Model
==============================================================================
                 coef    std err          t      P>|t|      95.0% Conf. Int.
------------------------------------------------------------------------------
mu          -1.2677e-03  9.703e-04     -1.306      0.191 [-3.169e-03,6.340e-04]
                                Volatility Model
==============================================================================
                 coef    std err          t      P>|t|      95.0% Conf. Int.
------------------------------------------------------------------------------
omega       3.4630e-05  6.304e-06      5.493  3.952e-08 [2.227e-05,4.699e-05]
alpha[1]        0.1000  3.453e-02      2.896  3.778e-03   [3.232e-02,  0.168]
beta[1]         0.8800  3.099e-02     28.397 2.187e-177   [  0.819,    0.941]
==============================================================================

Covariance estimator: robust
```

Here's an interpretation of the key components:

1. **Model Specification**: The model is a Constant Mean—GARCH(1,1) with a normal distribution for the error term. This means the model assumes a constant mean return and uses a GARCH(1,1) process to model the volatility.
2. **Optimization**: The optimization terminated successfully, meaning that the model's parameters have been estimated using maximum likelihood estimation. The negative log-likelihood (-2863.19) is the value of the objective function at the optimal solution, which is used to compare the fit of different models. Lower (more negative) values indicate a better fit.
3. **Mean Model**: The "mu" parameter represents the average (constant) return of BTC in the sample. The estimate is -0.0012677, but it is not statistically significant at the 5% level ($P > |t| = 0.191$), which means that the average return is not significantly different from zero.
4. **Volatility Model**: The GARCH(1,1) model has three parameters: omega, alpha[1], and beta[1].

    - omega: This is the constant term in the GARCH equation, representing the long-run average volatility. The estimate is $3.4630e-05$, and it is

statistically significant ($P < 0.001$), indicating that there is a non-zero long-run average volatility.
- alpha[1]: This parameter represents the ARCH effect, capturing the impact of past squared errors on current volatility. The estimate is 0.1000, and it is statistically significant ($P = 0.0038$), suggesting that there is a positive relationship between past shocks and current volatility. In other words, large shocks in the past are associated with higher volatility today.
- beta[1]: This parameter represents the GARCH effect, capturing the persistence of volatility over time. The estimate is 0.8800, and it is statistically significant ($P < 0.001$), indicating that past volatility has a positive impact on current volatility. The higher the beta[1] value, the more persistent the volatility.

5. Model Fit: The model fit can be assessed using the Akaike Information Criterion (AIC) and the Bayesian Information Criterion (BIC). Lower values for both criteria indicate a better model fit. In this case, the AIC is -5718.38, and the BIC is -5696.99.

The GARCH(1,1) model shows that there is a significant ARCH and GARCH effect in the BTC returns, indicating that past shocks and volatility play an important role in determining current volatility. The average return is not statistically significant, suggesting that the mean return is close to zero. The model fit can be assessed using the AIC and BIC values, which can be compared to alternative models to select the best-fitting model.

Once you have the GARCH(1,1) model fitted for the BTC returns, you can use it to forecast the volatility, future returns and prices. However, it's important to note that the GARCH model is designed to predict volatility, not returns directly. To predict the future price of BTC, you can combine the GARCH model's volatility forecasts with an appropriate mean model for returns.

```
# Forecast the next day's volatility
forecast = garch_fit.forecast(start=len(btc_returns) - 1,
horizon=1)
next_day_volatility = np.sqrt(forecast.variance.values[-1, :])
# Calculate the historical average return
mean_return = btc_returns.mean()

# For demonstration purposes, we'll use the historical average
return as the predicted return
next_day_return = mean_return
# Calculate the predicted next day price
current_price = data["BTC"].iloc[-1]
next_day_price = current_price * np.exp(next_day_return)

print("Predicted next day Volatility:", next_day_volatility)
print("Current BTC price:", current_price)
print("Predicted next day BTC price:", next_day_price)
```

Time series analysis is indispensable for understanding the dynamics of accounting and financial data, uncovering hidden patterns, and making informed decisions in a constantly changing economic landscape. Time series models enable them to analyze the behavior of financial variables, forecast future trends, identify potential risk factors, and evaluate the effectiveness of financial policies and strategies. The ability to model and predict financial time series is crucial for risk management, performance evaluation, and the development of trading strategies. By understanding the statistical properties of financial data, accounting and finance professionals can better anticipate market fluctuations, manage risk, and optimize asset allocation.

25

Panel Data

Panel data, also known as cross-sectional time series data, is a type of data that combines both cross-sectional and time series dimensions. In accounting and finance research, panel data typically consists of observations on multiple entities (such as firms, countries, or individuals) over multiple time periods. This data structure enables researchers to study the dynamics of relationships between variables and analyze the effects of various factors over time.

Panel data plays a crucial role in accounting and finance research as it allows researchers to control for unobserved heterogeneity between entities, which can lead to more accurate estimates and inferences. It also provides a richer source of information by combining both cross-sectional variation and time series variation, which can help in identifying causal relationships. Furthermore, panel data enables researchers to study dynamic relationships and analyze the impact of past events on current outcomes.

The structure of panel data includes two dimensions: cross-sectional and time series. The cross-sectional dimension refers to the individual entities being observed, such as companies or countries, while the time series dimension refers to the observations of these entities over different time periods. In a balanced panel, each entity has the same number of observations across all time periods. In contrast, an unbalanced panel has an unequal number of observations for different entities or time periods.

Working with panel data offers several benefits, such as the ability to control for unobserved fixed effects, which can reduce omitted variable bias and improve the accuracy of estimates. Moreover, it exploits both within-entity and between-entity variations, which can improve the power of statistical tests and increase the precision of estimates. Panel data also provides an opportunity to analyze dynamic relationships and causal effects, offering valuable insights into the underlying mechanisms of accounting and finance phenomena.

However, there are also some challenges associated with panel data. The presence of missing data, which can arise from non-random attrition, measurement errors, or incomplete data collection, may lead to biased estimates. The potential for endogeneity issues, such as reverse causality or simultaneity, can complicate the identification of causal relationships. Increased complexity of statistical models and estimation techniques may require specialized knowledge and computational resources.

Types of Panel Data Models

In panel data analysis, there are several types of models that researchers can employ to study the relationships between variables. The most common models used in accounting and finance research are fixed effects, random effects, and pooled OLS models. Each model is based on different assumptions and is suited for different types of research questions and data structures.

Fixed effects models assume that there are unobserved time-invariant factors that can affect the dependent variable, and these factors are unique to each entity. By controlling for these entity-specific fixed effects, this model accounts for unobserved heterogeneity and can provide unbiased estimates of the relationship between the independent and dependent variables. Fixed effects models are particularly useful when the primary interest is in the within-entity variation, rather than the between-entity variation.

Random effects models, on the other hand, assume that the unobserved time-invariant factors are random and uncorrelated with the independent variables. In this model, the unobserved heterogeneity is treated as a random effect, and the model estimates the average relationship between the independent and dependent variables across all entities. Random effects models are suitable when the primary interest is in both within-entity and between-entity variations, and when the unobserved heterogeneity is not systematically related to the independent variables.

Pooled OLS models treat panel data as a simple cross-sectional dataset by pooling all observations and estimating an ordinary least squares regression. This model assumes that there is no unobserved heterogeneity across entities or time periods, and that the relationships between variables are constant over time. Pooled OLS models can provide efficient estimates when these assumptions hold, but they may suffer from omitted variable bias and other issues when unobserved heterogeneity is present.

The choice between fixed effects and random effects models depends on the underlying assumptions and the research question being addressed. The Hausman test is a widely used statistical test that helps researchers decide between these two models. The test compares the estimates obtained from both models and evaluates the null hypothesis that the random effects model is consistent and efficient. If the Hausman test fails to reject the null hypothesis, the random effects model is preferred, as it provides more efficient estimates. However, if the test rejects the null hypothesis, it suggests that the random effects model is inconsistent, and the fixed effects model should be used instead. In Python, Hausman statistic can be calculated as under:

```python
import pandas as pd
import numpy as np
import scipy.stats as stats
from linearmodels import PanelOLS, RandomEffects

# Load your data
data = pd.read_csv('D:/Data/my_data4.csv')

# Make sure your 'Date' column is in datetime format
data['Date'] = pd.to_datetime(data['Date'])

# Set the index using both 'Date' and 'Year' variables
data = data.set_index(['Year', 'Date'])

# Dependent and independent variables
dependent_var = 'BTC_returns'
independent_vars = ['ETH_returns', 'BNB_returns', 'ADA_returns']

# Fixed Effects Model
fixed_effects_model = PanelOLS(data[dependent_var],
data[independent_vars], entity_effects=True).fit()

# Random Effects Model
random_effects_model = RandomEffects(data[dependent_var],
data[independent_vars]).fit()

# Perform the Hausman test manually
chi2_stat = (fixed_effects_model.params -
random_effects_model.params).T @
np.linalg.inv(fixed_effects_model.cov -
random_effects_model.cov) @ (fixed_effects_model.params -
random_effects_model.params)
df = len(fixed_effects_model.params)
p_value = 1 - stats.chi2.cdf(chi2_stat, df)

print(f"Hausman test statistic: {chi2_stat: .4f}")
print(f"Hausman test p-value: {p_value: .4f}")
```

This code calculates the Hausman test statistic manually using the parameter estimates and covariance matrices of the fixed effects and random effects models. It then calculates the *p*-value using the chi-squared distribution from the **scipy.stats** library. If the *p*-value is smaller than your chosen significance level (e.g., 0.05), you would reject the null hypothesis and prefer the fixed effects model. If the *p*-value is larger than the significance level, you would not reject the null hypothesis, and the random effects model might be more appropriate.

Pooled OLS Models

Pooled Ordinary Least Squares (OLS) Models are a relatively simple approach to panel data analysis. They treat the panel data as if it were a large cross-sectional dataset, ignoring the panel structure and any potential time or entity-specific effects. The equation for the Pooled OLS model can be written as follows:

$$Y\_it = \alpha + \beta X_{it} + \varepsilon_{it},$$

where Y_{it} represents the dependent variable, X_{it} is a matrix of independent variables, α is the intercept term, β is a vector of coefficients for the independent variables, ε_{it} is the error term, and i and t are the entity and time indexes, respectively. In this model, the same coefficients are estimated for all entities and time periods.

Despite its simplicity, the pooled OLS model has some advantages. It can provide a straightforward approach for analyzing relationships between variables when there is no reason to believe that entity or time-specific effects are present. Pooled OLS models can be easily implemented and interpreted, which makes them appealing for researchers who are just starting to work with panel data. However, there are several disadvantages of using pooled OLS models. One of the most important drawbacks is the potential for omitted variable bias, which can arise when unobserved heterogeneity is present in the data. Unobserved heterogeneity refers to the presence of factors that affect the dependent variable but are not included in the model. In the context of panel data, these factors may be constant across time within an entity or constant across entities within a time period. Ignoring these factors in the pooled OLS model can lead to biased estimates of the coefficients and incorrect inferences. Also, pooled OLS models may not efficiently use the information in the panel data. As these models ignore the panel structure, they do not account for any correlations between observations on the same entity or during the same time period. This can lead to inefficient estimation and may adversely affect hypothesis testing.

Pooled OLS models offer a simple and intuitive approach to panel data analysis. However, researchers should be cautious when using them, as they may not provide accurate results when unobserved heterogeneity or correlations between observations are present in the data. Alternative methods, such as fixed effects or random effects models, may be more suitable in these cases and should be considered as part of a robust panel data analysis.

Fixed Effect Models

The intuition behind the fixed effects model lies in the assumption that each entity in the panel data has its own unique, time-invariant characteristics that may impact the dependent variable. These unique characteristics can be unobservable factors or omitted variables, which, if not accounted for, may lead to biased estimates. The fixed effects model accounts for these unobserved factors by including entity-specific intercepts in the regression model, thereby capturing the time-invariant heterogeneity across entities.

Two common estimation methods for the fixed effects model are the within-group estimator and the least squares dummy variable (LSDV) approach. The within-group estimator works by subtracting the mean of each entity's variables from their respective values, essentially removing the entity-specific effects. The LSDV approach, on the other hand, introduces a set of dummy variables for each entity, except for one reference entity. This allows the model to capture the unique intercept for each entity, thus controlling for the unobserved time-invariant heterogeneity.

Advantages of the fixed effects model include its ability to control for omitted variables that are constant over time, providing unbiased estimates in the presence of unobserved time-invariant heterogeneity. However, the fixed effects model has some limitations. It cannot estimate the effects of time-invariant variables, as these variables will be absorbed by the entity-specific intercepts. Moreover, the fixed effects model may be less efficient than other models, such as the random effects model, when the unobserved heterogeneity is not correlated with the independent variables.

Python implementation of a fixed effect model using LDSV approach will be:

```python
import pandas as pd
import statsmodels.api as sm
import statsmodels.formula.api as smf

# Read the data
data = pd.read_csv("D:/Data/my_data4.csv")

# Set the index to 'Date'
data.set_index('Date', inplace=True)

# Estimate the fixed effects model using the LSDV approach
model = smf.ols('BTC ~ ETH + BNB + ADA + C(Year)',
data=data).fit()

# Print the results
print(model.summary())
```

The result will look like:

```
                            OLS Regression Results
==============================================================================
Dep. Variable:                    BTC   R-squared:                       0.932
Model:                            OLS   Adj. R-squared:                  0.932
Method:                 Least Squares   F-statistic:                     3048.
Date:                Mon, 08 May 2023   Prob (F-statistic):               0.00
Time:                        07:25:13   Log-Likelihood:                -15273.
No. Observations:                1553   AIC:                         3.056e+04
Df Residuals:                    1545   BIC:                         3.060e+04
Df Model:                           7
Covariance Type:            nonrobust
====================================================================================
                       coef    std err          t      P>|t|      [0.025      0.975]
------------------------------------------------------------------------------------
Intercept           7819.7514    502.000     15.577      0.000    6835.078    8804.424
C(Year)[T.2018]    -2237.1805    529.427     -4.226      0.000   -3275.653   -1198.708
C(Year)[T.2019]    -1469.0608    546.860     -2.686      0.007   -2541.728    -396.394
C(Year)[T.2020]     1753.3742    536.550      3.268      0.001     700.930    2805.818
C(Year)[T.2021]     2.267e+04    776.277     29.208      0.000    2.12e+04    2.42e+04
ETH                    3.9007      0.456      8.555      0.000       3.006       4.795
BNB                   17.9787      2.549      7.052      0.000      12.978      22.979
ADA                 -463.7062    537.346     -0.863      0.388   -1517.711     590.298
==============================================================================
Omnibus:                      253.932   Durbin-Watson:                   0.040
Prob(Omnibus):                  0.000   Jarque-Bera (JB):              616.046
Skew:                           0.899   Prob(JB):                    1.69e-134
Kurtosis:                       5.508   Cond. No.                     1.46e+04
==============================================================================

Notes:
[1] Standard Errors assume that the covariance matrix of the
errors is correctly specified.
[2] The condition number is large, 1.46e+04. This might indicate
that there are strong multicollinearity or other numerical
problems.
```

The model includes fixed effects for the **Year** variable, which represents the time series dimension of the panel data. The fixed effects are represented by the **C(Year)** terms in the model formula. The results indicate that the model has a high R-squared value of 0.932, suggesting a strong relationship between the independent variables and the dependent variable. The coefficients for the **ETH** and **BNB** variables are both statistically significant, with p-values of less than 0.001. However, the coefficient for the **ADA** variable is not statistically significant, with a *p*-value of 0.388. This suggests that **ADA** may not have a significant effect on **BTC**. The fixed effects coefficients suggest that **Year** has a significant effect on **BTC**, with the largest coefficient for the **Year 2021**.

Random Effect Models

The random effects model is another commonly used panel data model. It assumes that the individual-specific effects are uncorrelated with the explanatory variables, which can be a reasonable assumption if the individual-specific effects are randomly distributed. In contrast to the fixed effects model, the random effects model treats the individual-specific effects as random variables that are drawn from a common distribution.

The intuition behind the random effects model is that it estimates the average effect of the explanatory variables across all entities, while allowing the individual-specific effects to vary randomly. This approach can be useful when the goal is to estimate the average effect of the explanatory variables while controlling for the unobserved heterogeneity between entities.

The within-group estimator and the least squares dummy variable (LSDV) approach are commonly used methods to estimate the random effects model. The within-group estimator is based on the idea of taking the difference between each observation and its group mean, which eliminates the individual-specific effects. The resulting model is then estimated using ordinary least squares (OLS). The LSDV approach, on the other hand, involves adding a set of dummy variables for each entity to the model. These dummy variables capture the individual-specific effects and allow the model to estimate the random effects.

Advantages of the random effects model include its ability to estimate the average effect of the explanatory variables across all entities, while controlling for the unobserved heterogeneity between entities. Additionally, it can be more efficient than the fixed effects model if the individual-specific effects are randomly distributed. However, random effects model assumes that the individual-specific effects are uncorrelated with the explanatory variables, which may not hold in some cases. Moreover, it assumes that the individual-specific effects are randomly distributed, which may not be true if there are systematic differences between entities. Random effects model may also suffer from the incidental parameter problem, which occurs when the number of entities is relatively small compared to the number of time periods.

A Python implementation of random effect model can be done as follows:

```python
import pandas as pd
import statsmodels.formula.api as smf

# Read in the data
data = pd.read_csv('D:/Data/my_data4.csv')

# Fit the random effects model
# Specify the formula for the model: BTC as the dependent
variable, and ETH, BNB, and ADA as the explanatory variables
model = smf.mixedlm("BTC ~ ETH + BNB + ADA", data,
groups=data['Date'], re_formula="~1")
result = model.fit()

# Print the model summary
print(result.summary())
```

The estimation method used in the **mixedlm** function is the Maximum Likelihood Estimation (MLE) approach. The MLE method estimates the parameters by maximizing the likelihood function, which is a measure of how likely the observed data is, given the parameters of the model. By using MLE, the **mixedlm** function provides consistent and efficient estimates of the fixed and random effects in the presence of unobserved group-specific heterogeneity.

The result will look like:

```
           Mixed Linear Model Regression Results
=================================================================
Model:              MixedLM    Dependent Variable:   BTC
No. Observations:   1553       Method:               REML
No. Groups:         1553       Scale:                20647469.9795
Min. group size:    1          Log-Likelihood:       -15803.9195
Max. group size:    1          Converged:            Yes
Mean group size:    1.0
-----------------------------------------------------------------
              Coef.      Std.Err.     z      P>|z|   [0.025    0.975]
-----------------------------------------------------------------
Intercept    6726.135    195.821   34.348   0.000  6342.332  7109.938
ETH             5.911      0.037  160.321   0.000     5.839     5.983
BNB            40.393      2.221   18.185   0.000    36.039    44.747
ADA          3750.929    603.463    6.216   0.000  2568.163  4933.695
Group Var 20647469.979
=================================================================
```

The model is estimated using the mixed linear model regression method, with random effects for each entity in the panel data. Intercept term estimates the expected value of the dependent variable (BTC) when all the explanatory variables are equal to zero. In this model, the intercept is 6726.135, which means that when all the explanatory variables are equal to zero, the expected value of BTC is around 6726.135. The coefficients for the explanatory variables (ETH, BNB, and ADA) estimate the expected change in the dependent variable (BTC) for a one-unit increase in each of the explanatory variables, holding all other variables constant. For example, a one-unit increase in ETH is associated with a 5.911 increase in BTC, on average, while holding BNB and ADA constant. Similarly, a one-unit increase in BNB is associated with a 40.393 increase in BTC, on average, while holding ETH and ADA constant. The standard errors of the coefficients indicate the amount of variation in the estimated coefficients due to sampling error. The z-scores measure the number of standard errors away from the null hypothesis of zero effect and are used to test the statistical significance of the coefficients. In this model, all the explanatory variables are highly statistically significant, as indicated by their low p-values and high z-scores. Group variance estimate of 20,647,469.979 indicates the amount of variation in the dependent variable that is explained by the random effects for each entity in the panel data. This value is important for assessing the relative importance of the random effects versus the fixed effects in explaining the variation in the dependent variable.

Python implementation using GLS for the same data would be:

```
import pandas as pd
from linearmodels import PanelOLS
from linearmodels.panel import RandomEffects

# Read in the data
data = pd.read_csv('D:/Data/my_data4.csv')

# Add an index column as a unique identifier
data['UniqueIdentifier'] = data.index

# Set the index as a MultiIndex with Date as the first level and
the unique identifier as the second level
data = data.set_index(['Date', 'UniqueIdentifier'])

# Create the dependent variable and independent variables
dependent_variable = data['BTC']
explanatory_variables = data[['ETH', 'BNB', 'ADA']]

# Fit the random effects model using GLS
model = RandomEffects(dependent_variable, explanatory_variables)
result = model.fit()

# Print the model summary
print(result.summary())
```

Dynamic Panel Data Models

Dynamic Panel Data Models are a more advanced approach to analyzing panel data, particularly when the relationships between variables are dynamic or change over time. These models can be especially useful in accounting and finance research, as they allow researchers to study the impact of past values of variables on current outcomes, as well as capture time-varying effects and unobserved heterogeneity. Dynamic panel data models have been used to analyze a wide range of topics, such as corporate governance, financial performance, risk management, and capital structure decisions.

The key difference between static and dynamic panel data models lies in the inclusion of lagged dependent variables as explanatory variables. While static panel data models only consider the contemporaneous relationships between variables, dynamic panel data models incorporate the past values of the dependent variable to capture the persistence or inertia in the relationships. The general equation for a linear dynamic panel data model can be written as follows:

$$Y_{it} = \alpha + \beta_1 Y_{\{i,t-1\}} + \beta_2 X_{it} + \mu_i + \lambda_t + \varepsilon_{it},$$

where Y_{it} is the dependent variable, $Y_{\{i,\ t-1\}}$ is the lagged dependent variable, X_{it} is a matrix of independent variables, α is the intercept term, β_1 and β_2 are vectors of coefficients, μ_i and λ_t are the unobserved entity-specific and time-specific effects, respectively, and ε_{it} is the error term. i and t represent the entity and time indexes, respectively.

There are several types of dynamic panel data models, each with distinct features and suited for different research scenarios. Some of the most common dynamic panel data models include:

1. Linear Dynamic Panel Data Models: These models assume a linear relationship between the lagged dependent variable, independent variables, and the current value of the dependent variable. They are the most basic form of dynamic panel data models. Linear dynamic panel data models can capture the persistence or inertia in the relationships between variables, allowing researchers to study both short-term and long-term effects

and can be estimated using methods such as the Arellano-Bond Generalized Method of Moments (GMM) or the Arellano-Bover/Blundell-Bond System GMM. A Python implementation of these techniques is provided below:

```python
import pandas as pd
import xarray as xr
from linearmodels.panel import PanelOLS, FirstDifferenceOLS, compare
from linearmodels.iv import IV2SLS

# Load the data
data = pd.read_csv("D:/Data/my_data4.csv", parse_dates=["Date"], index_col=["Date"])

# Create the panel data structure using xarray
data = data.reset_index()
data = data.set_index(['Year', 'Date'])
data = data.sort_index()
data = xr.Dataset.from_dataframe(data)

# Convert the data to a DataFrame
df = data.to_dataframe()

# Arellano-Bond GMM Estimator
mod_ab_gmm = FirstDifferenceOLS(df['BTC'], df[['ETH', 'BNB', 'ADA']])
res_ab_gmm = mod_ab_gmm.fit(cov_type="kernel", kernel="bartlett", bandwidth=4)

# Arellano-Bover/Blundell-Bond System GMM Estimator
mod_sys_gmm = PanelOLS(df['BTC'], df[['ETH', 'BNB', 'ADA']],
entity_effects=True, time_effects=True, drop_absorbed=True)
res_sys_gmm = mod_sys_gmm.fit(cov_type="kernel", kernel="bartlett", bandwidth=4)

# Compare the results
print(compare({"Arellano-Bond GMM": res_ab_gmm, "System GMM": res_sys_gmm}))
```

Here, Arellano-Bond Generalized Method of Moments (GMM) estimator is implemented using the **FirstDifferenceOLS** function from the **linearmodels** package, with the dependent variable 'BTC' and independent variables 'ETH', 'BNB', and 'ADA'. The model is fitted, and the covariance matrix is specified as "kernel" with a Bartlett kernel and a bandwidth of 4. Similarly, the Arellano-Bover/Blundell-Bond System GMM estimator is implemented using the PanelOLS function with entity and time effects, and the same variables and covariance matrix specification as in the Arellano-Bond GMM.

Finally, the results of both models are compared and printed using the compare function from linearmodels.

2. Nonlinear Dynamic Panel Data Models: Nonlinear Dynamic Panel Data Models are a type of dynamic panel data model that allows for non-linear relationships between the dependent variable and the explanatory variables. These models can provide a more flexible framework for capturing complex dynamics in the data, such as threshold effects, regime-switching behavior, or interactions among variables. Non-linear dynamic panel data models can be particularly useful in accounting and finance research when linear models fail to accurately represent the underlying relationships.

Some common types of Nonlinear Dynamic Panel Data Models include:

- Dynamic Probit Models: These models are used when the dependent variable is binary, representing two possible outcomes. The dynamic probit model includes lagged values of the dependent variable as explanatory variables, capturing persistence or state-dependence in the binary outcomes. For instance, a dynamic probit model could be used to analyze corporate bankruptcy predictions, where the binary outcome could represent bankruptcy (1) or no bankruptcy (0).

A Python implementation dynamic probit model on D:/Data/my_data4.csv is as follows:

```python
import pandas as pd
from linearmodels.panel import PooledOLS
import statsmodels.api as sm

# Load your data
df = pd.read_csv('D:/Data/my_data4.csv')

# Ensure Date is in datetime format
df['Date'] = pd.to_datetime(df['Date'])

# Dummy entity
df['Year'] = 'Year_1'  # replace with actual entity column if available

# Set multi-index with Entity and Date
df.set_index(['Year', 'Date'], inplace=True)

# Create lagged dependent variable
df['BTC_lag'] = df.groupby('Year')['BTC'].shift(1)

# Drop the missing values generated by lag operation
df.dropna(inplace=True)

# Specify model (Note: this is a pooled OLS model, not a dynamic probit model)
exog = sm.add_constant(df[['ETH', 'BNB', 'ADA', 'BTC_lag']])
endog = df['BTC']
model = PooledOLS(endog, exog)
pooled_ols_res = model.fit()

# Print model summary
print(pooled_ols_res)
```

- The **linearmodels** package expects the data to be structured as a multi-index DataFrame, with the first level of the index being the entity (for example, a company or individual identifier) and the second level being the time dimension. As the data used here does not have a multi-index, a dummy **Year_1** has been introduced, which is not needed if you are dealing with a data with multi-level index.
- Dynamic Logit Models: Similar to probit models, dynamic logit models are used for binary dependent variables. However, they assume a logistic distribution for the error term, as opposed to the normal distribution assumed in the probit model. The choice between a logit and probit model typically depends on the specific research context and theoretical considerations.

Python's statsmodels and linearmodels libraries don't support dynamic logit models directly. However, you can create a workaround by including a lagged dependent variable in a standard Logit model.

```python
import pandas as pd
import statsmodels.api as sm

# Load your data
df = pd.read_csv('D:/Data/my_data4.csv')

# Ensure Date is in datetime format
df['Date'] = pd.to_datetime(df['Date'])

# Dummy entity
df['Year'] = df['Date'].dt.year

# Set index with Date
df.set_index('Date', inplace=True)

# Create lagged dependent variable
df['BTC_binary_lag'] = df.groupby('Year')['BTC_binary'].shift(1)

# Drop the missing values generated by lag operation
df.dropna(inplace=True)

# Specify dependent variable
endog = df['BTC_binary']

# Specify independent variables
exog = sm.add_constant(df[['ETH_returns', 'BNB_returns',
'ADA_returns', 'BTC_binary_lag']])

# Estimate the logit model
model = sm.Logit(endog, exog)
logit_res = model.fit()

# Print model summary
print(logit_res.summary())
```

- Dynamic Tobit Models: These models are used when the dependent variable is censored, meaning that it is only observed within a certain range. The dynamic Tobit model incorporates past values of the dependent variable to capture dynamic relationships in the censored outcomes. This could be useful, for example, in analyzing the impact of past financial performance on current dividend payouts, where the dividend payout cannot fall below zero.

Python libraries like **statsmodels** or **scikit-learn** do not directly support dynamic Tobit models. However, we can manually implement a Tobit model using a maximum likelihood estimation (MLE) approach. Below is a Python script that demonstrates how to implement a simple Tobit model.

```python
import pandas as pd
import numpy as np
from scipy import stats
from scipy.optimize import minimize

# Load the data
df = pd.read_csv('D:/Data/my_data4.csv')

# Trim the data
df = df[df['BTC_returns'] >= 0]

# Define the dependent and independent variables
y = df['BTC_returns']
X = df[['ETH_returns', 'BNB_returns', 'ADA_returns']]
X['lagged_BTC_returns'] = df['BTC_returns'].shift(1)   # Add the
lagged dependent variable
X = X.fillna(0)   # Handle any NaN values resulting from the lag
operation

# Add a constant to the independent variables matrix
X = sm.add_constant(X)

# Define the likelihood function
def likelihood(params):
    beta = params[:-1]
    sigma = params[-1]
    cdf_arg = (0 - np.dot(X, beta)) / sigma
    pdf_arg = (y - np.dot(X, beta)) / sigma
    ll = np.sum(stats.norm.logcdf(cdf_arg)) + np.sum(stats.norm.logpdf(pdf_arg))
    return -ll

# Initial guess for the parameters (betas and sigma)
init_params = np.ones(X.shape[1] + 1)

# Run the optimizer
res = minimize(likelihood, init_params, method='Nelder-Mead')

# Print the estimated parameters
print('Estimated parameters: ', res.x)
```

3. **Dynamic Panel Data Models with Time-Varying Coefficients**: These represent a particular class of dynamic panel models that can capture more complex temporal relationships. As the name suggests, these models allow the coefficients of the explanatory variables to change over time. This feature can be especially useful in situations where the relationships between variables are not constant, but instead evolve over time due to structural changes, policy shifts, or changing market conditions.

The general form of a dynamic panel data model with time-varying coefficients can be written as follows:

$$Y_{it} = \alpha_t + \beta_t * Y_{\{i,t-1\}} + \gamma_{t'} * X_{it} + \mu_i + \lambda_t + \varepsilon_{it},$$

where Y_{it} is the dependent variable, $Y_{\{i, t-1\}}$ is the lagged dependent variable, X_{it} is a matrix of independent variables, α_t, β_t, and γ_t are time-varying intercepts and coefficients, μ_i and λ_t are the unobserved entity-specific and time-specific effects, respectively, and ε_{it} is the error term. i and t represent the entity and time indexes, respectively. In these models, the coefficients β_t and γ_t are allowed to vary over time, capturing the temporal changes in the relationships between the dependent variable and the lagged dependent variable or independent variables. This allows the model to capture more complex dynamics, such as non-stationarity, structural breaks, or regime changes in the data.

Estimation of dynamic panel data models with time-varying coefficients can be challenging, as it requires the estimation of a larger number of parameters and involves more complex assumptions. Common estimation techniques include Maximum Likelihood, Bayesian methods, and Kalman Filtering. These methods often rely on strong assumptions about the process governing the time-varying coefficients, such as linearity or stationarity, and may require large amounts of data for reliable estimation.

4. Dynamic Panel Threshold Models: These are a type of dynamic panel data model that is particularly useful for examining non-linear relationships in panel data. These models allow for different regimes or states in the data, which can capture discontinuities or threshold effects in the relationships between variables. In a Dynamic Panel Threshold Model, the relationship between the dependent variable and the independent variables can change depending on the value of a threshold variable. The threshold variable can be one of the independent variables or an entirely separate variable. The model assumes that the data switches between different regimes when the threshold variable crosses a certain threshold value.

The general form of a Dynamic Panel Threshold Model can be written as follows:

$$Y_{it} = \alpha_1 + \beta_1 Y_{\{i,t-1\}} + \gamma_1' X_{it} + \varepsilon_{it} \text{ if } Z_{it} \leq \tau,$$

$$Y_{it} = \alpha_2 + \beta_2 Y_{\{i,t-1\}} + \gamma_2' X_{it} + \varepsilon_{it} \text{ if } Z_{it} > \tau,$$

where Y_it is the dependent variable, $Y_{\{i,\,t-1\}}$ is the lagged dependent variable, X_{it} is a matrix of independent variables, α_1, α_2, β_1, β_2, γ_1, and γ_2 are parameters to be estimated, Z_{it} is the threshold variable, τ is the threshold value, and ε_{it} is the error term. i and t represent the entity and time indexes, respectively.

The key feature of Dynamic Panel Threshold Models is their ability to capture different dynamics in different regimes, which can help uncover more complex and nuanced relationships in the data. For example, these models can show how the effect of a financial shock or policy change may differ above and below a certain threshold. Estimating Dynamic Panel Threshold Models can be quite challenging due to the presence of endogeneity and autocorrelation issues, as well as the need to determine the threshold value. The estimation usually involves iterative procedures, where the threshold value and the model parameters are estimated simultaneously. Hansen (1999) developed an estimator for static panel threshold models, and this approach can be extended to dynamic panel threshold models.

5. Dynamic Panel Data Models with Spatial Dependence: These are a sophisticated type of dynamic panel data models that account for spatial interactions or dependencies between the entities under study. These models are increasingly used in fields such as regional science, environmental economics, urban planning, and other areas where spatial relationships are crucial. In the context of accounting and finance research, dynamic panel data models with spatial dependence could be used to analyze phenomena such as the diffusion of corporate practices across firms, the impact of geographical proximity on investment decisions, or the spillover effects of financial shocks across countries or regions.

Spatial dependence in a dynamic panel data model implies that the dependent variable for one entity at a particular time period could be influenced not just by its own past values or the current and past values of other variables, but also by the values of the dependent variable for other entities. This spatial interaction is typically modeled through a spatial weight matrix that captures the relationship or proximity between entities.

The general form of a spatial dynamic panel data model can be written as follows:

$$Y_{it} = \rho W Y_{it} + \alpha + \beta_1 Y_{\{i,t-1\}} + \beta_2 X_{it} + \mu_i + \lambda_t + \varepsilon_{it}.$$

In this equation, WY_{it} is the spatially lagged dependent variable, where W is the spatial weight matrix and Y_{it} is the matrix of the dependent variable. ρ is the spatial autoregressive coefficient that measures the strength of the spatial dependence.

Estimating dynamic panel data models with spatial dependence can be challenging due to issues related to endogeneity, autocorrelation, and the proper specification of the spatial weight matrix. Several estimation methods have been developed to address these issues, including the spatial autoregressive model (SAR), the spatial error model (SEM), and the spatial Durbin model (SDM). The SAR model addresses spatial dependence in the dependent variable, the SEM model deals with spatial correlation in the error terms, and the SDM model accommodates spatial lags in both the dependent variable and independent variables. These models employ different strategies to achieve consistent and efficient estimation, such as Maximum Likelihood, Generalized Method of Moments, or instrumental variables techniques.

6. Dynamic Panel Data Models with Heterogeneous Coefficients: These are a type of panel data analysis that allows for the coefficients of the lagged dependent variable and independent variables to vary across entities or groups of entities. This feature enables researchers to study cross-sectional heterogeneity in the relationships between variables, such as differences in the impact of a policy or a financial shock across firms or countries. In contrast to standard dynamic panel data models, where the coefficients are assumed to be constant across all entities, DPD models allow for the possibility that the relationships between the variables differ across entities. This can be useful for capturing the unique characteristics of each entity and identifying factors that drive the variation in the relationships between the variables.

The equation for the DPD model can be written as follows:

$$Y_{\{it\}} = \alpha_i + \beta_i Y_{\{i,t-1\}} + X'_{\{it\}} \gamma_i + \varepsilon_{\{it\}},$$

where $Y_{\{it\}}$ is the dependent variable, $Y_{\{i, t-1\}}$ is the lagged dependent variable, $X'_{\{it\}}$ is a matrix of independent variables, α_i is the entity-specific intercept term, β_i and γ_i are the entity-specific coefficients for the lagged dependent variable and independent variables, respectively, and $\varepsilon_{\{it\}}$ is the error term. i and t represent the entity and time indexes, respectively.

Estimating DPD models can be challenging, as they involve a large number of parameters to be estimated and may require specialized estimation techniques. One approach to estimating DPD models is the Mean Group (MG) estimator, which assumes that the coefficients are different across entities but common over time. Another approach is the Pooled Mean Group (PMG) estimator, which assumes that the coefficients vary both across entities and over time but are constrained by a common mean. In addition to the MG and PMG estimators, researchers can also use other advanced techniques, such as the Least-Squares Dummy Variables (LSDV) estimator, which allows for the inclusion of entity-specific fixed effects to control for unobserved heterogeneity, or the Bayesian estimation approach, which can incorporate prior information and improve the precision of the estimates.

Panel Data Model Diagnostics

To ensure the reliability of the model results, it is important to perform diagnostic tests to assess the validity of the assumptions and the model fit.

Some common diagnostic tests for panel data models include:

- Heteroskedasticity tests: Heteroskedasticity refers to the presence of unequal variances in the error terms across entities or time periods. This can lead to biased and inefficient estimates of the coefficients. Diagnostic tests for heteroskedasticity include the Breusch–Pagan test, the White test, and the clustered standard errors approach. The code for these tests would be:

```
# Breusch-Pagan test
from statsmodels.stats.diagnostic import het_breuschpagan
bp_test = het_breuschpagan(model.resid, model.model.exog)
print('Breusch-Pagan test p-value:', bp_test[1])

# White test
from statsmodels.stats.diagnostic import het_white
white_test = het_white(model.resid, model.model.exog)
print('White test p-value:', white_test[1])

# Clustered standard errors
clustered_se = model.get_robustcov_results(cov_type='cluster',
use_correction=True, groups=data['Year'])
print('Clustered standard errors:\n', clustered_se.summary())
```

This is the code for the fixed effect model already discussed earlier in this chapter. Adaptation of this code needs to be included at the end of any regression code.

- Autocorrelation tests: Autocorrelation refers to the presence of correlation between the error terms across time periods within an entity. This can lead to biased and inefficient estimates of the coefficients. Diagnostic tests for autocorrelation include the Durbin–Watson test, the Breusch–Godfrey test, and the Arellano–Bond test.

```
# Durbin-Watson test
from statsmodels.stats.stattools import durbin_watson

dw_test = durbin_watson(model.resid)
print('Durbin-Watson test statistic:', dw_test)

# Breusch-Godfrey test
from statsmodels.stats.diagnostic import acorr_breusch_godfrey

bg_test = acorr_breusch_godfrey(model, nlags=1)
print('Breusch-Godfrey test p-value:', bg_test[1])
```

For Durbin–Watson test statistic, a value close to 2 suggests no significant autocorrelation, below 2 indicates positive autocorrelation and value above 2 indicates negative autocorrelation for residuals. A significant Breusch–Godfrey test p-value suggests autocorrelation in the residuals. Interpretation of the Arellano–Bond test involves comparing the test statistic against critical values at a chosen significance level to assess the presence of serial correlation and endogeneity.

- Multicollinearity tests: Multicollinearity refers to the high correlation between two or more independent variables, which can lead to instability and imprecision in the coefficient estimates. Diagnostic tests for multicollinearity include the variance inflation factor (VIF) test and the condition number test.

```
# VIF test
from statsmodels.stats.outliers_influence import
variance_inflation_factor
vif = pd.DataFrame({'VIF':
[variance_inflation_factor(model.model.exog, i) for i in
range(model.model.exog.shape[1])]},
index=model.model.exog_names)
print('VIF:\n', vif)

# Condition number test
from numpy.linalg import cond
cond_num = cond(model.model.exog)
print('Condition number:', cond_num)
```

VIF values below 1 suggest no multicollinearity, while values between 1 and 5 indicate low to moderate multicollinearity. VIF values between 5 and 10 suggest a moderate level of multicollinearity, while values above 10 indicate a high level of multicollinearity.

- Outliers and influential observations tests: Outliers and influential observations can have a disproportionate effect on the model estimates and lead to biased results. Diagnostic tests for outliers and influential observations include the Cook's distance test and the leverage plot.

```
# Cook's distance test
from statsmodels.stats.outliers_influence import OLSInfluence
import matplotlib.pyplot as plt

influence = OLSInfluence(model)
cooks_dist = pd.Series(influence.cooks_distance[0],
index=data.index, name="Cook's Distance")
print('Cook\'s distance:\n', cooks_dist)

# Leverage plot
from statsmodels.graphics.regressionplots import
plot_leverage_resid2
fig, ax = plt.subplots(figsize=(10, 6))
plot_leverage_resid2(model, ax=ax)
plt.show()
```

Cook's distance is a measure of the influence of a data point on a regression model. A value less than 1 indicates negligible influence, between 1 and 4 suggests moderate influence, and greater than 4 signifies a significant influence that may require careful consideration or potential removal from the analysis.

- Normality tests: Normality assumptions are often made about the error terms in panel data models. Diagnostic tests for normality include the Jarque–Bera test, the Shapiro–Wilk test, and the Kolmogorov–Smirnov test.

```python
# Jarque-Bera test
from scipy.stats import jarque_bera
jb_test = jarque_bera(model.resid)
print('Jarque-Bera test statistic:', jb_test[0])
print('Jarque-Bera test p-value:', jb_test[1])

# Shapiro-Wilk test
from scipy.stats import shapiro
shapiro_test = shapiro(model.resid)
print('Shapiro-Wilk test statistic:', shapiro_test[0])
print('Shapiro-Wilk test p-value:', shapiro_test[1])

# Kolmogorov-Smirnov test
from scipy.stats import kstest
ks_test = kstest(model.resid, 'norm')
print('Kolmogorov-Smirnov test statistic:', ks_test[0])
print('Kolmogorov-Smirnov test p-value:', ks_test[1])
```

For all these tests, a rejection of null hypothesis, as indicated by the test statistic/p values, indicates that the data is not normally distributed.

- Model specification tests: Model specification refers to the choice of functional form, the inclusion/exclusion of variables, and the selection of lag lengths. Diagnostic tests for model specification include the Ramsey Regression Equation Specification Error Test (RESET) and the Hausman test. RESET can be implemented as below:

```python
# RESET test
from statsmodels.stats.diagnostic import linear_reset
reset_test = linear_reset(model)
print('RESET test p-value:', reset_test.pvalue)
```

In the context of the RESET test, a significant p-value suggests that the observed relationship between the dependent variable and the independent variables cannot be adequately explained by the current model, indicating a potential functional form misspecification. Hausman test has already been discussed earlier.

Panel data analysis is a powerful and widely used method in academic research, particularly in the fields of accounting and finance. Panel data, also known as longitudinal or panel longitudinal data, refers to data that contains observations on multiple entities (such as firms or individuals) over time. This data structure enables researchers to analyze both cross-sectional and time series variations, providing valuable insights into various phenomena.

In academic accounting and finance research, panel data analysis has proven to be invaluable in investigating complex relationships and addressing key research questions. One major advantage of panel data is its ability to control for unobserved heterogeneity across entities. By including entity-specific fixed effects or random effects in the analysis, panel data models can account for individual characteristics that remain constant over time but may influence the dependent variable of interest. This helps mitigate omitted variable bias and provides more accurate estimates of the relationships under study.

Panel data analysis also allows researchers to examine dynamic relationships and explore the effects of changes over time. By capturing both within-entity and between-entity variations, panel data models provide insights into the short-term and long-term dynamics of various financial and accounting phenomena. Researchers can assess how factors such as corporate governance practices, financial reporting quality, capital structure decisions, and market conditions evolve and affect firm performance over time.

Moreover, panel data analysis enables researchers to study the impact of policy changes, regulatory interventions, or other exogenous shocks on accounting and financial outcomes. By comparing entities before and after a specific event or policy change, panel data models help identify the causal effects and evaluate the effectiveness of various interventions. This is particularly relevant in accounting and finance research where policy implications and regulatory changes have profound implications for financial markets, corporate behavior, and investor protection.

Panel data also allows for more efficient use of data and improves statistical power. By utilizing information from multiple entities and time periods, panel data analysis increases the effective sample size, thus enhancing the precision and reliability of the estimates. This is particularly beneficial in accounting and finance research, where datasets are often limited and it is crucial to maximize the information available.

In summary, panel data analysis is an essential tool in academic accounting and finance research. It offers unique advantages by accounting for unobserved heterogeneity, capturing dynamic relationships, assessing policy effects, and improving statistical efficiency. The application of panel data analysis enhances our understanding of financial and accounting phenomena, supports evidence-based policymaking, and contributes to the advancement of knowledge in these fields. As such, it plays a vital role in shaping the academic discourse and informing decisions in accounting and finance.

26

Special Techniques in Multivariate Analysis

Multivariate analysis has gained recognition in the accounting and finance research field as an indispensable tool, particularly for its ability to uncover patterns and relationships among multiple variables simultaneously. Unlike univariate and bivariate analysis, which focus on isolated variables or pairwise associations, multivariate analysis captures the intricate complexities inherent in financial phenomena. This comprehensive approach is particularly vital in studying phenomena such as corporate bankruptcy, where variables like liquidity ratios, profitability, leverage, and firm size collectively contribute to bankruptcy likelihood. By employing multivariate regression and other appropriate techniques, researchers can quantitatively assess the combined impact of these variables, providing a holistic understanding of the phenomenon. Moreover, multivariate analysis allows researchers to address confounding variables and establish causal relationships, a crucial aspect in many accounting and finance research studies. In this chapter, we will discuss some useful techniques of multivariate analysis.

Principal Component Analysis

Principal Component Analysis (PCA) is a widely used statistical technique in the field of academic research, particularly in the disciplines of accounting and finance. The method is rooted in the concept of dimension reduction, which seeks to simplify the complexity of multivariate data by extracting a set of uncorrelated variables, known as principal components. These components are linear combinations of the original variables that capture the maximum variance in the data, providing a comprehensive view of the dataset's structure.

One of the primary reasons for the ubiquity of PCA in accounting and finance research is its power in dealing with multicollinearity among variables. Multicollinearity is a common issue in these fields due to the interconnected nature of financial indicators. For instance, variables such as revenue, net income, and cash flow are often highly correlated, which can undermine the interpretability and reliability of regression models. PCA addresses this problem by creating principal components that are orthogonal to each other, thereby eliminating the issue of multicollinearity. Furthermore, PCA is instrumental in the visualization of high-dimensional data. As research in accounting and finance often involves dealing with numerous variables, it becomes challenging to visualize and understand the relationships among them. By reducing the dimensionality of the data, PCA allows researchers to visualize these relationships in two or three dimensions, thus providing insights that may not be readily apparent in the higher-dimensional space. In terms of interpretability, PCA can also help identify underlying latent factors that contribute to the variability in the data. For example, in finance, a few principal components might capture market-wide factors such as interest rates, inflation, and market sentiment. These components can then be used in subsequent analyses, providing a simpler, more interpretable model while retaining most of the information from the original dataset.

However, despite these advantages, PCA is not without its limitations. One significant drawback is that the principal components are often difficult to interpret. While they are linear combinations of the original variables, they do not always map directly onto intuitive concepts or factors. This lack of interpretability can make it challenging to communicate results and can limit the usefulness of PCA in certain contexts. Moreover, PCA assumes linear relationships among variables, which might not always hold in real-world financial data. This assumption can lead to the misrepresentation of underlying patterns and relationships, particularly when non-linear relationships exist. Furthermore, PCA, by design, gives more importance to variables with larger variances, which can sometimes lead to biased results if the variables are not appropriately scaled or if the variance is not necessarily indicative of the variable's importance.

Here is an example of a data D:/Data/acctg.csv in which gvkey is the cross sectional identifier and fyear is the time variable. The independent variable is ue_ce and rest all variables, eight in number, are dependent variables. PCA implementation, assuming there are no missing values and all variables are numerical, is as follows:

```
import pandas as pd
from sklearn.preprocessing import StandardScaler
from sklearn.decomposition import PCA

# Load your dataset
df = pd.read_csv('D:/Data/acctg.csv')

# Define the features and the target
features = df.drop(['gvkey', 'fyear', 'ue_ce'], axis=1)
target = df['ue_ce']

# Standardize the features to have mean=0 and variance=1
scaler = StandardScaler()
features_scaled = scaler.fit_transform(features)

# Apply PCA
pca = PCA(n_components=3)   # Reduce dimensionality to 3
features_pca = pca.fit_transform(features_scaled)

# Convert the results back to a pandas dataframe
pca_df = pd.DataFrame(data = features_pca, columns = ['Principal Component 1', 'Principal Component 2', 'Principal Component 3'])

# Add the target variable back into the DataFrame
pca_df['ue_ce'] = target

print(pca_df.head())
```

The resultant data will consist of 4 variables i.e., dependent variable and three components. The result will look like:

	Principal Component 1	Principal Component 2	Principal Component 3	ue_ce
0	-0.949901	-0.575974	-0.053595	0.016199
1	-0.778554	0.913777	0.329411	-0.019262
2	0.248846	-0.455588	-0.063551	0.019104
3	-1.022333	-0.022026	-0.178154	-0.002423
4	-0.994182	-0.230897	-0.12789	0.016183

Determining the optimal number of components in Principal Component Analysis (PCA) is a significant decision that affects the balance between maximizing explained variance and minimizing information loss due to dimensionality reduction. Several methods have been proposed in the literature to address this question, including:

1. **Eigenvalue One Criterion (Kaiser Criterion):** This rule suggests retaining all principal components with eigenvalues greater than one. The intuition behind this rule is that any principal component with an eigenvalue less than one explains less variance than a single original variable and thus may not be worth retaining.
2. **Scree Plot:** A scree plot is a simple line segment plot that shows the fraction of total variance in the data as explained or represented by each component. This is more of a visual aid, wherein the 'elbow' or 'knee' of the plot (the point of inflection on the curve) is often considered the appropriate number of dimensions.
3. **Cumulative Explained Variance Rule:** This rule involves deciding the number of components necessary to reach a certain cumulative explained variance threshold, commonly set at 95% or 99%. One starts by calculating the total variance explained by each component (in descending order), then adds these variances until the cumulative variance exceeds the pre-set threshold.
4. **Cross-Validation Methods:** This method involves running PCA on different subsets of data and identifying the number of components that minimize the cross-validation error. One popular approach for cross-validation in PCA is the "n-fold" method, where the dataset is divided into 'n' subsets, PCA is performed on 'n-1' subsets, and the model is validated on the remaining subset.
5. **Parallel Analysis:** Parallel analysis is a sophisticated method that compares the explained variance of each component to the explained variance of components from random data. Only components that explain more variance than the components from the random data are retained.

Here's a Python code that implements Kaiser Criterion, Scree Plot, and Cumulative Explained Variance Rule on the data mentioned earlier:

```python
import pandas as pd
import numpy as np
from sklearn.preprocessing import StandardScaler
from sklearn.decomposition import PCA
import matplotlib.pyplot as plt

# Load your dataset
df = pd.read_csv('D:/Data/acctg.csv')

# Define the features and the target
features = df.drop(['gvkey', 'fyear', 'ue_ce'], axis=1)

# Standardize the features to have mean=0 and variance=1
scaler = StandardScaler()
features_scaled = scaler.fit_transform(features)

# Apply PCA
pca = PCA()
features_pca = pca.fit(features_scaled)

# Kaiser Criterion
eigenvalues = pca.explained_variance_
n_components_kaiser = len(eigenvalues[eigenvalues > 1])

# Scree plot
plt.plot(np.arange(1, len(eigenvalues) + 1), eigenvalues, 'ro-',
linewidth=2)
plt.title('Scree Plot')
plt.xlabel('Principal Component')
plt.ylabel('Eigenvalue')
plt.show()

# Cumulative explained variance
cumulative_explained_variance =
np.cumsum(pca.explained_variance_ratio_)
n_components_95 =
len(cumulative_explained_variance[cumulative_explained_variance
<= 0.95])
n_components_99 =
len(cumulative_explained_variance[cumulative_explained_variance
<= 0.99])

# Print the results
print(f"Number of components according to Kaiser criterion:
{n_components_kaiser}")
print(f"Number of components to explain 95% variance:
{n_components_95}")
print(f"Number of components to explain 99% variance:
{n_components_99}")
```

k-fold cross-validation for different numbers of components, where k is specified by **n_folds** on the same data can be implemented as below:

```python
from sklearn.model_selection import cross_val_score

# Number of folds
n_folds = 5

# Maximum number of components to test
max_components = min(len(df), len(features.columns))

# List to store cross-validated scores
cv_scores = []

for n in range(1, max_components + 1):
    pca = PCA(n_components=n)
    scores = cross_val_score(pca, features_scaled, cv=n_folds)
    cv_scores.append(np.mean(scores))

# Plot the scores
plt.plot(range(1, max_components + 1), cv_scores, 'ro-',
linewidth=2)
plt.title('Cross-Validated Scores')
plt.xlabel('Number of Components')
plt.ylabel('Cross-Validated Score')
plt.show()

# The optimal number of components is the one that gives the
highest cross-validated score
optimal_components_cv = np.argmax(cv_scores) + 1
print(f"Optimal number of components (cross-validation): 
{optimal_components_cv}")
```

The cross-validated scores are then plotted against the number of components. The optimal number of components is the one that gives the highest cross-validated score.

Parallel analysis using Kaiser–Meyer–Olkin (KMO) measure, which is a measure of sampling adequacy recommended for factor analysis, is demonstrated below. The code performs parallel analysis by comparing the eigenvalues from the PCA with those from a random dataset. The optimal number of components is the one for which the eigenvalue is greater than the corresponding random value.

```
import numpy as np
from factor_analyzer.factor_analyzer import calculate_kmo

# Calculate the Kaiser-Meyer-Olkin (KMO) measure
kmo_all, kmo_model = calculate_kmo(features)

# Apply PCA again to compute eigenvalues
pca = PCA()
pca.fit(features_scaled)
eigenvalues = pca.explained_variance_

# Perform parallel analysis
parallel_analysis_results = pd.DataFrame({'eigenvalues':
eigenvalues})
random_val = []

for i in range(len(features.columns), 0, -1):

random_val.append(np.percentile(np.random.random_sample([1000,
i]).std(axis=0), 95))

parallel_analysis_results['random_val'] = random_val

# The optimal number of components is the one for which the
eigenvalue is greater than the corresponding random value
optimal_components_pa =
len(parallel_analysis_results[parallel_analysis_results['eigenva
lues'] > parallel_analysis_results['random_val']])

print(f"Optimal number of components (parallel analysis):
{optimal_components_pa}")
```

Factor Analysis

Factor Analysis is a multifaceted, statistical procedure predominantly utilized in the behavioral and social sciences, but its utility extends to the domain of accounting and finance research as well. It is a technique to identify latent variables or 'factors' from a set of observed variables. By doing so, it can condense the information contained in the original variables into a smaller set of factors with minimum loss of information.

Factor Analysis can be of profound importance in accounting and finance research. In many scenarios, the data under study in these fields contain numerous variables, which often may be correlated. Factor Analysis aids in simplifying the data structure by reducing the number of variables and detecting structure in the relationships between variables, which are used to classify variables. Consequently, it assists researchers in identifying the underlying dimensions of data, such as identifying the underlying factors that explain the patterns of correlations within a set of observed variables. For example, in finance research, Factor Analysis can be utilized to identify the key factors driving investment returns, while in accounting, it can help identify the major factors that influence a firm's financial performance or risk.

The application of Factor Analysis provides several advantages. It simplifies data interpretation by reducing the dimensionality. By dealing with fewer factors, as opposed to numerous individual variables, researchers can gain a more lucid understanding of the data. It also helps to remove the redundancy or duplication in the dataset due to correlated variables, leading to a more parsimonious model. This simplification assists in the creation of models that are less complex, more interpretable, and more generalizable. However, Factor Analysis assumes linear relationships among the variables, which may not always hold true. Also, the interpretability of the factors can sometimes be challenging, given that they are latent constructs. The Factor Analysis procedure is also sensitive to outliers, and thus, the data requires careful preprocessing before the analysis. Moreover, the results of the analysis can be complex and require a certain level of subjectivity in interpretation, which can lead to different researchers drawing different conclusions from the same data.

Despite these limitations, the utility of Factor Analysis in accounting and finance research is undeniable. By enabling the extraction of maximum common variance from all variables and putting them into a common score, Factor Analysis helps in ascertaining the most crucial variables in multivariate data. It provides an effective method to examine the interrelationships among a large number of variables and to explain these variables in terms of their common underlying dimensions.

Here's a Python code for Factor Analysis on the acctg.csv data we mentioned earlier:

```python
import pandas as pd
from factor_analyzer import FactorAnalyzer
import matplotlib.pyplot as plt

# Loading data
df = pd.read_csv('D:/Data/acctg.csv')

# Drop unnecessary columns if any
# df.drop(['gvkey', 'fyear'], axis=1, inplace=True)

# Dropping missing values rows
df.dropna(inplace=True)

# Checking adequacy
from factor_analyzer.factor_analyzer import
calculate_bartlett_sphericity
chi_square_value, p_value = calculate_bartlett_sphericity(df)
print(chi_square_value, p_value)

# Kaiser-Meyer-Olkin (KMO) Test measures the suitability of data
for factor analysis.
from factor_analyzer.factor_analyzer import calculate_kmo
kmo_all, kmo_model = calculate_kmo(df)

# Create factor analysis object and perform factor analysis
fa = FactorAnalyzer(rotation=None)
fa.fit(df)

# Check Eigenvalues
ev, v = fa.get_eigenvalues()

# Create scree plot using matplotlib
plt.scatter(range(1, df.shape[1]+1), ev)
plt.plot(range(1, df.shape[1]+1), ev)
plt.title('Scree Plot')
plt.xlabel('Factors')
plt.ylabel('Eigenvalue')
plt.grid()
plt.show()

# Perform factor analysis using the optimal number of factors
(based on scree plot)
fa = FactorAnalyzer(n_factors=3, rotation="varimax")   # Change
n_factors based on scree plot
fa.fit(df)

# Get the loadings
loadings = fa.loadings_
print(loadings)

# Get the factor variance
variance = fa.get_factor_variance()
print(variance)
```

You should adjust the number of factors (n_factors) based on the scree plot and your research questions. Note that the factor loadings represent the correlation between the observed variables and the factors.

The output will look like:

```
[[-4.49057623e-02 -4.23845990e-02 -4.26311450e-02]
 [-8.14329929e-03 -8.52487380e-03 -1.10104268e-02]
 [ 9.50644054e-01  1.05927663e-01 -5.29583997e-04]
 [ 9.25010472e-01  5.02712859e-02  2.48504104e-03]
 [ 1.83790709e-03 -1.80669699e-02  3.13259709e-01]
 [ 4.28985583e-01  8.77196326e-01  1.42728078e-02]
 [-3.52560937e-03  1.59205441e-02  1.01581817e+00]
 [-1.68251854e-01  9.02744690e-01  3.57748456e-02]
 [ 1.89518324e-02  3.57814760e-02  5.86684951e-01]
 [ 5.84235636e-02  3.61491311e-01 -4.26781119e-02]
 [ 1.61684290e-02 -3.13142174e-02  4.05019334e-01]]
(array([1.97783836, 1.73355511, 1.64350816]), array([0.17980349,
0.15759592, 0.14940983]), array([0.17980349, 0.33739941,
0.48680924]))
```

The first matrix shown is the 'Factor Loadings' matrix, which reflects the correlation between each variable and the inferred factors. Each row represents a variable, and each column represents a factor. A larger absolute value for a loading indicates a stronger association between the variable and the factor. Loadings can be positive or negative, indicating the direction of the relationship between a variable and a factor.

The second array shown in the output is the 'Factor Variance' matrix. This matrix contains three rows. The first row shows the 'SS Loadings', which is the sum of squared loadings for each factor (the sum of the square of each column in the loadings matrix). It represents the variance in all the variables accounted for by each factor. The second row shows the 'Proportion Var', which is the proportion of the total variance accounted for by each factor. The third row shows the 'Cumulative Var', which is the cumulative proportion of the total variance accounted for by the current and all preceding factors.

The interpretation of these results is largely contingent on the context of data and the research questions you are trying to answer. Your task as a researcher is to assign meaningful interpretations to the factors based on the loadings of the variables on the factors.

Cluster Analysis

Cluster analysis, a fundamental technique in the domain of multivariate analysis, holds substantial importance in academic research within accounting and finance. This analytical technique strives to identify homogeneous subgroups within larger heterogeneous datasets. It offers a versatile methodology for exploratory data analysis, hypothesis generation, and pattern recognition, facilitating an effective distillation of the essence from large volumes of data.

26 Special Techniques in Multivariate Analysis

In the broader context of accounting and finance research, cluster analysis can be employed for a multitude of tasks. For instance, it is extensively utilized for market segmentation, where investors or markets with similar characteristics are clustered together. This could aid in identifying distinct groups of investors who react differently to financial information, thereby providing valuable insights into investor behavior. Similarly, cluster analysis can be instrumental in identifying patterns in financial ratios or performances of companies, enabling more precise categorization of firms for comparative analysis. Cluster analysis can also be used for portfolio management and risk analysis. By clustering assets based on their returns and volatility, finance researchers can construct well-diversified portfolios that optimize the risk-return trade-off. In risk analysis, cluster analysis can be employed to identify groups of firms with similar risk profiles, thus facilitating a more efficient risk assessment and management process.

While cluster analysis offers a myriad of advantages, it is not without its limitations. A key challenge lies in the determination of the optimal number of clusters. Although various statistical criteria and rules of thumb exist, there is no universally applicable solution, making this largely a subjective decision. Another disadvantage is the sensitivity of cluster analysis to the initial setup and the specific algorithm employed, which can influence the final results. This necessitates careful consideration and validation of the clustering process. Further, cluster analysis is vulnerable to the curse of dimensionality. As the number of variables increases, the volume of the data space expands exponentially, leading to sparser data distribution. This can result in less meaningful clusters, thereby diminishing the effectiveness of the analysis. Additionally, cluster analysis requires that the data be of good quality. Erroneous or missing data can significantly distort the clustering process, leading to misleading results.

It is crucial to remember that cluster analysis is fundamentally an exploratory descriptive technique. It provides a useful way to classify observations, but it does not infer causal relationships. Thus, the results of a cluster analysis must be interpreted with caution, keeping in mind the broader research context and the limitations of the method.

A Python implementation of cluster analysis using k-means with three clusters on acctg.csv data is as follows:

```python
import pandas as pd
from sklearn.cluster import KMeans
from sklearn.preprocessing import StandardScaler

# Load the data
data = pd.read_csv('D:/Data/acctg.csv')

# Identify the variables
cross_sectional_var = 'gvkey'
time_var = 'fyear'
dependent_var = 'ue_ce'
independent_vars = data.columns.drop([cross_sectional_var,
time_var, dependent_var])

# Standardize the independent variables
scaler = StandardScaler()
data_scaled = scaler.fit_transform(data[independent_vars])

# Apply k-means clustering
# Determine number of clusters you want (let's say 3 for this
example)
k = 3
kmeans = KMeans(n_clusters=k, random_state=0)
data['cluster'] = kmeans.fit_predict(data_scaled)

# Now, 'data' contains an additional column 'cluster' which
indicates the cluster each row belongs to.
```

Please replace '3' with the desired number of clusters and adapt the variable names to match your data. This script will add a new column **cluster** to your DataFrame, indicating the cluster to which each row belongs.

Determining the optimal number of clusters for a clustering analysis is a critical aspect and involves selecting a suitable number of groups that effectively capture the underlying structure within the data. Several methods exist to aid in this decision-making process. In this section, we will discuss two commonly used approaches: the elbow method and the silhouette method. We will illustrate the implementation of these techniques using the acctg.csv dataset mentioned earlier in Python.

1. **The Elbow Method:**

The elbow method aims to identify the number of clusters by evaluating the within-cluster sum of squares (WCSS) for different values of **k**. The WCSS represents the sum of the squared distances between each data point and its corresponding cluster centroid. As the number of clusters increases, the WCSS tends to decrease, as more clusters can better fit the data. However, beyond a certain point, the benefit of adding more clusters diminishes, resulting in a less significant reduction in WCSS. This optimal number of clusters is often denoted as the "elbow point" or "knee point" in the plot of WCSS against **k**.

```
import pandas as pd
import matplotlib.pyplot as plt
from sklearn.cluster import KMeans
from sklearn.preprocessing import StandardScaler

# Load the data
data = pd.read_csv('D:/Data/acctg.csv')

# Identify the variables
independent_vars = data.columns.drop(['gvkey', 'fyear',
'ue_ce'])

# Standardize the independent variables
scaler = StandardScaler()
data_scaled = scaler.fit_transform(data[independent_vars])

# Perform the elbow method
wcss = []
max_clusters = 10  # Maximum number of clusters to consider

for k in range(1, max_clusters + 1):
    kmeans = KMeans(n_clusters=k, random_state=0)
    kmeans.fit(data_scaled)
    wcss.append(kmeans.inertia_)

# Plot the WCSS values
plt.plot(range(1, max_clusters + 1), wcss)
plt.xlabel('Number of Clusters')
plt.ylabel('Within-Cluster Sum of Squares (WCSS)')
plt.title('Elbow Method')
plt.show()
```

The output will be a plot similar to the one below:

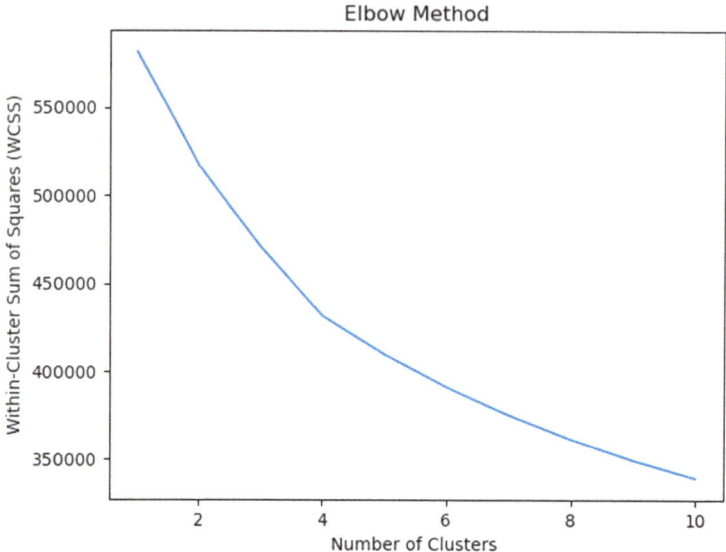

In the resulting plot, we observe the decreasing trend of WCSS as the number of clusters (**k**) increases. The optimal number of clusters can be determined by visually identifying the "elbow point" where the rate of decrease in WCSS significantly diminishes, 4 in this case.

2. **The Silhouette Method:**

The silhouette method provides a quantitative measure of the quality and separation of clusters. It considers the average silhouette coefficient across all data points. The silhouette coefficient for each data point compares the cohesion within its assigned cluster (a measure of similarity to other points in the cluster) to the separation from points in neighboring clusters. Higher silhouette coefficients indicate better-defined and well-separated clusters.

```python
import pandas as pd
import matplotlib.pyplot as plt
from sklearn.cluster import KMeans
from sklearn.metrics import silhouette_score
from sklearn.preprocessing import StandardScaler

# Load the data
data = pd.read_csv('D:/Data/acctg.csv')

# Identify the variables
independent_vars = data.columns.drop(['gvkey', 'fyear', 'ue_ce'])

# Standardize the independent variables
scaler = StandardScaler()
data_scaled = scaler.fit_transform(data[independent_vars])

# Perform the silhouette analysis
silhouette_scores = []
max_clusters = 10  # Maximum number of clusters to consider

for k in range(2, max_clusters + 1):
    kmeans = KMeans(n_clusters=k, random_state=0)
    cluster_labels = kmeans.fit_predict(data_scaled)
    silhouette_avg = silhouette_score(data_scaled, cluster_labels)
    silhouette_scores.append(silhouette_avg)

# Plot the silhouette scores
plt.plot(range(2, max_clusters + 1), silhouette_scores)
plt.xlabel('Number of Clusters')
plt.ylabel('Average Silhouette Score')
plt.title('Silhouette Method')
plt.show()
```

The output will be a plot similar to the one shown below:

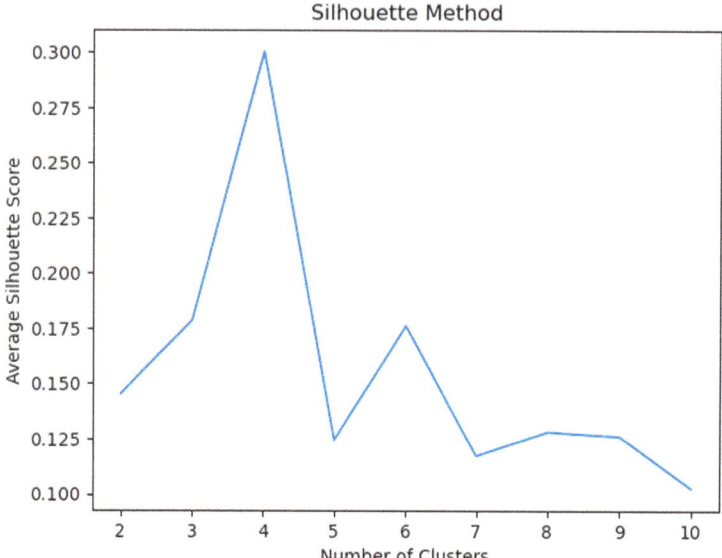

In the plot, the average silhouette score is plotted against the number of clusters (**k**). Higher silhouette scores indicate better clustering results, where data points within clusters are more similar to each other than to those in other clusters. The optimal number of clusters can be determined by selecting the value of **k** that maximizes the average silhouette score (4 in this case).

In addition to the widely used k-means clustering algorithm, several other methods have been developed for clustering analysis. These methods offer alternative approaches to identifying clusters within datasets, each with its own underlying assumptions and characteristics.

1. Hierarchical Clustering:

Hierarchical clustering aims to create a hierarchical structure of clusters by iteratively merging or splitting clusters based on their similarity. This method does not require the predetermined number of clusters but instead generates a dendrogram that represents the hierarchical relationships between data points. Hierarchical clustering can be agglomerative, starting with each data point as an individual cluster and progressively merging them, or divisive, beginning with all data points in a single cluster and subsequently splitting them into smaller clusters.

2. Density-Based Clustering:

Density-based clustering algorithms, such as DBSCAN (Density-Based Spatial Clustering of Applications with Noise), identify clusters based on the density of data points. Instead of assuming spherical clusters like k-means, density-based methods can discover clusters of arbitrary shapes. These algorithms classify dense regions of data points as clusters and identify outliers or noise as points with low density.

3. Model-Based Clustering:

Model-based clustering approaches assume that the data points are generated from a mixture of probability distributions. These methods estimate the parameters of the underlying distributions and assign data points to clusters based on their likelihood of belonging to a specific distribution. Examples of model-based clustering algorithms include Gaussian Mixture Models (GMM) and Finite Mixture Models (FMM).

4. Fuzzy Clustering:

Fuzzy clustering allows data points to belong to multiple clusters with varying degrees of membership. Unlike traditional hard clustering algorithms that assign each data point to a single cluster, fuzzy clustering assigns membership values indicating the degree to which each point belongs to different clusters. Fuzzy C-Means (FCM) is a well-known algorithm for fuzzy clustering.

5. Spectral Clustering:

Spectral clustering combines techniques from linear algebra and graph theory to partition data points into clusters. It leverages the spectral properties of the data similarity matrix and performs dimensionality reduction using eigenvectors. Spectral clustering can effectively handle complex datasets with non-linear structures and has shown promising results in various applications.

Canonical Correlation Analysis

Canonical Correlation Analysis (CCA) is a statistical technique employed to explore the relationship between two sets of variables by identifying the underlying linear combinations that exhibit maximum correlation. This method is particularly valuable when dealing with multiple dependent and independent variables, as it allows researchers to investigate their joint relationship. By identifying the canonical variables, CCA facilitates the understanding of shared variation and provides insights into the underlying data structure.

In the domains of accounting and finance research, CCA has found extensive application. In financial economics, researchers have utilized CCA to examine the relationship between financial market variables, such as stock prices, interest rates, and exchange rates. This analysis aids in understanding the interactions between different financial markets and evaluating the impact of macroeconomic factors on asset prices. In accounting research, CCA has been employed to explore associations between financial statement variables, such as profitability, liquidity, and firm-specific characteristics. Furthermore, researchers have used CCA to investigate the relationship between accounting measures and market-related variables, such as stock returns and market value.

CCA enables researchers to uncover latent constructs or underlying dimensions that explain the relationships between sets of variables. By identifying the canonical variables, researchers can develop more robust theories and hypotheses by better understanding the complex associations between financial and accounting variables. Moreover, CCA provides a comprehensive framework for assessing the joint significance of multiple dependent and independent variables. This is particularly valuable in accounting and finance, where researchers often analyze large datasets with multiple interrelated variables. CCA also facilitates the examination of the relative importance of different variables in explaining shared variation, assisting in variable selection and model refinement.

CCA accommodates a wide range of research questions by allowing for the exploration of complex relationships between two sets of variables. Its flexibility makes it applicable to diverse research contexts, thereby enhancing its utility in the field. CCA also provides a robust statistical framework that accounts for the multivariate nature of variables, ensuring more accurate and reliable results and improving the validity of findings. However, assumes linearity between the sets of variables, and thus, may not adequately capture underlying associations in cases of nonlinearity. CCA requires a relatively large sample size to obtain stable and reliable estimates. In situations where sample sizes are limited, alternative techniques may be more appropriate.

A Python implementation of CCA on acctg.csv data we mentioned earlier is as follows:

```
import pandas as pd
from sklearn.cross_decomposition import CCA

# Load the data from the CSV file
data = pd.read_csv('D:/Data/acctg.csv')

# Separate the dependent variable (ue_ce) and independent
variables
dependent_variable = data['ue_ce']
independent_variables = data[['ue_gp', 'ldue_chgp', 'ue_sga',
'ldue_chsga', 'ue_cog', 'ldue_chcog']]

# Perform Canonical Correlation Analysis
cca = CCA(n_components=1)   # Set the number of canonical
variables to 1
cca.fit(independent_variables, dependent_variable)

# Retrieve the canonical variables
canonical_vars_X, canonical_vars_Y =
cca.transform(independent_variables,
dependent_variable.values.reshape(-1, 1))

# Print the canonical correlation coefficient
canonical_corr_coef = cca.score(independent_variables,
dependent_variable)
print("Canonical Correlation Coefficient:", canonical_corr_coef)
```

In Canonical Correlation Analysis (CCA), the number of canonical correlations and their associated canonical variables depends on the number of variables in each set (independent and dependent). Specifically, the number of canonical correlations is equal to the minimum of the number of variables in each set. The canonical correlation coefficient measures the strength and direction of the linear relationship between the canonical variables of the two sets of variables. The canonical correlation coefficient ranges from 0 to 1, where a value of 1 indicates a perfect positive correlation, 0 indicates no correlation, and -1 indicates a perfect negative correlation.

Discriminant Analysis

Discriminant Analysis (DA) as a multivariate analysis is a method aimed at achieving optimal discrimination among multiple groups based on a set of independent variables by identifying the variables that exhibit the highest discriminatory power among the groups and construct a linear combination of these variables that maximizes the separation between the groups. DA is primarily used for classification tasks with categorical or discrete target variables. In accounting and finance research, DA is extensively used for financial distress prediction, credit scoring, risk assessment, and fraud detection. It enables the differentiation of financially distressed firms from non-distressed ones, assists in evaluating credit risks, and aids in detecting fraudulent transactions or activities. By identifying key financial indicators, DA facilitates the development of predictive models and enhances the effectiveness of fraud prevention and investigation strategies.

Discriminant Analysis holds notable significance in academic research within accounting and finance due to its ability to analyze complex datasets with multiple variables, providing insights into group relationships and differences that may not be apparent when examining variables individually. It facilitates the investigation of independent variables' discriminative power in classifying observations into predefined groups, offering valuable insights into the underlying factors driving group differences. Moreover, Discriminant Analysis allows for the evaluation of the overall predictive accuracy of the model, enabling researchers to assess its practical utility and potential application in real-world scenarios through measures like classification accuracy, sensitivity, and specificity.

However, researchers should be mindful of the assumptions and limitations of Discriminant Analysis. Violation of the linearity assumption, which assumes linearity between independent variables and the discriminant function, can lead to suboptimal discrimination outcomes. Deviations from the assumption of multivariate normality within each group can impact model accuracy. Furthermore, the assumption of equal covariance matrices across groups, known as homoscedasticity, is assumed and violation of this assumption, referred to as heteroscedasticity, can affect group discrimination accuracy. It is important to test for homogeneity of covariance matrices and consider alternative methods, such as quadratic discriminant analysis, when heteroscedasticity is present. Discriminant Analysis requires a priori defined groups, and its effectiveness relies on the existence of distinct groups based on prior knowledge or theory. If the groups are ill-defined or exhibit overlapping categories, the discriminant model may yield less meaningful results.

A Python implementation of DA on D:/Data/acctg2.csv where ue_ce_cat is the dependent variable and ue_gp, ldue_chgp, ue_sga, ldue_chsga, ue_cog, ldue_chcog, and MBE are independent variables is provided below:

```python
import pandas as pd
import statsmodels.api as sm

# Load the dataset
data_path = "D:/Data/acctg2.csv"
df = pd.read_csv(data_path)

# Separate the independent variables (features) and dependent
variable (target)
X = df[['ue_gp', 'ldue_chgp', 'ue_sga', 'ldue_chsga', 'ue_cog',
'ldue_chcog']]
y = df['ue_ce_cat']

# Add a constant term to the independent variables
X = sm.add_constant(X)

# Create and fit the Logistic Regression model
logit_model = sm.Logit(y, X)
logit_result = logit_model.fit()

# Create a DataFrame to store the results
results_df = pd.DataFrame({
    'Variable': X.columns[1:],
    'Coefficient': logit_result.params.values[1:],
    't-statistic': logit_result.tvalues.values[1:],
    'p-value': logit_result.pvalues.values[1:]
})

# Set the index of the DataFrame
results_df.set_index('Variable', inplace=True)

# Print the results table
print(results_df)
```

The output will be a table with DA coefficients, *t*-stats, and *p*-values for all independent variables.

Multivariate analysis holds an indispensable role in the sphere of accounting and finance research, illuminating the interconnectedness of multiple variables and their collective impact on financial phenomena. The utilization of these sophisticated statistical techniques allows researchers to apprehend the intricate nature of financial markets, business operations, and economic systems. It provides the ability to scrutinize the complex relationships among financial variables that would otherwise remain obscured in univariate or bivariate analyses. The diverse suite of multivariate analysis methods, including but not limited to Principal Component Analysis, Cluster Analysis, and Discriminant Analysis, etc. ensures the flexibility and adaptability required to address the multifaceted research questions prevalent in the field.

Part VI

Advanced Topics

27

Deep Learning

In the ever-evolving field of accounting research, the integration of cutting-edge technologies has become increasingly crucial to gain insights from complex data patterns. Deep learning, an advanced subset of machine learning, has emerged as a powerful tool for addressing a wide range of applications in various domains, including finance and accounting. Deep learning is a subfield of machine learning that focuses on algorithms inspired by the structure and function of the human brain, known as artificial neural networks. These algorithms enable computers to learn and extract complex patterns and representations from vast amounts of data, often surpassing human capabilities in specific tasks.

The term "deep" in deep learning refers to the depth of the artificial neural networks, which consist of multiple layers of interconnected nodes or neurons. Each layer processes and transforms the input data, gradually learning higher-level features and abstractions as information passes through successive layers. This hierarchical learning approach enables deep learning models to discover intricate structures and relationships within the data, making them highly effective for tasks such as image recognition, natural language processing, and various other applications.

Machine learning and deep learning are related but distinct concepts, with deep learning being a specialized subset of machine learning. Here are the key differences between the two:

1. Architecture: Machine learning encompasses a wide range of algorithms and techniques, such as linear regression, decision trees, and support vector machines, which are used to build models that can learn from and make predictions based on data. Deep learning, on the other hand, specifically refers to models that utilize artificial neural networks, particularly those with multiple hidden layers, which allow for the learning of complex patterns and representations.
2. Feature Extraction: In traditional machine learning, the selection and engineering of features (variables or attributes) often require significant domain knowledge and manual intervention. The success of a machine learning model largely depends on the quality of the chosen features. Deep learning models, however, are capable of automatically learning and extracting relevant features from raw data through their layered architectures, reducing the need for manual feature engineering.
3. Data Requirements: Deep learning models typically require larger amounts of data compared to traditional machine learning models. This is because the complex architectures of deep learning models have a higher capacity to learn and generalize, but they also need more data to avoid overfitting and to achieve better performance. Traditional machine learning techniques can often perform well with smaller datasets and may be more suitable in situations where data is limited.
4. Computational Resources: Deep learning models, due to their complex architectures and the large datasets they require, generally demand more computational power and resources, such as Graphics Processing Units (GPUs) or specialized hardware like Tensor Processing Units (TPUs), for training and inference. Traditional machine learning models, in contrast, are usually less resource-intensive and can often be trained on standard CPUs.
5. Interpretability: Traditional machine learning models, such as linear regression or decision trees, are often more interpretable and easier to understand, as they provide explicit relationships between input features and the output. Deep learning models, with their intricate architectures and numerous parameters, are considered "black boxes" in comparison, making it challenging to interpret or explain their predictions and decisions.

Deep learning has found extensive applications across various domains due to its capacity to learn complex patterns and representations from large datasets. In image recognition and computer vision, deep learning models like Convolutional Neural Networks (CNNs) have demonstrated outstanding performance in tasks such as object detection, image classification, and facial recognition. These applications have significant implications for industries like security, automotive, health care, and retail.

In the realm of natural language processing (NLP), deep learning techniques, including Recurrent Neural Networks (RNNs), Long Short-Term Memory (LSTM) networks, and Transformer models, have revolutionized tasks like machine translation, sentiment analysis, text summarization, and chatbots. Deep learning has also been employed to develop highly accurate speech recognition systems, enabling voice assistants to understand and respond to spoken commands, and generating realistic human-like speech for text-to-speech systems. Furthermore, deep learning models have been utilized for creating personalized recommendations in industries such as e-commerce, entertainment, and social media. In health care, deep learning has been applied to various applications, including medical image analysis, drug discovery, and patient monitoring.

Deep learning is extremely useful for academic research in various fields, including accounting and finance. Its ability to handle large and complex datasets and learn intricate patterns makes it particularly suitable for addressing research questions in these areas. One of the key applications of deep learning in accounting and finance research is financial statement analysis. Deep learning models can be employed to analyze financial statements, extracting valuable information that may help predict future financial performance, evaluate solvency, and assess credit risk.

Another area where deep learning has made a significant impact is fraud detection. By leveraging deep learning techniques, researchers can detect fraudulent activities in financial data, identifying anomalies, irregularities, or suspicious patterns that may indicate financial fraud or earnings manipulation. Additionally, deep learning models have proven valuable in predicting the likelihood of corporate bankruptcy or financial distress by analyzing various financial ratios, market data, and other relevant features.

Furthermore, deep learning has found applications in algorithmic trading and portfolio management. Researchers use these advanced models to develop trading strategies and optimize portfolio management by predicting asset prices, modeling market sentiment, and identifying patterns in financial time series data.

An example of application of deep learning in accounting and finance research is to predict corporate bankruptcy. Predicting corporate bankruptcy is an essential task in accounting research, as it helps investors, financial institutions, and other stakeholders assess the financial health of companies and make informed decisions. To develop a deep learning model that can effectively predict the likelihood of bankruptcy based on historical financial data, researchers must first collect financial data of companies, including both bankrupt and non-bankrupt firms. This data can be obtained from sources like financial statements, stock market data, and economic indicators, with relevant features such as financial ratios, market capitalization, and stock price volatility.

Once the data has been collected, it needs to be cleaned and preprocessed to address missing values, outliers, and scaling issues. The dataset is then divided into training, validation, and testing sets to facilitate model training, tuning, and evaluation. Researchers can choose a suitable deep learning architecture for the problem, such as a feedforward neural network, recurrent neural network (RNN), or long short-term memory (LSTM) network. These architectures can effectively model the relationships between financial features and bankruptcy likelihood.

During the model training phase, the chosen deep learning model is trained on the training dataset using a suitable loss function and optimization algorithm. Researchers can experiment with various hyperparameters to optimize the model's performance. Regular evaluation of the model's performance on the validation dataset during training helps monitor its learning progress and prevent overfitting. Researchers can use techniques like early stopping, dropout, or L1/L2 regularization to improve generalization and adjust hyperparameters as necessary to achieve the best performance on the validation dataset.

After the model is trained and tuned, its performance is evaluated on the test dataset using relevant metrics such as accuracy, precision, recall, F1-score, and area under the ROC curve (AUC-ROC). This assessment helps determine the model's effectiveness in predicting bankruptcy. Finally, researchers can analyze the results and, if possible, identify the features that the deep learning model found most important in predicting bankruptcy. This information can provide valuable insights into the key factors driving the likelihood of corporate bankruptcy and help stakeholders make informed decisions. By following this process, they can develop an effective model that assists investors, financial institutions, and other stakeholders in assessing the financial health of companies and making better decisions.

Neuron

In the context of artificial neural networks (ANN) and deep learning, a neuron, also referred to as a node or artificial neuron, is a fundamental computational unit that mimics the basic functioning of biological neurons in the human brain. It receives input from other neurons or external sources, processes the information, and produces an output based on the input and its internal activation function.

$$f\left(b + \sum_{i=1}^{n} x_i w_i\right)$$

An example of a neuron showing the input $(x_1–x_n)$, their corresponding weights $(w_1–w_n)$, a bias (b) and the activation function f applied to weighted sum of the inputs.[1]

An artificial neuron typically consists of the following components:

1. Inputs: These are the incoming connections from other neurons or data sources. Each input is associated with a weight, which represents the strength or importance of the connection.
2. Weighted Sum: The neuron computes the weighted sum of its inputs by multiplying each input by its corresponding weight and then summing the results. This sum is also known as the neuron's net input or pre-activation value.
3. Bias: A bias term is added to the weighted sum to shift the activation function, enabling the neuron to learn more complex relationships between inputs and outputs. The bias acts as a constant offset and is also a learnable parameter.
4. Activation Function: The activation function is applied to the weighted sum (plus bias) to transform the neuron's output, introducing non-linearity to the network. Common activation functions include the sigmoid, hyperbolic tangent (tanh), ReLU (rectified linear unit), and softmax functions.
5. Output: The result of applying the activation function is the neuron's output, which can be sent to other neurons in the network or used as the final output for the model.

During training, the weights and biases of neurons in the network are adjusted through a process called backpropagation, which involves minimizing a loss function using optimization algorithms like gradient descent. This process allows the artificial neural network to learn complex patterns and relationships in the input data.

---

[1] Adapted from https://medium.com/hyunjulie/activation-functions-a-short-summary-8450c1b1d426.

Deep Learning Techniques and Architectures

Deep learning encompasses a variety of techniques and architectures designed to handle different types of data and problems. Some of the most prominent deep learning techniques include:

1. **Feedforward Neural Networks (FNN):** Also known as Multi-Layer Perceptrons (MLPs), these are the simplest form of artificial neural networks, where information flows in one direction from input to output through multiple hidden layers. They are primarily used for classification and regression tasks.

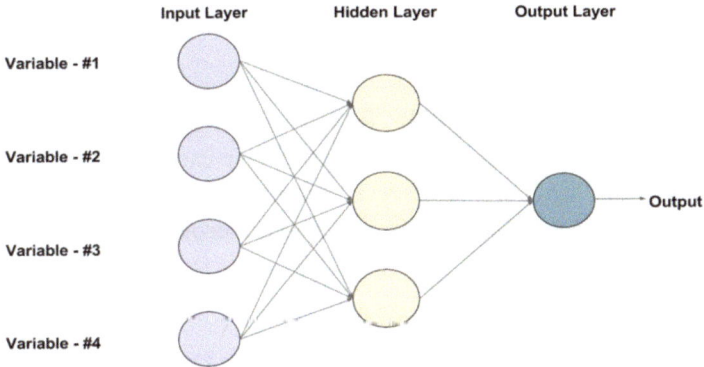

An example of a FFN with one Hidden Layer.[2]

2. **Convolutional Neural Networks (CNN):** Designed to handle grid-like data, such as images, CNNs employ convolutional layers to scan and learn spatial hierarchies of features. They are widely used for image recognition, object detection, and other computer vision tasks.

---

[2] Adapted from https://learnopencv.com/understanding-feedforward-neural-networks/.

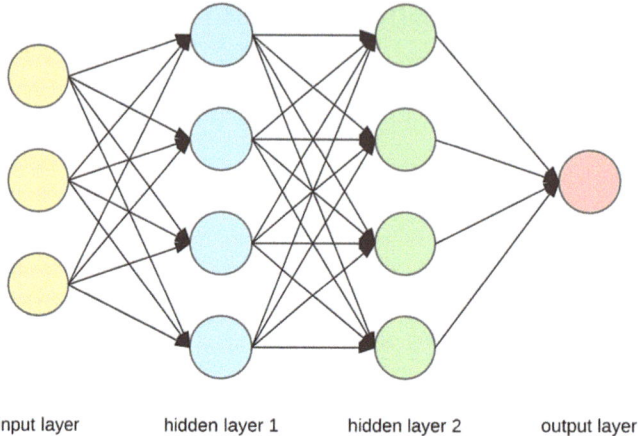

An example of a CNN with two Hidden Layers.[3]

3. **Recurrent Neural Networks (RNN)**: RNNs are designed to process sequential data by maintaining an internal state that can capture information from previous time steps. They are commonly used in natural language processing, time series analysis, and speech recognition tasks.

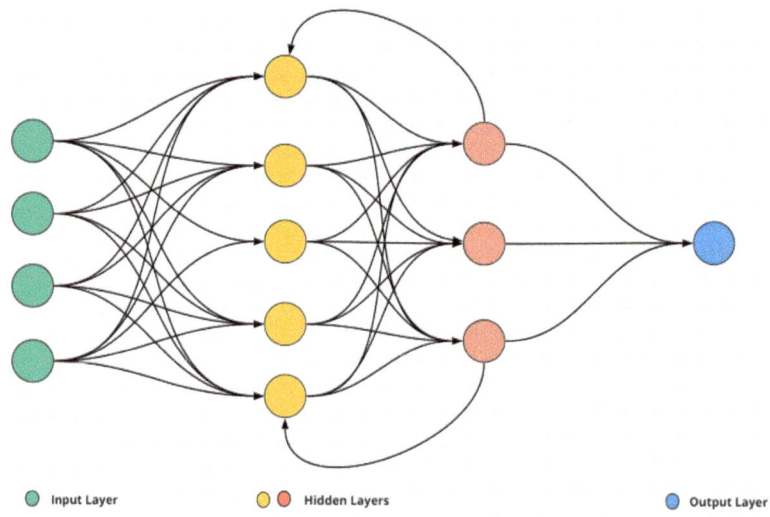

---

[3] Adapted from https://medium.com/@eyeq/convolutional-neural-network-cnn-overview-33026f15dd28.

An example of a RNN with Recurrent Loops.[4]

RNNs can process sequential data, making them suitable for tasks involving text data, such as sentiment analysis or financial news classification. They can capture the dependencies in the text by maintaining an internal state.

4. **Long Short-Term Memory (LSTM)**: LSTM is a specific type of RNN that addresses the vanishing gradient problem, allowing the network to learn long-term dependencies in sequential data more effectively. LSTM models consist of LSTM cells that have input gate, forget gate, and output gate. Forget gate allows these cells to discard information that is not useful in the long term. They are used in various sequence-to-sequence tasks, such as machine translation and text generation.

   LSTMs are a type of RNN designed to learn long-term dependencies in sequential data more effectively. They can be employed in various NLP tasks, including text classification, machine translation, and text summarization, which are relevant to accounting and finance research.

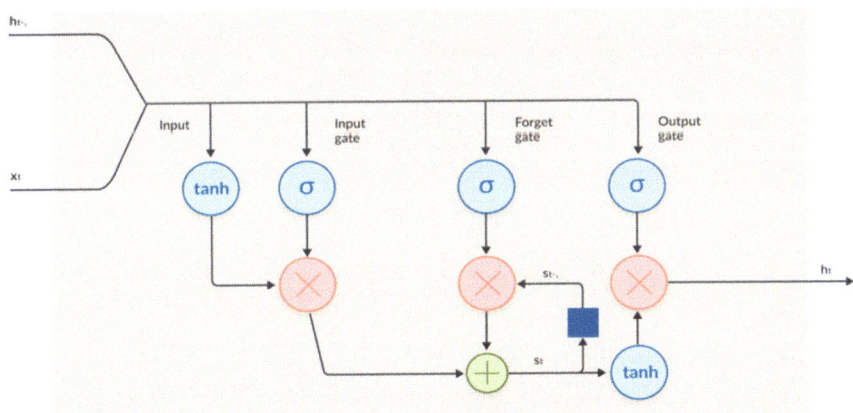

An example of a LSTM cell.[5]

5. **Gated Recurrent Units (GRU)**: Similar to LSTM, GRU is another RNN variant that addresses the vanishing gradient problem with a simpler architecture. They are used in tasks like language modeling, sentiment analysis, and speech recognition.

---

[4] Adapted from https://dataaspirant.com/how-recurrent-neural-network-rnn-works/.
[5] Adapted from https://indiantechwarrior.com/all-about-deep-learning-long-short-term-memory-lstm-networks/.

6. **Transformer Models**: Transformers are a more recent deep learning architecture designed to handle sequential data without recurrent connections. They rely on self-attention mechanisms to process input data in parallel rather than sequentially, leading to improved training efficiency and scalability. Transformers have been used in various NLP tasks, such as document classification, information extraction, and topic modeling, which can be beneficial in accounting and finance research. Transformers are widely used in advanced natural language processing tasks, such as BERT, GPT, and T5 models.
7. **Autoencoders**: These are unsupervised deep learning models used for tasks like dimensionality reduction, feature learning, and data denoising. They consist of an encoder that compresses input data into a lower-dimensional representation and a decoder that reconstructs the original data from this representation.

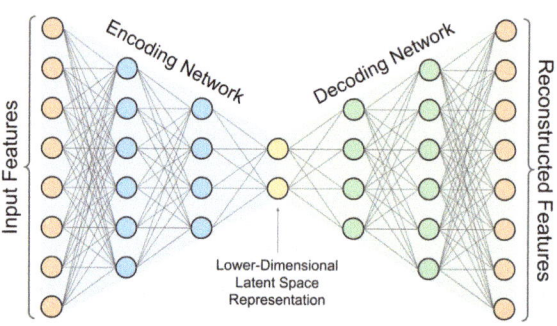

Network Architecture for an Ordinary Autoencoder.[6]

8. **Variational Autoencoders (VAE)**: VAEs are a type of generative model that extends autoencoders by introducing a probabilistic layer, allowing them to generate new data samples. They are used in tasks like image generation, data augmentation, and unsupervised feature learning.
9. **Generative Adversarial Networks (GAN)**: GANs are a class of generative models that consist of two neural networks, a generator, and a discriminator, that compete against each other in a zero-sum game. They are used for generating new data samples, such as images, and enhancing the quality of existing data.

---

[6] Adapted from https://www.assemblyai.com/blog/introduction-to-variational-autoencoders-using-keras/.

Implementation of Deep Learning in Accounting and Finance

The best architecture or technique for accounting and finance text data depends on various factors such as the size and complexity of the dataset, the specific task at hand, and the available resources. However, some commonly used architectures and techniques for NLP tasks in accounting and finance include recurrent neural networks (RNNs) such as Long Short-Term Memory (LSTM) and Gated Recurrent Unit (GRU), convolutional neural networks (CNNs), transformer-based models such as BERT and GPT. Ultimately, the choice of architecture or technique depends on the specific problem you are trying to solve and the characteristics of your dataset. It is recommended to experiment with multiple architectures and techniques and evaluate their performance on your dataset before deciding on the best one for your task.

Transformer Models

In terms of deep learning architectures, a common choice for text classification tasks is the transformer-based models such as BERT, GPT-2, and T5. These models have achieved state-of-the-art performance on many natural language processing tasks, including sentiment analysis. However, these models can be computationally expensive and require significant computing resources, especially for large documents like 10Ks. If you Here's a step-by-step demonstration of implementation of deep learning in the context of sentiment analysis, using Financial PhraseBank dictionary, on a 10-K statements text stored in D:/Data/10K0K, using BERT model (Transformer Model) is as follows:

1. Import required libraries and modules:

```
import torch
import pandas as pd
from sklearn.model_selection import train_test_split
from transformers import BertTokenizerFast, BertForSequenceClassification, Trainer, TrainingArguments
```

These imports include the necessary libraries and modules for handling data, creating the model, and training the model.

2. Load the data and preprocess it:

```
df = pd.read_csv("D:/python
codes/Repositories/FinancialPhraseBank/all-data.csv",
encoding='utf-8', header=None, names=['sentiment', 'text'],
dtype=str)
df['sentiment'] = df['sentiment'].map({'positive': 2,
'negative': 0, 'neutral': 1})
df['sentiment'] = df['sentiment'].astype(int)
train_df, val_df = train_test_split(df, test_size=0.2,
random_state=42)
train_df.reset_index(drop=True, inplace=True)
val_df.reset_index(drop=True, inplace=True)
```

Here, the data is loaded from the CSV file and the sentiment labels are converted to integer values. The integer values have to be within a range [0, C-1], where C is the number of classes. In our case, we have 3 classes: positive, negative, and neutral, so the labels should be in the range [0, 2]. The dataset is then split into training and validation sets. **test_size = 0.2** in the **train_test_split** function specifies that 20% of the dataset should be used as the validation set, and 80% should be used as the training set. **random_state = 42** sets the random seed used for shuffling the data before splitting it into the training and validation sets. By setting the random seed to a fixed value, you can ensure that the data will be split in the same way every time you run the code. This is useful for reproducibility purposes.

3. Create the tokenizer and the model:

```
tokenizer = BertTokenizerFast.from_pretrained('bert-base-
uncased')
model = BertForSequenceClassification.from_pretrained('bert-
base-uncased', num_labels=3)
```

The tokenizer is responsible for converting text into the format required by the BERT model. The model is an instance of BertForSequenceClassification, which is a BERT model fine-tuned for sequence classification tasks.

4. Set hyperparameters and define the Dataset class:

```
MAX_LEN = 128
BATCH_SIZE = 16
EPOCHS = 3

class SentimentDataset(torch.utils.data.Dataset):
    ...

# Create the dataset and dataloaders
train_dataset = SentimentDataset(train_df, tokenizer, MAX_LEN)
val_dataset = SentimentDataset(val_df, tokenizer, MAX_LEN)
```

This part sets the hyperparameters for the model and creates a custom PyTorch Dataset class to handle the input data. The dataset instances are then created for training and validation. **MAX_LEN**: Maximum length of input text sequences. Sequences that are longer than **MAX_LEN** will be truncated and sequences that are shorter will be padded with special tokens. **BATCH_SIZE** is number of training examples processed in one iteration during training. **EPOCHS** is number of times the entire training dataset is passed forward and backward through the neural network during training.

5. Define training arguments and create the Trainer:

```
training_args = TrainingArguments(
    ...
)

trainer = Trainer(
    model=model,
    args=training_args,
    train_dataset=train_dataset,
    eval_dataset=val_dataset
)
```

Here, the training arguments are set, and a Trainer instance is created to handle the training and evaluation of the model.

6. Train the model:

```
trainer.train()
```

This line starts the training process using the Trainer instance.

7. Create a function for making predictions:

```
def predict_sentiment(text, model, tokenizer):
    ...
```

This function takes a text input, preprocesses it using the tokenizer, and feeds it to the trained model to predict the sentiment.

8. Test the model with a new 10-K statement:

```
new_text = "Your new 10-K statement text here."
sentiment = predict_sentiment(new_text, model, tokenizer)

if sentiment == 1:
    print("The sentiment of the given 10-K statement is Positive.")
elif sentiment == -1:
    print("The sentiment of the given 10-K statement is Negative.")
else:
    print("The sentiment of the given 10-K statement is Neutral.")
```

Finally, this part uses the **predict_sentiment()** function to predict the sentiment of a new 10-K statement and prints the result. It is pertinent to mention that BERT model and tokenizer handle text preprocessing internally, which includes lowercasing and tokenization. As a result, there is no need for any additional normalization before using the tokenizer. BERT tokenizer expects input text in a specific format, and it will handle the necessary preprocessing steps internally. For example, the tokenizer will convert the text to lowercase (since it's using 'bert-base-uncased') and split it into tokens, including WordPiece tokens for out-of-vocabulary words.

A sample of detailed code is:

```python
import torch
import os
import pandas as pd
from sklearn.model_selection import train_test_split
from transformers import BertTokenizerFast,
BertForSequenceClassification, Trainer, TrainingArguments

# Load the data
df = pd.read_csv("D:/python
codes/Repositories/FinancialPhraseBank/all-data.csv",
encoding='utf-8', header=None, names=['sentiment', 'text'],
dtype=str)

# Replace the sentiment labels with numeric values
df['sentiment'] = df['sentiment'].map({'positive': 2,
'negative': 0, 'neutral': 1})

# Convert the sentiment column to integers
df['sentiment'] = df['sentiment'].astype(int)

# Split the dataset into training and validation sets
train_df, val_df = train_test_split(df, test_size=0.2,
random_state=42)
train_df.reset_index(drop=True, inplace=True)
val_df.reset_index(drop=True, inplace=True)

# Tokenizer and model
tokenizer = BertTokenizerFast.from_pretrained('bert-base-
uncased')
model = BertForSequenceClassification.from_pretrained('bert-
base-uncased', num_labels=3)

# Hyperparameters
MAX_LEN = 128
BATCH_SIZE = 16
EPOCHS = 3

# Dataset class
class SentimentDataset(torch.utils.data.Dataset):
    def __init__(self, dataframe, tokenizer, max_len):
        self.tokenizer = tokenizer
        self.data = dataframe
        self.text = dataframe.text
        self.sentiment = dataframe.sentiment
        self.max_len = max_len

    def __len__(self):
        return len(self.text)
```

```python
    def __getitem__(self, index):
        text = str(self.text[index])
        text = " ".join(text.split())

        inputs = self.tokenizer.encode_plus(
            text,
            None,
            add_special_tokens=True,
            max_length=self.max_len,
            padding='max_length',
            return_token_type_ids=True,
            truncation=True
        )

        return {
            'input_ids': torch.tensor(inputs['input_ids'], dtype=torch.long),
            'attention_mask': torch.tensor(inputs['attention_mask'], dtype=torch.long),
            'labels': torch.tensor(self.sentiment[index], dtype=torch.long)
        }

# Create the dataset and dataloaders
train_dataset = SentimentDataset(train_df, tokenizer, MAX_LEN)
val_dataset = SentimentDataset(val_df, tokenizer, MAX_LEN)

# Training arguments
training_args = TrainingArguments(
    output_dir='./results',
    num_train_epochs=EPOCHS,
    por_device_train_batch_size=BATCH_SIZE,
    per_device_eval_batch_size=BATCH_SIZE,
    warmup_steps=500,
    weight_decay=0.01,
    logging_dir='./logs',
)

# Trainer
trainer = Trainer(
    model=model,
    args=training_args,
    train_dataset=train_dataset,
    eval_dataset=val_dataset
)

# Start training
trainer.train()
```

```python
# Prediction function
def predict_sentiment(text, model, tokenizer):
    inputs = tokenizer.encode_plus(
        text,
        None,
        add_special_tokens=True,
        max_length=MAX_LEN,
        padding="max_length",
        return_token_type_ids=True,
        truncation=True
    )

    input_ids = torch.tensor([inputs['input_ids']], dtype=torch.long)
    attention_mask = torch.tensor([inputs['attention_mask']], dtype=torch.long)

    with torch.no_grad():
        outputs = model(input_ids, attention_mask=attention_mask)
        logits = outputs.logits
        sentiment = torch.argmax(logits, dim=1).item()

    return sentiment


# Directory containing text files
dir_path = "D:/Data/10K"

# Iterate over all text files in the directory
for file_name in os.listdir(dir_path):
    if file_name.endswith(".txt"):
        # Read the text file
        with open(os.path.join(dir_path, file_name), "r") as f:
            new_text = f.read()

        # Get the sentiment prediction for the text
        sentiment = predict_sentiment(new_text, model, tokenizer)


        # Print the sentiment label
        if sentiment == 2:
            print(f"{file_name}: The sentiment of the given text is Positive.")
        elif sentiment == 0:
            print(f"{file_name}: The sentiment of the given text is Negative.")
        else:
            print(f"{file_name}: The sentiment of the given text is Neutral.")
```

The output will look like:

```
20110223_10-K_edgar_data_1326355_0001144204-11-010380.txt: The
sentiment of the given text is Neutral.
20110223_10-K_edgar_data_1331875_0000950123-11-017206.txt: The
sentiment of the given text is Positive.
20110223_10-K_edgar_data_1335103_0001144204-11-010387.txt: The
sentiment of the given text is Neutral.
20110223_10-K_edgar_data_1338749_0001193125-11-043305.txt: The
sentiment of the given text is Negative.
20110223_10-K_edgar_data_1357615_0001193125-11-043559.txt: The
sentiment of the given text is Positive.
```

However, since 10K statements are big, a good approach is to use a combination of topic modeling and sentiment analysis. Topic modeling can be used to identify the main themes or topics in a large document like a 10K, which can then be used to inform the sentiment analysis. For example, if the topic is about positive earnings, then it's likely that the sentiment will also be positive. This approach can help to reduce the amount of noise in the data and improve the accuracy of the sentiment analysis. A code that performs LDA on all the text files and then calculates the sentiment score as an aggregate of sentiments for all the topics is provided below. This code creates 20 topics. Determining the optimum number of topics has already been discussed in section on Topic Modeling.

```python
import os
import torch
import pandas as pd
import gensim
from gensim import corpora
from gensim.models import LdaModel
from sklearn.model_selection import train_test_split
from transformers import BertTokenizerFast,
BertForSequenceClassification, Trainer, TrainingArguments

# Add this function to preprocess the text data
def preprocess_data(text):
    return gensim.utils.simple_preprocess(text)

# Load the data
df = pd.read_csv("D:/python
codes/Repositories/FinancialPhraseBank/all-data.csv",
encoding='utf-8', header=None, names=['sentiment', 'text'],
dtype=str)

# Replace the sentiment labels with numeric values
df['sentiment'] = df['sentiment'].map({'positive': 2,
'negative': 0, 'neutral': 1})

# Convert the sentiment column to integers
df['sentiment'] = df['sentiment'].astype(int)

# Split the dataset into training and validation sets
train_df, val_df = train_test_split(df, test_size=0.2,
random_state=42)
train_df.reset_index(drop=True, inplace=True)
val_df.reset_index(drop=True, inplace=True)

# Tokenizer and model
tokenizer = BertTokenizerFast.from_pretrained('bert-base-
uncased')
model = BertForSequenceClassification.from_pretrained('bert-
base-uncased', num_labels=3)

# Hyperparameters
MAX_LEN = 128
BATCH_SIZE = 16
EPOCHS = 3

# Dataset class
class SentimentDataset(torch.utils.data.Dataset):
    def __init__(self, dataframe, tokenizer, max_len):
        self.tokenizer = tokenizer
        self.data = dataframe
        self.text = dataframe.text
        self.sentiment = dataframe.sentiment
        self.max_len = max_len
```

```python
    def __len__(self):
        return len(self.text)

    def __getitem__(self, index):
        text = str(self.text[index])
        text = " ".join(text.split())

        inputs = self.tokenizer.encode_plus(
            text,
            None,
            add_special_tokens=True,
            max_length=self.max_len,
            padding='max_length',
            return_token_type_ids=True,
            truncation=True
        )

        return {
            'input_ids': torch.tensor(inputs['input_ids'], dtype=torch.long),
            'attention_mask': torch.tensor(inputs['attention_mask'], dtype=torch.long),
            'labels': torch.tensor(self.sentiment[index], dtype=torch.long)
        }

# Create the dataset and dataloaders
train_dataset = SentimentDataset(train_df, tokenizer, MAX_LEN)
val_dataset = SentimentDataset(val_df, tokenizer, MAX_LEN)

# Training arguments
training_args = TrainingArguments(
    output_dir='./results',
    num_train_epochs=EPOCHS,
    per_device_train_batch_size=BATCH_SIZE,
    per_device_eval_batch_size=BATCH_SIZE,
    warmup_steps=500,
    weight_decay=0.01,
    logging_dir='./logs',
)

# Trainer
trainer = Trainer(
    model=model,
    args=training_args,
    train_dataset=train_dataset,
    eval_dataset=val_dataset
)

# Start training
trainer.train()
```

```python
# LDA Model
def get_lda_model(texts, num_topics=20, passes=20):
    dictionary = corpora.Dictionary(texts)
    corpus = [dictionary.doc2bow(text) for text in texts]
    lda_model = LdaModel(corpus, num_topics=num_topics,
id2word=dictionary, passes=passes)
    return lda_model, dictionary

# Directory containing text files
dir_path = "D:/Data/10K"

# Iterate over all text files in the directory and preprocess
them
texts = []
file_list = []
for file_name in os.listdir(dir_path):
    if file_name.endswith(".txt"):
        with open(os.path.join(dir_path, file_name), "r") as f:
            file_text = f.read()
            texts.append(preprocess_data(file_text))
            file_list.append(file_name)

# Create LDA model
num_topics = 20
lda_model, dictionary = get_lda_model(texts, num_topics)

# Function to calculate sentiment scores for the topics
def sentiment_score_for_topics(lda_model, text, model,
tokenizer):
    bow = dictionary.doc2bow(text)
    topic_distribution = lda_model.get_document_topics(bow)
    sentiment_score = 0


    for topic, prob in topic_distribution:
        topic_terms = lda_model.show_topic(topic)
        terms = " ".join([term for term, _ in topic_terms])
        sentiment = predict_sentiment(terms, model, tokenizer)
        sentiment_score += sentiment * prob

    return sentiment_score

# Iterate over all text files in the directory and predict
sentiment score
for file_text, file_name in zip(texts, file_list):
    # Get the sentiment score for the topics in the text
    sentiment_score = sentiment_score_for_topics(lda_model,
file_text, model, tokenizer)
    print(sentiment_score)
```

10K is a very lengthy and complex document. This code can also be applied to any accounting and financial text data.

FNN Models

BERT is computationally very intensive and it may take several days to get output, depending on the size of your data. FNN, which is simple and computationally less intensive as compared to BERT can give good results when combined with topic modeling such as LDA. However, FNN model's performance might not be as good as BERT. A sample code of FNN with topic modeling is provided below. This code is similar in all respects to the above code, the only difference being FNN instead of BERT.

```python
import os
import torch
import torch.nn as nn
import torch.optim as optim
import pandas as pd
import gensim
from gensim import corpora
from gensim.models import LdaModel
from sklearn.model_selection import train_test_split
from torch.utils.data import DataLoader

# Add this function to preprocess the text data
def preprocess_data(text):
    return gensim.utils.simple_preprocess(text)

# Load the data
df = pd.read_csv("D:/python
codes/Repositories/FinancialPhraseBank/all-data.csv",
encoding='utf-8', header=None, names=['sentiment', 'text'],
dtype=str)

# Replace the sentiment labels with numeric values
df['sentiment'] = df['sentiment'].map({'positive': 2,
'negative': 0, 'neutral': 1})

# Convert the sentiment column to integers
df['sentiment'] = df['sentiment'].astype(int)

# Split the dataset into training and validation sets
train_df, val_df = train_test_split(df, test_size=0.2,
random_state=42)
train_df.reset_index(drop=True, inplace=True)
val_df.reset_index(drop=True, inplace=True)

# Tokenizer
tokenizer = gensim.utils.simple_preprocess

# Hyperparameters
MAX_LEN = 128
BATCH_SIZE = 16
EPOCHS = 3
EMBEDDING_DIM = 100
HIDDEN_DIM = 64
NUM_CLASSES = 3

# Dataset class
class SentimentDataset(torch.utils.data.Dataset):
    def __init__(self, dataframe, tokenizer, max_len):
        self.tokenizer = tokenizer
        self.data = dataframe
        self.text = dataframe.text
        self.sentiment = dataframe.sentiment
```

```python
        self.max_len = max_len

    def __len__(self):
        return len(self.text)

    def __getitem__(self, index):
        text = str(self.text[index])
        tokens = self.tokenizer(text)
        padded_tokens = tokens[:self.max_len] + [''] * (self.max_len - len(tokens))

        # Create word indices
        indices = [word2index[word] for word in padded_tokens]

        return {
            'input_ids': torch.tensor(indices, dtype=torch.long),
            'labels': torch.tensor(self.sentiment[index], dtype=torch.long)
        }

# Create the dataset and dataloaders
train_dataset = SentimentDataset(train_df, tokenizer, MAX_LEN)
val_dataset = SentimentDataset(val_df, tokenizer, MAX_LEN)

train_dataloader = DataLoader(train_dataset, batch_size=BATCH_SIZE, shuffle=True)
val_dataloader = DataLoader(val_dataset, batch_size=BATCH_SIZE)

# Create a word index mapping
words = set()
for text in train_df.text:
    words.update(tokenizer(text))
word2index = {word: idx for idx, word in enumerate(words)}

# FNN model
class FNN(nn.Module):
    def __init__(self, vocab_size, embedding_dim, hidden_dim, num_classes):
        super(FNN, self).__init__()
        self.embedding = nn.Embedding(vocab_size, embedding_dim)
        self.fc1 = nn.Linear(embedding_dim, hidden_dim)
        self.fc2 = nn.Linear(hidden_dim, num_classes)

    def forward(self, x):
        x = self.embedding(x)
        x = torch.mean(x, dim=1)
        x = torch.relu(self.fc1(x))
        x = self.fc2(x)
        return x
```

```python
# Initialize the model
vocab_size = len(words)
model = FNN(vocab_size, EMBEDDING_DIM, HIDDEN_DIM, NUM
# Create the optimizer and the loss function
optimizer = optim.Adam(model.parameters(), lr=0.001)
loss_function = nn.CrossEntropyLoss()

# Training loop
for epoch in range(EPOCHS):
    model.train()
    for batch in train_dataloader:
        optimizer.zero_grad()
        input_ids = batch['input_ids']
        labels = batch['labels']
        outputs = model(input_ids)
        loss = loss_function(outputs, labels)
        loss.backward()
        optimizer.step()

    # Validation
    model.eval()
    total_val_loss = 0
    correct_predictions = 0
    with torch.no_grad():
        for batch in val_dataloader:
            input_ids = batch['input_ids']
            labels = batch['labels']
            outputs = model(input_ids)
            loss = loss_function(outputs, labels)
            total_val_loss += loss.item()

            _, predicted = torch.max(outputs, 1)
            correct_predictions += (predicted ==
labels).sum().item()

    val_loss = total_val_loss / len(val_dataloader)
    val_accuracy = correct_predictions / len(val_dataset)
    print(f"Epoch {epoch + 1}/{EPOCHS}, Validation Loss:
{val_loss}, Validation Accuracy: {val_accuracy}")

# LDA Model
def get_lda_model(texts, num_topics=20, passes=20):
    dictionary = corpora.Dictionary(texts)
    corpus = [dictionary.doc2bow(text) for text in texts]
    lda_model = LdaModel(corpus, num_topics=num_topics,
id2word=dictionary, passes=passes)
    return lda_model, dictionary

# Directory containing text files
dir_path = "D:/Data/10K"
```

```python
# Iterate over all text files in the directory and preprocess
them
texts = []
file_list = []
for file_name in os.listdir(dir_path):
    if file_name.endswith(".txt"):
        with open(os.path.join(dir_path, file_name), "r") as f:
            file_text = f.read()
            texts.append(preprocess_data(file_text))
            file_list.append(file_name)

# Create LDA model
num_topics = 20
lda_model, dictionary = get_lda_model(texts, num_topics)

# Function to calculate sentiment scores for the topics
def sentiment_score_for_topics(lda_model, text, model):
    bow = dictionary.doc2bow(text)
    topic_distribution = lda_model.get_document_topics(bow)
    sentiment_score = 0

    for topic, prob in topic_distribution:
        topic_terms = lda_model.show_topic(topic)
        terms = " ".join([term for term, _ in topic_terms])

        # Tokenize and pad the terms
        tokens = tokenizer(terms)
        padded_tokens = tokens[:MAX_LEN] + [''] * (MAX_LEN -
len(tokens))
        indices = [word2index.get(word, 0) for word in
padded_tokens]

        sentiment = model(torch.tensor([indices]))
        _, predicted = torch.max(sentiment, 1)
        sentiment_score += predicted.item() * prob

    return sentiment_score

# Iterate over all text files in the directory and predict
sentiment score
for file_text, file_name in zip(texts, file_list):
    # Get the sentiment score for the topics in the text
    sentiment_score = sentiment_score_for_topics(lda_model,
file_text, model)
    print(f"Sentiment score for {file_name}: {sentiment_score}")
```

The code first imports the necessary libraries and reads the labeled dataset for training the FNN model. It then preprocesses the dataset and tokenizes the text, splitting it into training and validation sets. A custom PyTorch dataset class, **SentimentDataset**, is created to manage the input–output pairs from the preprocessed dataset. Then, the FNN model is defined and trained on the dataset using the Adam optimizer and cross-entropy loss. The performance of the model is evaluated on the validation set after each training epoch. After training the FNN model, the code focuses on applying LDA to a set of text documents in a directory. Each document is read, preprocessed, and added to a list of texts. Then, the LDA model is trained on this list of preprocessed texts to generate topic distributions for each document.

A custom function, **sentiment_score_for_topics**, calculates the sentiment scores of the topics within each document using the trained FNN model. This function tokenizes and pads the terms extracted from the topics, and the FNN model predicts the sentiment of these terms. The sentiment scores are then weighted by the probability of the topics in the document and summed to produce the final sentiment score for each document. Finally, the sentiment scores are calculated for all text files in the directory and printed. This code demonstrates how topic modeling can be combined with sentiment analysis to assess the sentiment of a collection of documents based on their topics.

LSTM Models

LSTM, a type of RNN, modification of the code to get sentiment score after creating topic through LDA is as follows:

```python
import os
import torch
import pandas as pd
import gensim
from gensim import corpora
from gensim.models import LdaModel
from sklearn.model_selection import train_test_split
from torch import nn
from torch.utils.data import DataLoader
from torch.optim import AdamW
from torch.optim.lr_scheduler import StepLR

# Preprocessing and LDA functions remain the same

# RNN Model (LSTM)
class LSTMModel(nn.Module):
    def __init__(self, vocab_size, embed_dim, hidden_dim, num_layers, num_classes, dropout=0.5):
        super(LSTMModel, self).__init__()
        self.embedding = nn.Embedding(vocab_size, embed_dim)
        self.lstm = nn.LSTM(embed_dim, hidden_dim, num_layers, batch_first=True, dropout=dropout)
        self.fc = nn.Linear(hidden_dim, num_classes)

    def forward(self, x):
        x = self.embedding(x)
        _, (hidden, _) = self.lstm(x)
        out = self.fc(hidden[-1])
        return out

# Hyperparameters
MAX_LEN = 128
BATCH_SIZE = 16
EPOCHS = 3
LEARNING_RATE = 1e-4
LR_STEP = 1
GAMMA = 0.95
EMBED_DIM = 100
HIDDEN_DIM = 256
NUM_LAYERS = 2

# Dataset class
class SentimentDataset(torch.utils.data.Dataset):
    def __init__(self, dataframe, tokenizer, max_len):
        self.tokenizer = tokenizer
        self.data = dataframe
        self.text = dataframe.text
        self.sentiment = dataframe.sentiment
        self.max_len = max_len

    def __len__(self):
        return len(self.text)
```

```python
    def __getitem__(self, index):
        text = str(self.text[index])
        text = " ".join(text.split())

        inputs = self.tokenizer.encode_plus(
            text,
            None,
            add_special_tokens=True,
            max_length=self.max_len,
            padding='max_length',
            return_token_type_ids=True,
            truncation=True
        )

        return {
            'input_ids': torch.tensor(inputs['input_ids'], dtype=torch.long),
            'attention_mask': torch.tensor(inputs['attention_mask'], dtype=torch.long),
            'labels': torch.tensor(self.sentiment[index], dtype=torch.long)
        }
# Create the dataset and dataloaders
train_dataset = SentimentDataset(train_df, tokenizer, MAX_LEN)
val_dataset = SentimentDataset(val_df, tokenizer, MAX_LEN)
train_loader = DataLoader(train_dataset, batch_size=BATCH_SIZE, shuffle=True)
val_loader = DataLoader(val_dataset, batch_size=BATCH_SIZE, shuffle=False)

# Initialize the model
vocab_size = len(tokenizer.vocab)
num_classes = 3
model = LSTMModel(vocab_size, EMBED_DIM, HIDDEN_DIM, NUM_LAYERS, num_classes)

# Loss function and optimizer
criterion = nn.CrossEntropyLoss()
optimizer = AdamW(model.parameters(), lr=LEARNING_RATE)
scheduler = StepLR(optimizer, step_size=LR_STEP, gamma=GAMMA)

# Training loop
for epoch in range(EPOCHS):
    model.train()
    for i, batch in enumerate(train_loader):
        optimizer.zero_grad()
        input_ids, attention_mask, labels = batch['input_ids'], batch['attention_mask'], batch['labels']
        outputs = model(input_ids)
        loss = criterion(outputs, labels)
        loss.backward()
        optimizer.step()
```

```python
    model.eval()
    total_loss, total_correct, total_count = 0, 0, 0
    for batch in val_loader:
        input_ids, attention_mask, labels = batch['input_ids'],
batch['attention_mask'], batch['labels']
        with torch.no_grad():
            outputs = model(input_ids)
            loss = criterion(outputs, labels)
            _, predicted = torch.max(outputs, 1)
            total_loss += loss.item()
            total_correct += (predicted == labels).sum().item()
            total_count += labels.size(0)

    scheduler.step()

    print(f"Epoch {epoch + 1}/{EPOCHS} - Validation Loss:
{total_loss / total_count:.4f} - Accuracy: {total_correct /
total_count:.4f}")

# Prediction function
def predict_sentiment(text, model, tokenizer):
    inputs = tokenizer.encode_plus(
        text,
        None,
        add_special_tokens=True,
        max_length=MAX_LEN,
        padding="max_length",
        return_token_type_ids=True,
        truncation=True
    )

    input_ids = torch.tensor([inputs['input_ids']],
dtype=torch.long)
    attention_mask = torch.tensor([inputs['attention_mask']],
dtype=torch.long)

    with torch.no_grad():
        outputs = model(input_ids,
attention_mask=attention_mask)
        logits = outputs.logits
        sentiment = torch.argmax(logits, dim=1).item()

    return sentiment

# LDA Model
def get_lda_model(texts, num_topics=10, passes=20):
    dictionary = corpora.Dictionary(texts)
    corpus = [dictionary.doc2bow(text) for text in texts]
    lda_model = LdaModel(corpus, num_topics=num_topics,
id2word=dictionary, passes=passes)
    return lda_model, dictionary
```

```python
# Function to calculate sentiment scores for the topics
def sentiment_score_for_topics(lda_model, text, model,
tokenizer):
    bow = dictionary.doc2bow(text)
    topic_distribution = lda_model.get_document_topics(bow)
    sentiment_score = 0

    for topic, prob in topic_distribution:
        topic_terms = lda_model.show_topic(topic)
        terms = " ".join([term for term, _ in topic_terms])
        sentiment = predict_sentiment(terms, model, tokenizer)
        sentiment_score += sentiment * prob

    return sentiment_score

# Directory containing text files
dir_path = "D:/Data/10K"

# Iterate over all text files in the directory and preprocess
them
texts = []
file_list = []
for file_name in os.listdir(dir_path):
    if file_name.endswith(".txt"):
        with open(os.path.join(dir_path, file_name), "r") as f:
            file_text = f.read()
            texts.append(preprocess_data(file_text))
            file_list.append(file_name)

# Create LDA model
num_topics = 20
lda_model, dictionary = get_lda_model(texts, num_topics)

# Iterate over all text files in the directory and predict
sentiment score
for file_text, file_name in zip(texts, file_list):
    # Get the sentiment score for the topics in the text
    sentiment_score = sentiment_score_for_topics(lda_model,
file_text, model, tokenizer)
    print(f"{file_name}: The sentiment score for the given text
is {sentiment_score:.2f}")
```

The code uses the PyTorch framework for building and training the model. The first part of the code imports the necessary libraries, including PyTorch, Pandas, Gensim, and Scikit-learn. It then defines the LSTM model class, which inherits from the PyTorch **nn.Module** class. The **LSTMModel** class takes as input several hyperparameters, such as the vocabulary size, embedding dimension, hidden dimension, number of layers, and number of output classes. The **forward()** method of the class processes the input data by first passing it through an embedding layer, followed by the LSTM layer, and finally through a linear layer to produce the output.

Next, the code defines the hyperparameters, such as the maximum sequence length, batch size, number of epochs, and learning rate. It also defines the dataset class and creates the training and validation datasets and data loaders. The model is then initialized with the hyperparameters, and the loss function and optimizer are defined. The code then trains the model for the specified number of epochs, evaluating the validation loss and accuracy at each epoch.

After training, the code defines a **predict_sentiment()** function that takes as input a text string and returns the predicted sentiment of the text using the trained model. The code then defines an LDA topic modeling function, which takes as input a list of texts and returns an LDA model and dictionary. The code also defines a **sentiment_score_for_topics()** function, which takes as input an LDA model, text, and the trained sentiment analysis model, and returns the sentiment score of the text by computing the sentiment scores of the individual topics in the text.

Finally, the code iterates over all the text files in a given directory, preprocesses them, and passes them through the LDA model and sentiment analysis model to compute the sentiment scores for the texts. The sentiment scores are then printed along with the corresponding file names. This code can be useful for analyzing the sentiment of large volumes of textual data and extracting insights from them.

RNN (LSTM) model might not perform as well as the BERT model in certain cases, and you may need to fine-tune the hyperparameters to achieve better performance. Fine-tuning a model involves adjusting its hyperparameters to achieve better performance. Some key steps include training for more epochs, adjusting the learning rate, tuning the learning rate scheduler, modifying the LSTM architecture, using pre-trained word embeddings, applying regularization techniques, optimizing the batch size, and performing model selection. It's essential to monitor the model's performance on a validation set during this process to avoid overfitting and ensure generalization to unseen data. Keep in mind that fine-tuning can be time-consuming, and it might require multiple iterations of experimentation to find the best combination of hyperparameters.

GRU Model

GRU model is RNN, similar to LSTM and implementation is similar to LSTM. There are only two changes required in the code above to change the model from LSTM to GRU.

1. Make a few modifications to the **LSTMModel** class definition. Here is the changed model class renamed as **GRUModel**:

```
# GRU Model
class GRUModel(nn.Module):
    def __init__(self, vocab_size, embed_dim, hidden_dim,
num_layers, num_classes, dropout=0.5):
        super(GRUModel, self).__init__()
        self.embedding = nn.Embedding(vocab_size, embed_dim)
        self.gru = nn.GRU(embed_dim, hidden_dim, num_layers,
batch_first=True, dropout=dropout)
        self.fc = nn.Linear(hidden_dim, num_classes)

    def forward(self, x):
        x = self.embedding(x)
        _, hidden = self.gru(x)
        out = self.fc(hidden[-1])
        return out
```

2. Update the model initialization to use the **GRUModel** instead of the **LSTMModel**:

```
# Initialize the model
vocab_size = len(tokenizer.vocab)
num_classes = 3
model = GRUModel(vocab_size, EMBED_DIM, HIDDEN_DIM, NUM_LAYERS,
num_classes)
```

Again, just like LSTM, the GRU model also need to be fine-tuned.

CNN Model

Convolutional Neural Networks can capture local patterns in the input data and are often used for image recognition tasks. They can also be applied to text data, with 1D convolutions, to detect local patterns in the text. CNNs generally require less training time compared to RNNs (like LSTM or GRU) but may not perform as well in capturing long-range dependencies in the text. Here is a 1D implementation of CNN in the context of sentiment analysis after LDA:

```python
import os
import torch
import pandas as pd
import gensim
from gensim import corpora
from gensim.models import LdaModel
from sklearn.model_selection import train_test_split
from torch import nn
from torch.utils.data import DataLoader
from torch.optim import AdamW
from torch.optim.lr_scheduler import StepLR

# Preprocessing and LDA functions remain the same

# CNN Model (1D Convolution)
class CNNModel(nn.Module):
    def __init__(self, vocab_size, embed_dim, num_classes, dropout=0.5):
        super(CNNModel, self).__init__()
        self.embedding = nn.Embedding(vocab_size, embed_dim)
        self.conv1 = nn.Conv1d(embed_dim, 128, kernel_size=3, padding=1)
        self.pool1 = nn.MaxPool1d(2)
        self.conv2 = nn.Conv1d(128, 256, kernel_size=3, padding=1)
        self.pool2 = nn.MaxPool1d(2)
        self.fc = nn.Linear(256 * 32, num_classes)
        self.dropout = nn.Dropout(dropout)

    def forward(self, x):
        x = self.embedding(x)
        x = x.transpose(1, 2)
        x = self.dropout(F.relu(self.conv1(x)))
        x = self.pool1(x)
        x = self.dropout(F.relu(self.conv2(x)))
        x = self.pool2(x)
        x = x.view(x.size(0), -1)
        out = self.fc(x)
        return out

# Hyperparameters
MAX_LEN = 128
BATCH_SIZE = 16
EPOCHS = 3
LEARNING_RATE = 1e-4
LR_STEP = 1
GAMMA = 0.95
EMBED_DIM = 100
```

```python
# Dataset class, train_dataset, val_dataset, train_loader, and
val_loader remain the same

# Initialize the model
vocab_size = len(tokenizer.vocab)
num_classes = 3
model = CNNModel(vocab_size, EMBED_DIM, num_classes)

# Loss function and optimizer
criterion = nn.CrossEntropyLoss()
optimizer = AdamW(model.parameters(), lr=LEARNING_RATE)
scheduler = StepLR(optimizer, step_size=LR_STEP, gamma=GAMMA)

# Training loop
for epoch in range(EPOCHS):
    model.train()
    for i, batch in enumerate(train_loader):
        optimizer.zero_grad()
        input_ids, attention_mask, labels = batch['input_ids'], batch['attention_mask'], batch['labels']
        outputs = model(input_ids)
        loss = criterion(outputs, labels)
        loss.backward()
        optimizer.step()

    model.eval()
    total_loss, total_correct, total_count = 0, 0, 0
    for batch in val_loader:
        input_ids, attention_mask, labels = batch['input_ids'], batch['attention_mask'], batch['labels']
        with torch.no_grad():
            outputs = model(input_ids)
            loss = criterion(outputs, labels)
            _, predicted = torch.max(outputs, 1)
            total_loss += loss.item()
            total_correct += (predicted == labels).sum().item()
            total_count += labels.size(0)

    scheduler.step()

    print(f"Epoch {epoch + 1}/{EPOCHS} - Validation Loss: {total_loss / total_count:.4f} - Accuracy: {total_correct / total_count:.4f}")

# Prediction function
def predict_sentiment(text, model, tokenizer):
    inputs = tokenizer.encode_plus(
        text,
        None,
        add_special_tokens=True,
        max_length=MAX_LEN,
```

```python
        padding="max_length",
        return_token_type_ids=True,
        truncation=True
    )
    input_ids = torch.tensor([inputs['input_ids']], dtype=torch.long)
    attention_mask = torch.tensor([inputs['attention_mask']], dtype=torch.long)

    with torch.no_grad():
        outputs = model(input_ids)
        _, predicted = torch.max(outputs, 1)
        sentiment = predicted.item()

    return sentiment

# LDA Model
def get_lda_model(texts, num_topics=20, passes=20):
    dictionary = corpora.Dictionary(texts)
    corpus = [dictionary.doc2bow(text) for text in texts]
    lda_model = LdaModel(corpus, num_topics=num_topics, id2word=dictionary, passes=passes)
    return lda_model, dictionary

# Function to calculate sentiment scores for the topics
def sentiment_score_for_topics(lda_model, text, model, tokenizer):
    bow = dictionary.doc2bow(text)
    topic_distribution = lda_model.get_document_topics(bow)
    sentiment_score = 0

    for topic, prob in topic_distribution:
        topic_terms = lda_model.show_topic(topic)
        terms = " ".join([term for term, _ in topic_terms])
        sentiment = predict_sentiment(terms, model, tokenizer)
        sentiment_score += sentiment * prob

    return sentiment_score

# Directory containing text files
dir_path = "D:/Data/10K"

# Iterate over all text files in the directory and preprocess them
texts = []
file_list = []
for file_name in os.listdir(dir_path):
    if file_name.endswith(".txt"):
        with open(os.path.join(dir_path, file_name), "r") as f:
            file_text = f.read()
            texts.append(preprocess_data(file_text))
            file_list.append(file_name)
```

```
# Create LDA model
num_topics = 20
lda_model, dictionary = get_lda_model(texts, num_topics)

# Iterate over all text files in the directory and predict
sentiment score
for file_text, file_name in zip(texts, file_list):
    # Get the sentiment score for the topics in the text
    sentiment_score = sentiment_score_for_topics(lda_model,
file_text, model, tokenizer)
    print(f"{file_name}: The sentiment score for the given text
is {sentiment_score:.2f}")
```

In the beginning, necessary libraries such as PyTorch, pandas, and Gensim are imported. The preprocessing and LDA functions are assumed to be already defined, as they remain unchanged. Two deep learning models, an LSTM-based and a 1D CNN-based model, are defined as separate classes. Both models share a common architecture: They start with an embedding layer, followed by layers specific to the model (either LSTM or 1D Convolution layers), and end with a fully connected layer for classification. Hyperparameters for the models, such as the maximum sequence length, batch size, learning rate, and embedding dimensions, are defined as well.

A custom PyTorch dataset class is created to handle the tokenization and padding of the input text. This class also takes care of mapping the input text to their corresponding sentiment labels. DataLoader objects are created for both the training and validation datasets, which are then used during the training process. The training loop iterates through the DataLoader objects, feeding the input text into the selected deep learning model, computing the loss using the CrossEntropyLoss criterion, and updating the model's parameters using the AdamW optimizer. The learning rate scheduler is also employed to adjust the learning rate during training. After each epoch, the model's performance is evaluated on the validation dataset to measure its accuracy and loss.

The **predict_sentiment** function is defined to use the trained model for predicting the sentiment of a given text. It tokenizes the input text, pads it, and passes it through the model. The model then returns the predicted sentiment. The LDA model is created using the Gensim library, and a function **sentiment_score_for_topics** is defined to calculate the sentiment score of topics in the text. This function obtains the topic distribution for a given text, predicts the sentiment for each topic, and calculates the weighted average of the sentiment scores based on the topic probabilities.

Finally, the code iterates through a directory of text files, preprocessing each document and creating an LDA model for the entire corpus. It then calculates the sentiment score for each document using the previously defined **sentiment_score_for_topics** function, which combines the LDA model and the deep learning model to generate a comprehensive sentiment score. The code outputs the sentiment score for each document, providing an insight into the overall sentiment of the corpus.

Autoencoder Model

Autoencoders are unsupervised learning models typically used for dimensionality reduction, feature extraction, or denoising data. They learn to reconstruct their input data through a process of encoding and decoding, where the encoded representation is usually a lower-dimensional representation of the original data. Autoencoders have been used in academic research across various domains, including accounting and finance. While they might not be as prevalent as other machine learning techniques, their applications in these fields can be valuable for tasks such as anomaly detection, feature extraction, and dimensionality reduction.

Here is an example of anomaly detection on topics after LDA using encoders:

```python
import os
import numpy as np
import pandas as pd
from sklearn.model_selection import train_test_split
from gensim import corpora
from gensim.models import LdaModel
from tensorflow.keras.layers import Input, Dense
from tensorflow.keras.models import Model

# Preprocessing and LDA functions
def preprocess_data(text):
    # Implement your preprocessing steps here
    pass

def get_lda_model(texts, num_topics=20, passes=20):
    dictionary = corpora.Dictionary(texts)
    corpus = [dictionary.doc2bow(text) for text in texts]
    lda_model = LdaModel(corpus, num_topics=num_topics, id2word=dictionary, passes=passes)
    return lda_model, dictionary, corpus

# Load text files from directory
dir_path = "D:/Data/10K"
file_list = [file_name for file_name in os.listdir(dir_path) if file_name.endswith(".txt")]

texts = []
for file_name in file_list:
    with open(os.path.join(dir_path, file_name), "r") as f:
        texts.append(preprocess_data(f.read()))

# Apply LDA on preprocessed texts
num_topics = 20
lda_model, dictionary, corpus = get_lda_model(texts, num_topics)

# Extract topic distributions for each document
topic_distributions = np.array([np.array([topic_prob for _, topic_prob in lda_model.get_document_topics(bow)]) for bow in corpus])

# Split data into train and test sets
train_data, test_data = train_test_split(topic_distributions, test_size=0.2, random_state=42)

# Define the autoencoder model
encoding_dim = 4
input_layer = Input(shape=(num_topics,))
encoded = Dense(encoding_dim, activation="relu")(input_layer)
decoded = Dense(num_topics, activation="sigmoid")(encoded)

autoencoder = Model(input_layer, decoded)
encoder = Model(input_layer, encoded)
```

```
# Compile and train the autoencoder model
autoencoder.compile(optimizer="adam",
loss="binary_crossentropy")
autoencoder.fit(train_data, train_data, epochs=100,
batch_size=256, shuffle=True, validation_data=(test_data,
test_data))

# Compute reconstruction errors for anomaly detection
train_data_encoded = encoder.predict(train_data)
train_data_decoded = autoencoder.predict(train_data)
train_data_reconstruction_error = np.mean(np.square(train_data -
train_data_decoded), axis=1)

test_data_encoded = encoder.predict(test_data)
test_data_decoded = autoencoder.predict(test_data)
test_data_reconstruction_error = np.mean(np.square(test_data -
test_data_decoded), axis=1)

# Set a threshold for anomaly detection based on the
reconstruction errors of the training data
threshold = np.percentile(train_data_reconstruction_error, 95)

# Identify anomalies in the test data
anomalies = test_data_reconstruction_error > threshold
print(f"Number of anomalies detected: {np.sum(anomalies)}")
```

In this code, we first preprocess the 10-K statement text files and apply Latent Dirichlet Allocation (LDA) to extract topics from the documents. By applying LDA, we transform the high-dimensional text data into a lower-dimensional representation based on the learned topic distributions, which allows us to focus on topic-level features for anomaly detection. We then construct an autoencoder model for anomaly detection on the topic distributions. Autoencoders are unsupervised neural networks that consist of an encoder and a decoder. The encoder maps the input data to a lower-dimensional latent space, and the decoder reconstructs the input data from the latent representation. In this context, we train the autoencoder to minimize the reconstruction error between the input topic distributions and the reconstructed topic distributions. After training the autoencoder, we calculate the reconstruction errors for both the training and test datasets. These errors represent the difference between the original topic distributions and their reconstructed counterparts. We set a threshold based on the 95th percentile of the training data's reconstruction errors to identify anomalies. Documents with reconstruction errors above the threshold are considered anomalous, indicating that their topic distributions deviate significantly from the norm.

The code outputs the number of anomalies detected in the test data, which can be interpreted as documents with unusual topic distributions.

The use of deep learning in accounting and finance has been steadily increasing as the field recognizes the potential of these powerful models to extract valuable insights from vast amounts of unstructured text data. By employing deep learning techniques, practitioners in accounting and finance can automate complex tasks, enhance decision-making processes, and gain a competitive edge in the industry.

Some of the most promising applications of deep learning in accounting and finance include fraud detection, risk management, sentiment analysis, and financial document analysis. By leveraging advanced models such as Convolutional Neural Networks (CNNs), Recurrent Neural Networks (RNNs), LSTMs, and autoencoders, researchers, and professionals can uncover hidden patterns and relationships in the data that traditional methods might overlook.

Moreover, the combination of deep learning with other machine learning techniques, such as Latent Dirichlet Allocation (LDA) for topic modeling, can provide a more comprehensive understanding of financial documents and their underlying structures. This integration enables the identification of anomalies and unusual patterns in the data, which can be crucial for early detection of potential risks and fraud.

As the volume of financial data continues to grow, the importance of deep learning in accounting and finance will only become more pronounced. By staying ahead of these advancements and understanding the latest developments in deep learning techniques, professionals in the field can ensure they are well-equipped to tackle the challenges of the future and drive innovation in accounting and finance.

With the knowledge gained from this chapter on various deep learning models and their applications, we hope to inspire further exploration and adoption of these techniques in the accounting and finance domain, leading to more efficient, accurate, and insightful analysis of financial data.

Most Common Errors and Solutions

- **SyntaxError**: This error is usually caused by incorrect Python syntax. It could be due to a missing colon at the end of a function or class definition, incorrect indentation, or a missing bracket.
  Solution: Review your code carefully for any syntax mistakes. Python's error message will often give a hint about where the error occurred.
- **IndentationError**: Python uses indentation to delimit blocks of code. If you have mismatched or inconsistent indentation, you'll encounter this error.
  Solution: Make sure that your code is correctly indented. As a rule of thumb, Python code within the same block should have the same level of indentation.
- **NameError**: This error is raised when a variable or function is used before it has been defined, or if it's not in the current namespace.
  Solution: Ensure that the variable or function you're trying to use has been defined before you use it, and that it's in the correct scope.
- **TypeError**: This error is raised when an operation is performed on a data type that does not support it, like trying to concatenate a string with an integer.
  Solution: Check the data types of your variables before performing operations on them. You can use the **type()** function to do this. If necessary, convert data types before performing operations.
- **ValueError**: This error is raised when a function receives an argument of the right data type, but an inappropriate value.

Solution: Make sure that the values you're using are appropriate for the context. For example, if you're trying to convert a string to an integer using **int()**, make sure the string represents a valid integer.
- **ImportError**: This error is raised when an **import** statement fails to find the module definition or when a **from ... import** fails to find a name that is to be imported.
Solution: Check the spelling and case of your module name. Also, ensure that the module is installed and accessible in the Python path.
- **ZeroDivisionError**: This error is raised when you try to divide a number by zero.
Solution: Add checks in your code to prevent division by zero. For example, you might add an **if** statement to check that the denominator is not zero before performing the division.
- **FileNotFoundError**: This error is raised when Python can't locate the file you're trying to open with built-in open() function.
Solution: Ensure the file exists at the path you're specifying. Also, check for any typos in the filename or path.
- **AttributeError**: This error occurs when you try to access an attribute or method that doesn't exist on an object.
Solution: Check the object you're trying to access the attribute or method on. Ensure that it is defined and that it is spelled correctly. Use the **dir()** function to see all the attributes and methods of an object.
- **KeyError**: This error is raised when a dictionary is accessed with a key that does not exist in the dictionary.
Solution: Before accessing a key, check if it exists in the dictionary using the **in** keyword or the dictionary's **get()** method, which returns **None** instead of raising an error if the key is not found.
- **IndexError**: This error is raised when you try to access an index that does not exist in a list, tuple, or string.
Solution: Make sure the index you're trying to access exists. You can use the **len()** function to check the length of your sequence.
- **ModuleNotFoundError**: This error occurs when the module you are trying to import cannot be found.
Solution: Check the spelling and case of the module name. Make sure the module is installed in your Python environment and is in your Python path.
- **MemoryError**: This error is raised when an operation runs out of memory.

Solution: Review your code to find places where you could reduce memory usage. This could involve deleting objects that aren't needed anymore, using generators instead of lists where possible, or optimizing your algorithms.

- **OverflowError**: This error occurs when the result of an arithmetic operation is too large to be expressed by Python's built-in numerical types.

    Solution: Use the **decimal** or **fractions** modules for more precise arithmetic, or use the **numpy** or **scipy** libraries if you're working with large numerical datasets.

- **StopIteration**: This error is raised by the **next()** function (or an **iterator** object's **__next__()** method) to signal that there are no further items produced by the iterator.

    Solution: If you're writing a function or method that uses **next()**, consider using the **default** parameter to avoid raising a **StopIteration** error. In Python 3.7 and later, you can use the **for** loop to iterate over the items in an iterator, which automatically handles the **StopIteration** error.

Index

A

Accessing Data from Other Sources 109
Accessing Data from SEC EDGAR 101
Accessing Data from WRDS 91
Advanced regressions 365–380
Algorithmic trading 462
Applications in accounting and finance 5, 374
ARMA, ARIMA, SARIMA and SARIMAX 395–398
Artificial neural networks 459–461
Assertions 51
Autocorrelation and partial autocorrelation 385–387
Autocorrelations 207
Autoencoder model 497–500
Autoencoders 468

B

Bag-of-words model 217
Bankruptcy prediction 462
Bar chart 158, 187
Basic descriptive statistics 200
Basic regression 303–308
BeautifulSoup 61
Benefits of Python in accounting and finance research 4
Box plot 161
Break, continue and pass statements 18
Building blocks of regex 33

C

Canonical correlation analysis 451–453
Character classes 39
Cleaning HTML tags 105–107, 129–132
Cluster analysis 444–451
Cluster Sampling 210
CNN model 492–497
Connecting to WRDS 93
Control flow statements 13
Convolutional neural networks (CNN) 465–466
Corpus 147
Correlation coefficient 206
Creating corpus 152–154

© The Editor(s) (if applicable) and The Author(s), under exclusive license to Springer Nature Switzerland AG 2024
S. Kumar, *Python for Accounting and Finance*, https://doi.org/10.1007/978-3-031-54680-8

D

Dask 72
Data access libraries 61
Data contained in webpages 110–112
Data from NOAA 116–118
Data from Twitter 118–121
Data handling 381–384
Data manipulation libraries 67
Data on cryptocurrencies 114–116
Data on Yahoo Finance 110–114
Data preprocessing 472–473
Data structures in Python 21
Data Types, Variables, and Operators 12
Data visualization 3, 7, 74, 157, 163, 167, 172, 178, 187, 196, 198
Data visualization libraries 74
Decision trees 259–260
Deep learning 263, 459–500
Deep Learning Model Implementation 469
Deep learning models 266
Descriptive statistics 199
Dictionaries 22
Dimensionality reduction 192
Discriminant analysis 453–455
Downloading data from Compustat 92–93
Downloading data from CRSP 95–96
Downloading data from IBES 96–97
Downloading data from Thomson Reuters 97–99
Downloading text files 101–107
Dynamic panel data models 419–429

E

ELMo 248
Escape sequences 45

Exponential moving average (EMA) 208
Exponential smoothing 208
Extracting useful parts 126–129

F

Factor analysis 441–444
FastText 247
Feedforward neural networks (FNN) 465
File handling in Python 23
Financial statement analysis 461
Fixed effects models 415–417
For and while loops 14
Fraud detection 462
Functions and modules 20

G

GARCH model 405–409
Gated recurrent units (GRU) 467
Generative adversarial networks (GAN) 468
GloVe 247
Google Trends data 121–124
Groups in regex 47
GRU model 491

H

Heatmap 160, 174
Histogram 158, 159

I

If, else, and elif statements 13
Input and output 22
Installing and setting up Python environment 8
Instrumental variables regression 376–380
IQR method 205

Index

K
Keras 85
Kurtosis 200

L
Latent Dirichlet Allocation (LDA) 213–224
Lemmatization 136–137
Lexicon-based methods 278–281
Limiting by date range 104–105
Line chart 158, 163
Linguistic rules 274–278
Lists 21
Literals 33
Logistic regression 298–300, 319–322
Logit regression 339–341
Long short-term memory (LSTM) 467
Lowercasing 134
LSTM model 485–491

M
Machine learning algorithms 282–300
Machine learning libraries 82
Matplotlib 158, 184
Matplotlib library 74
Mean 199
Median 199
Metacharacters 34
Mode 199
Modified Z-score method 205
Moving average 208
Multiple regression 309–314

N
Naive Bayes 250–255, 283–287
Negative binomial regression 374–376
Network graph 187

Non-negative matrix factorization (NMF) 224–228
Numerical data visualization 157
NumPy 70

O
Object Oriented Programming in Python 26
Outlier detection 204
Overview of python programming language 5

P
Pandas 67
Panel data 411–429
Panel data diagnostics 429–433
Pearson's correlation coefficient 206
Perplexity 217
Pie Chart 159, 169
Plotly 167
Poisson regression 373–374
Polynomial regression 343–355
Pooled OLS models 414–417
Portfolio management 462
Principal component analysis 436–441
Probabilistic latent semantic analysis (PLSA) 229–231
Probit and logit regression 329–341
Probit regression 329–335
PyMC3 80
PyTorch 86

Q
Quantifiers 37
Quantile regression 357–364

R
Random effects models 417–419
Random forests 261–264, 294–298

Recurrent neural networks (RNN) 466
re functions 32
Regression diagnostics 314–317
Regular expressions cheat sheet 56
Removing special characters 134
Renaming files 148–150
Requests 63
Rule-based methods 266–274

S

Sampling techniques 209
Scatter plot 158, 162, 164, 206
Scikit-learn 82
SciPy 76
Scrapy 65
Seaborn 163
Sentiment analysis 461
Sets 22
Simple Exponential Smoothing (SES) 208
Simple linear regression 306–308
Simple moving average (SMA) 208
Simple Random Sampling 209
Skewness 200, 201
Sorting files 150–154
Special considerations 142–145
Special techniques in multivariate analysis 435–455
Standard deviation 199
Stationarity 383–389
Statistical analysis libraries 76
StatModels 78
Stemming 135–136
Stop word removal 135
Stratified Sampling 209
Substitution or replacement metacharacters 49
Support vector machines (SVMs) 287–294
Surface Plot 168
Systematic Sampling 209

T

TensorFlow 84, 85
Text classification 32, 133, 134, 136, 147, 152, 224, 243, 244, 247, 249, 250, 252, 254–256, 259, 261, 263, 264, 467, 469
Text data visualization 179
Text extraction and cleaning 125
Text normalization 133, 134
Time series analysis 207, 381–409
Tobit regression 365–380
Tokenizing 134
Topic modeling 213, 214, 271, 277, 297, 477, 481
Transformer models 461
t-SNE 192
Tukey's Fences 206
Tuples 21
Two-stage least squares regression 378–380
Types of panel data models 412–414

U

UMAP 196, 197

V

Variance 200, 201
Variational autoencoders (VAE) 468
VAR model 398–401
VECM model 402–405

W

Word cloud 179, 185, 186
Word embeddings 192, 194, 196, 243–245, 247, 248, 250, 264, 491
Word2Vec 243–248

Z

Z-score method 204

MIX
Papier aus verantwortungsvollen Quellen
Paper from responsible sources
FSC® C105338

If you have any concerns about our products,
you can contact us on
ProductSafety@springernature.com

In case Publisher is established outside the EU,
the EU authorized representative is:
Springer Nature Customer Service Center GmbH
Europaplatz 3, 69115 Heidelberg, Germany

Printed by Libri Plureos GmbH
in Hamburg, Germany